The Essential Guide to Environmental Chemistry

Prof. Dr Georg Schwedt
Institute for Inorganic
and Analytical Chemistry
of the Technical University Clausthal
Paul-Ernst-Straße 4
38678 Clausthal-Zellerfeld
Germany

Colour table production:

Schreiber VIS
Joachim Schreiber
Karolinenstr. 26
64342 Seeheim

In cooperation with Werner Wildermuth, Stuttgart

Important Notice: This volume has been compiled by professionals. The user must know that simply dealing with chemicals and microorganisms introduces a latent risk. Theoretically, additional risks can occur due to incorrect amounts of materials.

The authors, the editor and the publisher have taken a great deal of care to see to it that the amounts used and the experimental protocols correspond to the state of the art when the volume was published. Still, the publisher cannot accept any liability for the accuracy of this information. Each user is asked to accept the responsibility to carefully verify whether the amounts of materials, the experimental protocols or other information are plausible based on the understanding of a natural scientist. In the event there is any doubt, the reader is strongly advised to consult with a competent colleague. Even the publisher is willing to offer support in clarifying questions which might arise. In spite of this, every application described in this volume is at the user's own risk.

The Essential Guide to Environmental Chemistry

Georg Schwedt
Technische Univerität, Clausthal

Translated by
Brooks Haderlie,
Brigham Young University - Idaho, Rexburg

JOHN WILEY & SONS, LTD
Chichester • New York • Weinheim • Brisbane • Singapore • Toronto

Baffins Lane, Chichester,
West Sussex PO19 1UD, UK

National 01243 779777
International (+44) 1243 779777

e-mail (for orders and customer service enquiries): cs-books@wiley.co.uk

Visit our Home Page on http://www.wiley.co.uk
or
http://www.wiley.com

Other Wiley Editorial Offices

John Wiley & Sons, Inc., 605 Third Avenue,
New York, NY 10158-0012, USA

Wiley-VCH Verlag GmbH, Pappelallee 3,
D-69469 Weinheim, Germany

John Wiley, Australia Ltd, 33 Park Road, Milton,
Queensland 4064, Australia

John Wiley & Sons (Asia) Pte Ltd, 2 Clementi Loop #02-01,
Jin Xing Distripark, Singapore 0512

John Wiley & Sons (Canada) Ltd, 22 Worcester Road,
Rexdale, Ontario M9W 1L1, Canada

Library of Congress Cataloging-in-Publication Data

Schwedt, Georg, 1943-
[Taschenatlas der Umweltchemie. English]
Essential Guide to Environmental Chemistry / Georg Schwedt.
p.cm.
Translation of: Taschenatlas der Umweltchemie.
Includes bibliographical references and index.
ISBN 0-471-89954-2
1. Environmental chemistry – Handbooks, manuals, etc. I. Title
TD193 .S3813 2001
623 – dc21 2001040143

British Library Cataloguing in Publication Data

A catalogue record for this book is available from the British Library

ISBN 0 471 89954 2

Typeset by Hilite Design & Reprographics Limited, Southampton
Printed and bound in Italy by Conti Tipocolor Arti Grafiche, Florence
This book is printed on acid-free paper responsibly manufactured from sustainable forestry, in which at least two trees are planted for each one used for paper production.

Contents

4 Soil

Preface

Many of the sciences have aided in our understanding of the complex environmental behaviour of chemical substances. These include chemistry and physics, meteorology and hydrology, biology and geology. Additional special disciplines within the analytical environmental sciences are microbiology, toxicology, biochemistry and soil science.

This handbook connects the most important principles of atmospheric chemistry, hydrochemistry and soil chemistry with the materials-oriented physicochemical, bio-ecological and special chemical– technical processes.

As an introduction to the topic, material cycles are discussed in particular detail because of the necessary holistic and therefore interdisciplinary approach. Individual facts in this and subsequent chapters are intentionally repeated in order to emphasise their significance for the various aspects of environmental chemistry.

In the other chapters on environmental chemistry in the atmosphere, the hydrosphere and the pedosphere, we include the biosphere as we discuss ecochemical interrelationships with increasing specialisation of the problems and environmental (protection) technology.

Chapters 5 and 6 deal with all four spheres and the environmental chemistry of selected xenobiotics and heavy metals, with problem- and effect-oriented environmental analysis, and with methods in ecotoxicology.

Graphical depictions and colour tables are especially well suited to demonstrate complex relationships even without exhaustive text descriptions which are reserved for textbooks. They give the users the opportunity to develop their own foundation for understanding the environmental chemistry extensively described here from the standpoint of their own level of knowledge and technological viewpoint. On the other hand, they cannot replace the aforementioned textbooks, which should be consulted for studying the material at a deeper level.

The author thanks all of his colleagues cited in the Bibliography who have compiled books dealing with this broad topic. With the figures in their books, they have provided the author with the foundations for the subsequent development and design of the colour tables by the graphical artist Joachim Schreiber. I would like to thank him as well for the fruitful cooperation, as with *The Essential Guide to Analytical Chemistry*.

The author welcomes critical comments on and suggestions for this first attempt at a comprehensive graphical depiction of the broadly laid-out topic of environment and chemistry, including technology.

Georg Schwedt
Clausthal
Spring 1996

Introduction

According to Friedhelm Korte, one of the founders of the field of ecological chemistry in Germany, by this term 'we are to understand, from a comprehensive viewpoint, according to the subject matter of both parts of the term, the chemical processes and chemical interactions and the consequences resulting therefrom in the ecosphere'. With the industrialisation which has greatly increased since the early 1960s, and as a result of the limiting of natural resources, ecological chemistry has been made the focal point of public interest as part of analytical environmental research. The complexity of the interrelationships and of the processes requires far-reaching interdisciplinary cooperation and a new networked thinking (Frederic Vester). The terms 'ecological chemistry', 'ecochemistry' and 'environmental chemistry' are used synonymously. Crucial contributions to ecological chemistry are provided by technical fields such as geoscience (especially geochemistry), biology, agriculture (especially soil science), biochemistry and toxicology (as ecotoxicology). Ecological chemistry is to chemistry as ecology is to biology. In contrast to ecological chemistry, chemical ecology is limited 'to natural substances, the natural field of application and control mechanisms (allelo chemicals) of the organisms in ecosystems' (F. Korte). However, the transitions between chemical ecology and ecological chemistry are fluid.

By 'environmental chemistry' (synonymously used with 'ecological chemistry' and "ecochemistry"), we mean a cross-sectional discipline extending across many fields and with close interactions with biology, ecology, ecotoxicology, chemistry, hydrology, meteorology, geochemistry and technology (Römpp, 1993: *Lexikon Umwelt*, Stuttgart: Thieme). Focal points of research are experiments dealing with mass transport, with the distribution and transformation of chemical substances (environmental chemicals) in the four spheres of atmosphere, hydrosphere, pedosphere and biosphere, and in the transitions between these spheres. Furthermore, environmental chemistry deals with the physical and chemical interactions among chemical substances, especially between environmental chemicals and the naturally occurring material components of the environment. Some of the most important processes include biogeochemical cycles, entry pathways of substances and material transport processes (distribution in environmental compartments, geo-accumulation and bio-accumulation, wet and dry deposition, volatility and mobility), as well as biotic and abiotic transformations (such as photochemical transformations, abiotic and biochemical–enzymic degradation, and biotransformation). Ecotoxicology and environmental analysis form a significant proportion of the advances made in environmental chemistry: new and improved analytical methods with high efficiency with respect to the separation of complex mixtures of materials and low limits of detection, plus new biological–biochemical methods, including problem-oriented concepts in ecotoxicology, have a high ranking in the field of ecochemistry. Physicochemical processes in the environment are being studied based on laboratory, field and model system analyses including mathematical modelling. Even in environmental (protection) technology, in addition to physicochemical processes, those processes that approximate the natural processes in the environment (e.g. biodegradation) are finding increased application.

1 Material Cycles

1.1 Geochemical cycles

1.1.1 Environmental areas

Our environment is divided into spheres. By 'biosphere', we mean all of the layers of the Earth inhabited by living organisms. We usually use the term 'ecosphere' today. The latter is in turn divided into different ecosystems. The atmosphere is the mass of air surrounding the earth with a fluid upper limit bordering on outer space. The ecosphere includes only inhabited space. Meteorologically the atmosphere is divided into troposphere and tropopause, stratosphere and stratopause, mesosphere and mesopause, and thermosphere, based on physical properties. The pedosphere represents the highest region of the soil which is inhabited by living organisms. It borders on the lithosphere, the outermost rock layer (geological foundation) down to a depth of some 100 km. The pedosphere is penetrated by the atmosphere and the hydrosphere (as groundwater in this case), i.e. by all other compartments (as individual parts of a complex ecosystem for the characterisation of material transformation and transport processes via interfaces) of terrestrial ecosystems.

1.1.2 Endogenous and exogenous material cycles

All geological processes on Earth can be summarily described as a cycle of the materials. Due to orogenesis or epeirogenesis, magma travels from the inner Earth up to the Earth's surface and hardens to form magmatic rock. Magma (Greek for 'kneaded mass, thick salve') is the name given to the glowing hot silicatic molten rock in the inner Earth. Orogenesis is a temporally and spatially limited mountainous formation process (also called tectogenesis). Epeirogenesis describes reversible, wide-ranging lifting and sagging of parts of the Earth's crust over long periods of geological time.

Metamorphosis includes the transformation of rocks. It takes place in the pressure and temperature fields of the Earth's crust, whereby mineral reactions (e.g. recrystallisation) take place as a result of changes in the physicochemical equilibrium. Depending on the origin of the starting material, metamorphic rocks are called orthorocks (former magmatic rocks) or para-rocks (former sedimentary rocks).

The magmatic and metamorphic rock which is exposed due to lifting is acted upon by exogenous forces – i.e. gravity, temperature, effects of water, ice and wind – to produce weathering products in solid or soluble form as well as soils. Chemical and biological processes in particular play a role in the formation of soils. Soil is the term used for the uppermost, inhabited weathering layer of the Earth's crust (see pedosphere). Weathering products and soils are displaced and deposited in other locations in the form of clastic or (bio)chemical sediments. Clastic sedimentary rocks are the products of mechanical weathering of stone, and are also called debris. Sedimentary loose rock such as dust, sand, claywash, mud ooze and peat are formed. With chemical and physical effects, in the course of diagenesis new sedimentary solid rocks are formed, such as sand, dolomite and limestone, shale clays and lignite.

Humans interfere with these natural cycles, in particular via the 'anthropogenic exploitation of natural resources', i.e. by the mining of ores and rocks, and by the use of water. This involves especially the processes of weathering and erosion, natural soil erosion due to the effects of wind or water, and the transport and redistribution of rocks or soil.

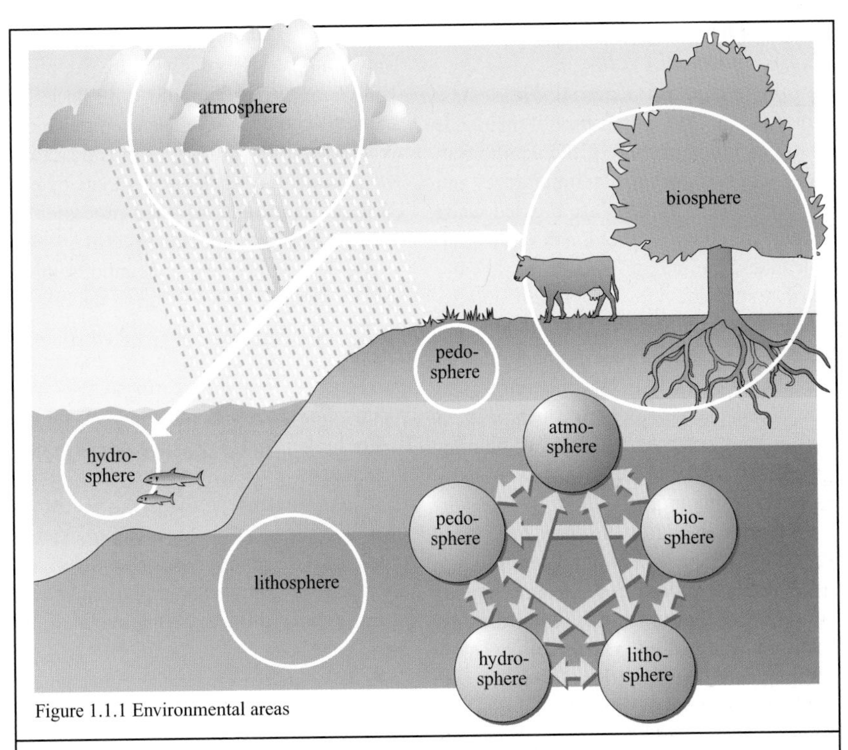

Figure 1.1.1 Environmental areas

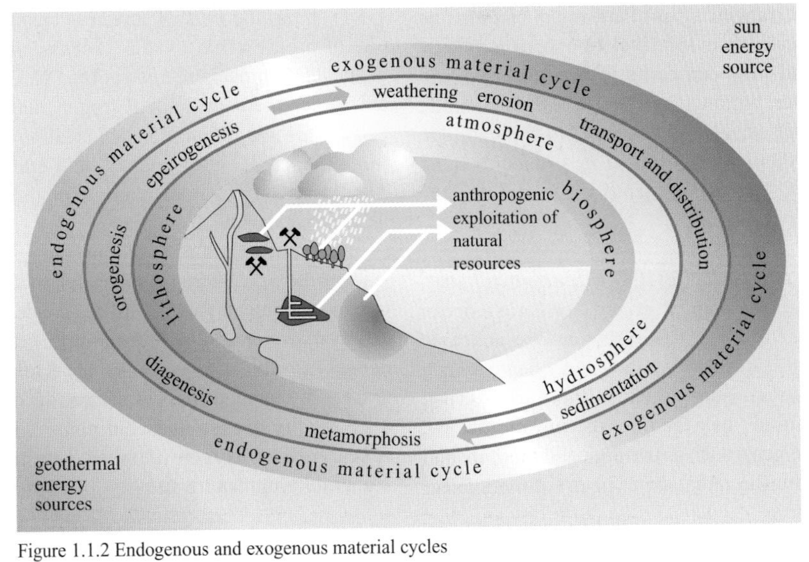

Figure 1.1.2 Endogenous and exogenous material cycles

1.1.3 The geological material cycle

For the geological material cycle, it is important that every transport of magma is based on a disturbance of the equilibrium from changes in the temperature and pressure ratios. Plutons are created when magma remains in the Earth's crust; in volcanoes, the magma finally reaches the Earth's surface. After the liquid magma phase, the actual hardening starts at about 1200 °C. The plutonic rocks are formed in the phases of early crystallisation (up to 900 °) and primary crystallisation (up to 600 °). They consist of silicates containing less silicic acid and the principal mass of lithogenous minerals.

In the geological cycle, new magmatic rocks can form via sinking to greater depths and the processes of metamorphosis, subsequent melting and re-solidifying upon being lifted up. On the other hand, sedimentary loose and solid rock can weather immediately after formation and again after being exposed, and it can be rearranged again. Para-rocks can also be created from solid rocks via metamorphosis, and they can come to the surface again even without melting. Finally, magmatic rock can also be converted to metamorphic orthorock even before it is made to stand out, and then enter again into the exogenous cycle.

1.1.4 The 'crust–ocean machine'

The 'geological mixer', or 'crust–ocean machine' of Garrels and Mackenzie (1971), starts with the energy centres: with the build-up of heat in the Earth's crust, caused by the decay of radioactive elements such as uranium and thorium, and with the radiation of heat via the nuclear processes taking place on the Sun. The endogenous processes are supplied with energy from the radioactive decay in the Earth's core. Gases and steam cause the weathering of the primary magmatic rocks. The sedimentary rocks and oceans of our Earth are the results of these processes in the form of a geological long-term effect.

The oceans represent a gigantic 'settling and evaporation vessel'. Due to the effects of gravity, the sedimentary rocks sink to greater depths, carried on by convection currents. At first they form wedges and prisms, but at these depths they are folded and/or broken and, depending on the depth, they undergo processes of diagenesis (solidification), metamorphosis (transformation) or melting (anatexis) – recycled rocks are produced. (Recycled) gases and water that are released during metamorphosis likewise return back to the weathering cycle on the Earth's surface.

In the course of the Earth's development during a period of 4000 million years, different spheres were formed (Section 1.1.1). First, the Earth's core was created out of iron-nickel. In the process, the primeval atmosphere of noble gases, ammonia, methane and hydrogen dissipated. This was followed by the creation of the three shells: core, mantle, and crust. The oxygen-free atmosphere contained gaseous materials from the inner Earth (water condensation, carbon dioxide, sulphur dioxide, hydrogen chloride). In the next developmental step, with the condensation of water and the formation of the oceans (as a result of cooling), the Earth's crust also changed its composition in the areas of the oceans and continents. The appearance of oxygen is linked with the higher development of life.

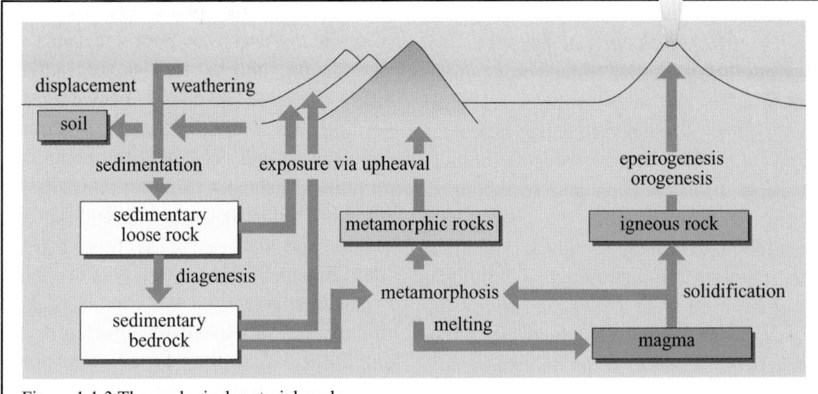

Figure 1.1.3 The geological material cycle

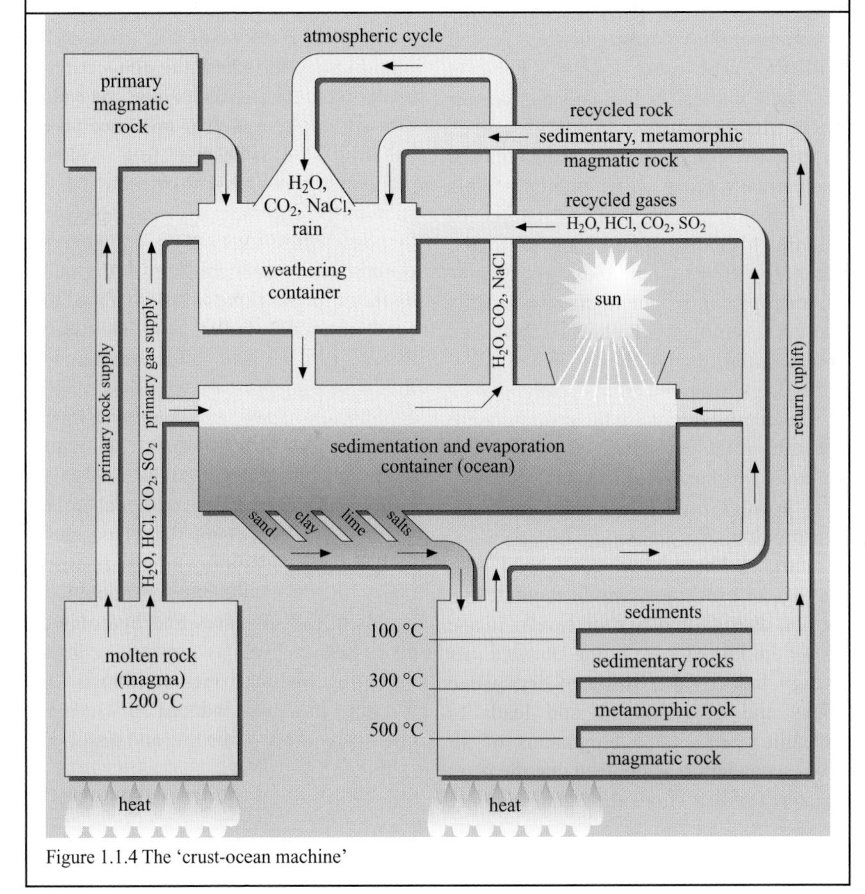

Figure 1.1.4 The 'crust-ocean machine'

1.1.5 *The Earth as a biogeochemical factory*

A view that is similar to the 'crust–ocean machine' is the analogy of a chemical factory that is reduced to a few processes (Sievers, 1974). Energy receives the heat machine from the Sun and in the inner Earth by means of the previously mentioned processes (Section 1.1.4). All of the processes are distributed to individual reactors. The heat generator drives the winds, ocean currents and the cycles of the water and the rock. Water is viewed as a means of transportation and as a chemical reagent. Over long periods of time, the reactions of the volcanic emissions (acids) with the basic rocks led to a constant composition of the oceans and to an atmosphere with a constant carbon dioxide level. Eruptive rock became soils, sediments and sedimentary rocks. After photosynthesis had become possible in the course of the Earth's development, and with the development of life, the biosphere became an 'entropy pump': due to the continuously radiating sunlight which was converted into thermal energy, the biosphere drove the biological and material cycles.

The oceans symbolise the central (main) reactor, which is connected to all other reactors. Mechanical erosion is depicted as the 'powder mill', the liquid extractor representing the chemical processes of weathering. The third reactor encompasses biological processes, which regulate the carbon dioxide and oxygen levels. If one views the Earth's core as the 'furnace', its energy induces the lifting of crystalline rocks and of sediments and leads to volcanic emissions. Components of all three secondary reactors feed into the main reactor. Detritus (from the Latin, meaning 'rubbed or ground off') is used in geology to indicate the rock debris created by weathering, and in biology to indicate finely dispersed materials – suspended and settling substances in waters – derived from the natural decay of dead plants and animals. These in turn contain organic residues such as lignin, cellulose and chitin, and they serve as food for the 'detritus-eaters'. The consumers are those organisms that feed on biomass, dead organisms, refuse such as foliage and excrement, as well as the accompanying microorganisms, and that then digest or mineralise these materials.

Dissolving processes take place in the liquid extractor, where the connection to the gas containers with oxygen and carbon dioxide emphasises their influence on the solubility of inorganic (e.g. calcium carbonate by CO_2) and organic (by O_2) substances. The water cycle is depicted by the distillation system, going from the oceans and to condensation in the liquid extractor. In this representation of the Earth as a chemical factory, in addition to the reactors we also differentiate among the following phases: gaseous, liquid, geologically sedimentary and crystalline.

All the processes which take place in the biogeochemical factory can be divided into biological/biochemical and geochemical/geophysical processes. The first group includes all metabolic processes (especially in the bioreactor). Geophysical/geochemical processes and hydrological processes, as well as erosion, sedimentation, geological metamorphosis and transport processes induced by winds, are principally those of melting and dissolving.

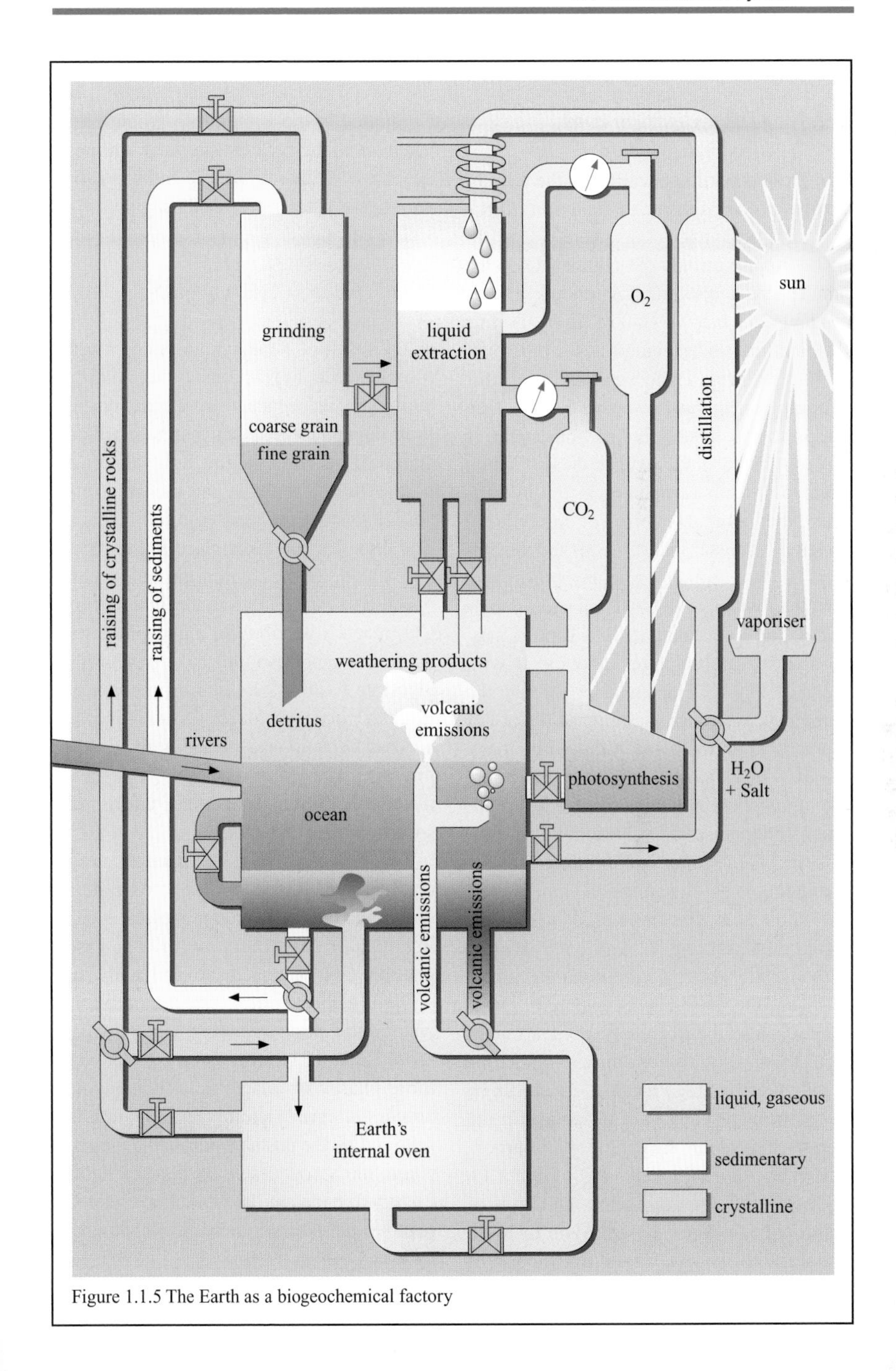

Figure 1.1.5 The Earth as a biogeochemical factory

1.2 The carbon cycle

1.2.1 Mineralisation and biosynthesis

The greatest portion of carbon in the form of carbon dioxide is stored in the oceans (3.8×10^{16} kg C) and in the atmosphere (7.2×10^{14} kg C). Approximately 15% of the CO_2 in the atmosphere is converted via photosynthesis in plants (see below). Half of this is used by the plants to produce biomass; the other half – starting from glucose – is used for acquiring energy and respiring back as CO_2 (respiration). Thus, the carbon cycle is directly linked to the oxygen cycle. Carbon reservoirs form carbonates in the hydrosphere (see inorganic forms of carbon), the biosphere (mussel shells, bones) and the lithosphere (lime, coral reefs in the hydrosphere boundary with 6×10^{15} kg C plus fossil fuels such as petroleum, natural gas, hard coal and lignite, and peat with 1.2×10^{15} kg C). Humans interfere with the carbon cycle through burning and the use of biomass. From the living and dead biomass, 6×10^{12} kg C (also as CO_2) is released into the atmosphere annually due to decomposition (mineralisation via microbial metabolism). The storage of biomass in the sediments (as fossilisation under the exclusion of air) removes carbon from the cycle (approximately 10^{11} kg C yr^{-1}). Another carbon cycle takes place between the atmosphere and the water: approximately 10^{14} kg C are exchanged between these compartments as CO_2 annually. Photosynthesis (by marine plankton with 65% of the total C taken up by the plant world) and the CO_2 uptake in the oceans represent sinks for the carbon. Currently approximately 4% of the CO_2 emitted into the atmosphere annually is of anthropogenic origin (combustion of fossil and non-fossil fuels, by the destruction of forests and soils and their outcomes). Compared to the amount of carbon that is taken out of the cycle by means of settling in oceanic sediment, the anthropogenic emission is greater by a factor of 50.

The equation

$$6CO_2 + 6H_2O \rightarrow C_6H_{12}O_6 + 6O_2$$

as the result of photosynthesis describes a complex chain of reactions. It takes place as two largely independent partial reactions, the light-dependent primary reaction (photochemical reaction in the light absorbing pigments, chlorophyll) and the light-independent secondary reaction (dark reaction). The dark reaction takes place in the stroma (framework) of the chloroplasts and requires $NADPH/H^+$ and ATP from the light reaction. CO_2 reacts with ribulose diphosphate with the formation of two molecules of phosphoglycerate (with three C atoms, a characteristic of C_3 plants), whereby CO_2 is reduced to a carbohydrate precursor and is fixed in the process. The photosynthesis system is disturbed by environmental toxins and specifically by herbicides.

The processes of mineralisation are in contrast to those of photosynthesis. Anaerobic decomposition processes, such as in a landfill (in compost in this case), generate CO_2 plus methane, and acids such as acetic acid and propionic acid. An aerobic phase (utilisation of atmospheric oxygen) shortly after the sedimentation is followed by three phases of anaerobic decomposition processes. An acidic fermentation is followed by the unstable and stable methane phase (methanogenesis). By photooxidation, carbon dioxide can be generated again for photosynthesis from methane via formaldehyde and carbon monoxide.

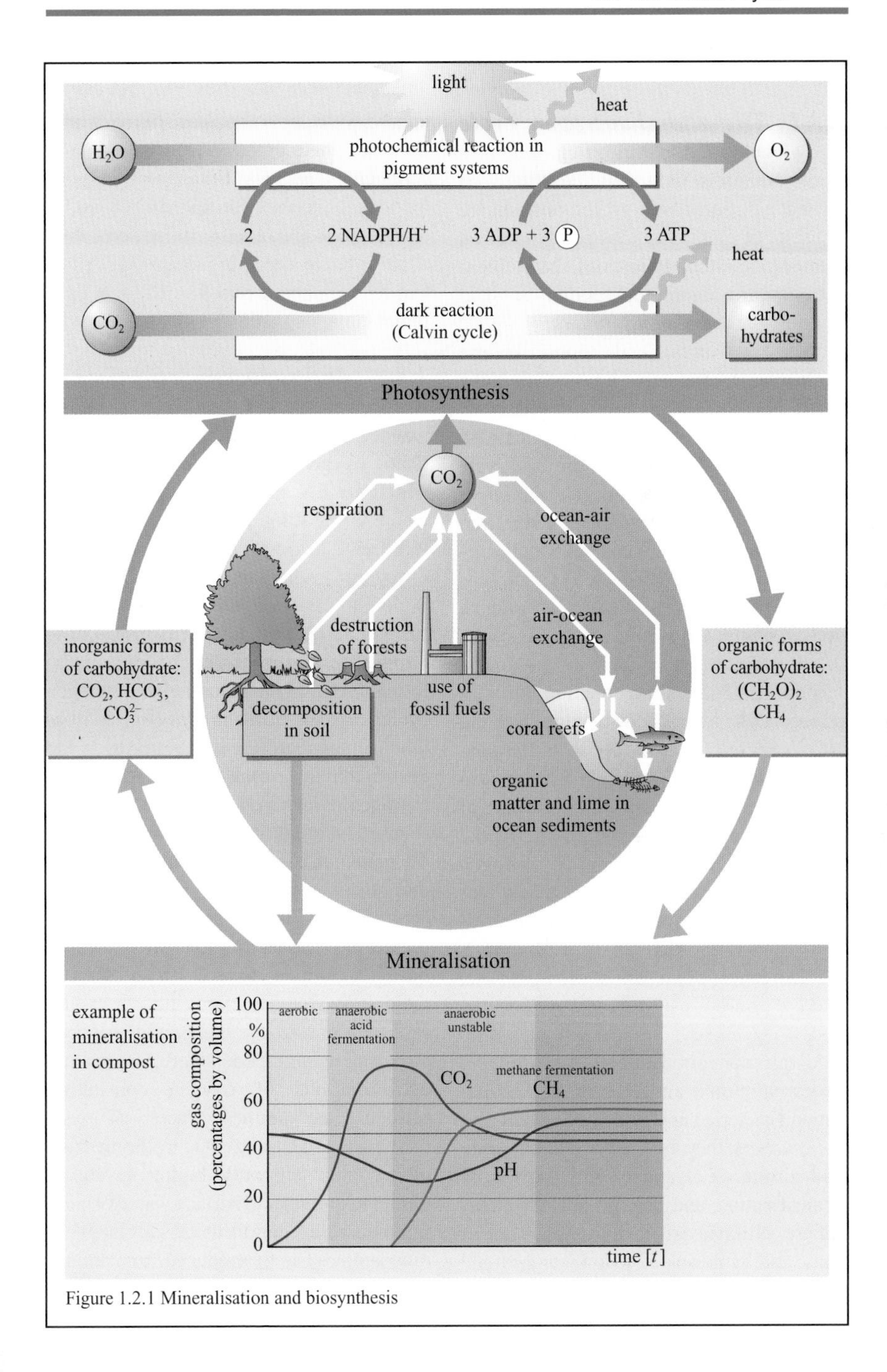

Figure 1.2.1 Mineralisation and biosynthesis

1.3 The nitrogen cycle

1.3.1 The global nitrogen cycle

The global nitrogen cycle is characterised by its numerous oxidation states between –3 and +5. Some 80% of the atmosphere consists of elemental nitrogen (oxidation number 0). Other species of the element nitrogen are organic compounds such as amino acids, proteins, amino sugars, amides and urea, and especially the inorganic substances NO_3^- (+5), NO_2^- (+3), NH_4^+ (–3) and nitrogen oxides NO (+2), NO_2 (+4), and N_2O (+1); see also Section 1.3.3. The biological nitrogen cycle is determined by nitrogen fixation, whereby atmospheric nitrogen enters into the hydrosphere and pedosphere and especially in biomass. Biological fixation is done by microorganisms and blue algae and in symbiosis of microorganisms with higher plants (e.g. rhizobia with legumes – legume bacteria) (nitrogen assimilation as a biocatalytic process). The oxides NO and NO_2 especially play a role in the atmospheric chemistry of nitrogen. Proteins can be formed from inorganic forms of nitrogen (ammonia assimilation). In the reverse process, organic nitrogen compounds can be converted back into ammonia by ammonification (by deaminating bacteria such as *Pseudomonas*). Microorganisms that cannot use light as an energy source use this to obtain the necessary energy: via oxidation, amino acids are converted into carbon dioxide, water, ammonia and energy; via nitrification (bacteria such as *Nitrosomonas* and *Nitrobacter*), they are converted into nitrite and nitrate. N_2O and N_2 are formed by denitrification, and they go into the atmosphere. Nitrates are easily washed out of soils, and by means of sediments they end up in the deep-sea floor or in the lithosphere. Volcanoes transport nitrogen into the atmosphere in the form of ammonia or nitrogen oxides. The lithosphere contains 0.2×10^{18} t N, the hydrosphere 23×10^{12} t, and the atmosphere 3.0×10^{15} t. Some 0.92×10^{12} t N are stored in biomass, 1.7×10^9 t in living biomass and 9×10^{11} t in dead biomass.

1.3.2 Anthropogenic effects

Nitrogen molecules are split in a natural manner in thunderstorms, and with atmospheric oxygen they form nitrogen oxides, which are a component of acid rain but which at the same time serve as nitrogen fertiliser. The growth of plants is limited by nitrogen, so humans disturb the equilibrium in the cycle with fertilisers (nitrate, ammonia and organic fertilisers). Using nitrogen fertilisers causes the same transformations as in the natural global nitrogen cycle: reactive nitrogen compounds from the atmosphere contribute to the destruction of ozone (see Section 2.2). Nitrogen oxides also enter into the atmosphere as a result of combustion processes. The intensification of the nitrogen cycle due to anthropogenic effects causes ecological problems such as the regionally increasing concentration of volatile nitrogen compounds (NO_x, N_2O, NH_3) in the troposphere and stratosphere, a greater exchange of nitrogen compounds between atmosphere and pedosphere, increasing concentrations of oxygen consuming nitrogen compounds such as urea $(NH_2)_2CO$, NH^+_4 and NO_2 in the hydrosphere, plus generally higher levels of nitrate in ground and surface waters. Under unfavourable conditions, carcinogenic nitrosamines can be produced from amines and nitrite.

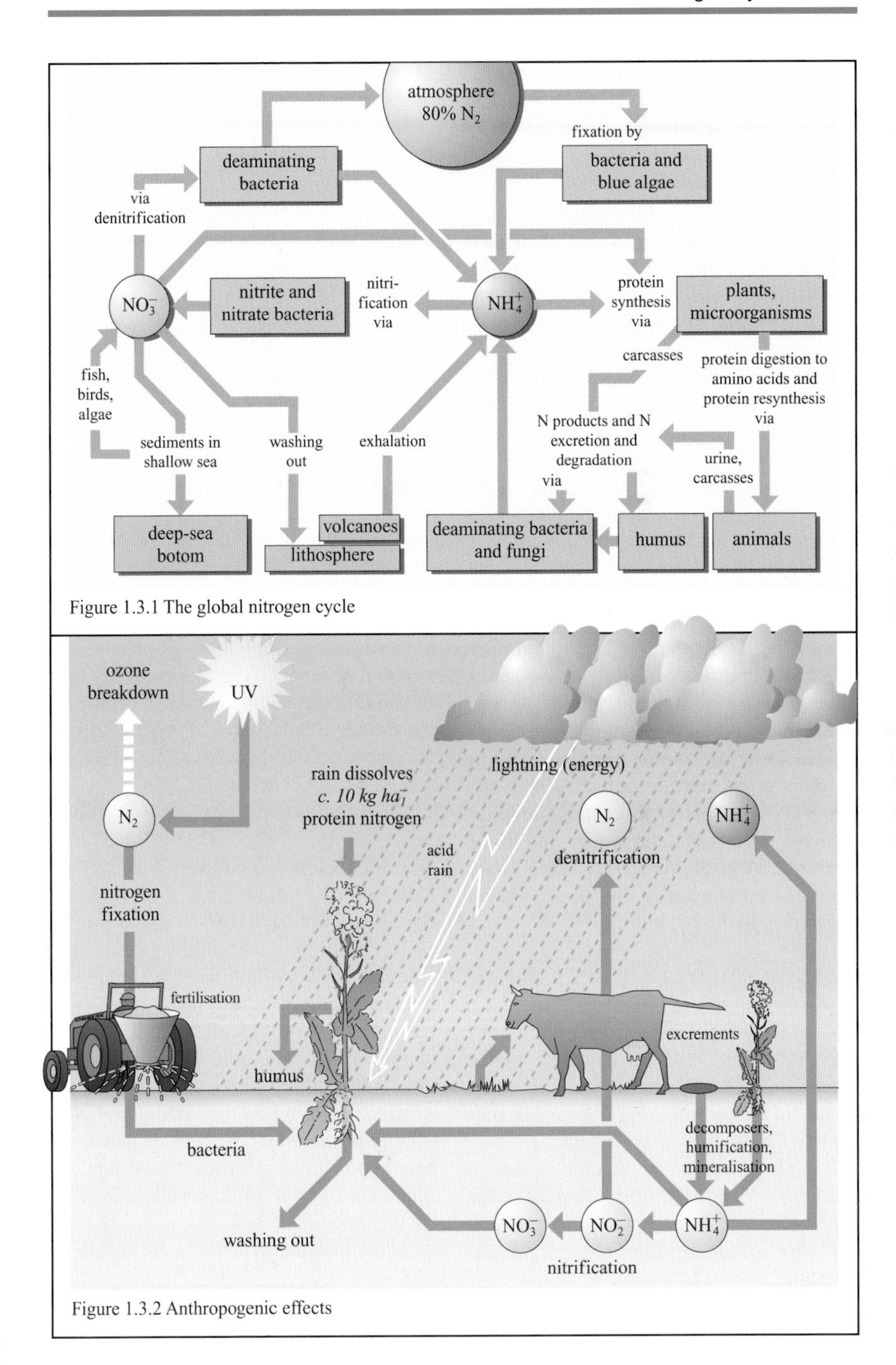

Figure 1.3.1 The global nitrogen cycle

Figure 1.3.2 Anthropogenic effects

1.3.3 Ammonification, nitrification, denitrification

The increase in nitrogen concentration in surface waters can be traced back to a number of different water-soluble nitrogen compounds with differing oxidation numbers between –3 and +5: urea as a fertiliser component and a metabolic product of certain animals whose main end product of protein metabolism is urea, e.g. sharks, terrestrial amphibians, some turtles and all mammals. Ammonia and ammonium salts are derived from fertilisation, from putrefaction and biological food chains, and from waste water. Nitrates are components of fertilisers and an end product of nitrogen species with low oxidation numbers. The processes of the microbial nitrogen cycle include redox and acid–base reactions, plus reaction mechanisms that lead to the formation or splitting of C–N bonds.

In ammonification, owing to nitrogen fixation (Section 1.3.1), atmospheric nitrogen is incorporated into organic molecules (proteins). Because of metabolic processes by organisms, proteins or amino acids are converted into the excretion products urea $(NH_2)_2CO$ and ammonia, which exists as the ammonium ion in the presence of acids. Ammonia can go back into the cycle for the formation of proteins. According to the equation

$$(NH_2)_2CO + 2H_2O \leftrightarrows NH_4^+ + NH_3 + HCO_3^-$$

the hydrolysis of urea catalysed by the enzyme urease leads to ammonia or the ammonium ion (depending on pH conditions). In the environment, ammonium salts go into the water cycle or the soil.

In nitrification, in aerobic zones a microbial oxidation of ammonium ions takes place where nitrite is converted to nitrate as processes of nitrification. Microorganisms of the *Nitrosomonas* genus oxidise ammonium ions to nitrite:

$$2NH_4^+ + 3O_2 + 2H_2O \rightarrow 2NO_2^- + 4H_3O^+$$

(molar reaction enthalpy: 260 kJ mol^{-1}). In the second stage, *Nitrobacter* species oxidise the nitrite to nitrate:

$$2NO_2^- + O_2 \rightarrow 2NO_3^-$$

(molar reaction enthalpy: 100 kJmol^{-1}). The overall equation for the nitrification processes is

$$NH_4^+ + 2O_2 + H_2O \rightarrow NO_3^- + 2H_3O^+$$

Denitrification, or the reduction of nitrate to atmospheric nitrogen, takes place in anoxic zones (anoxic = based on oxygen deficiency) due to metabolic processes by numerous facultative anaerobic, heterotrophic microorganisms. They use nitrate nitrogen as an electron acceptor for the oxidative breakdown of organic carbon compounds. Denitrification occurs in stages; {H} stands for H donors:

$$NO_3^- + 2\{H\} \rightarrow NO_2^- + H_2O$$
$$NO_2^- + 2\{H\} \rightarrow NO + H_2O$$
$$2NO + 2\{H\} \rightarrow N_2O + H_2O$$
$$N_2O + 2\{H\} \rightarrow N_2 + H_2O$$

The end product of nitrification is nitrogen; intermediates, which can end up in the environment if insufficient H donors (e.g. methanol or acetic acid from the carbon cycle) are available, also include the nitrogen oxides NO (nitrogen monoxide) and N_2O (dinitrogen monoxide). N_2O in particular enters the atmosphere from anthropogenic sources (as a result of fertilisation). In order for the nitrogen cycle to run as described in the soil, temperatures of 5–8°C are necessary.

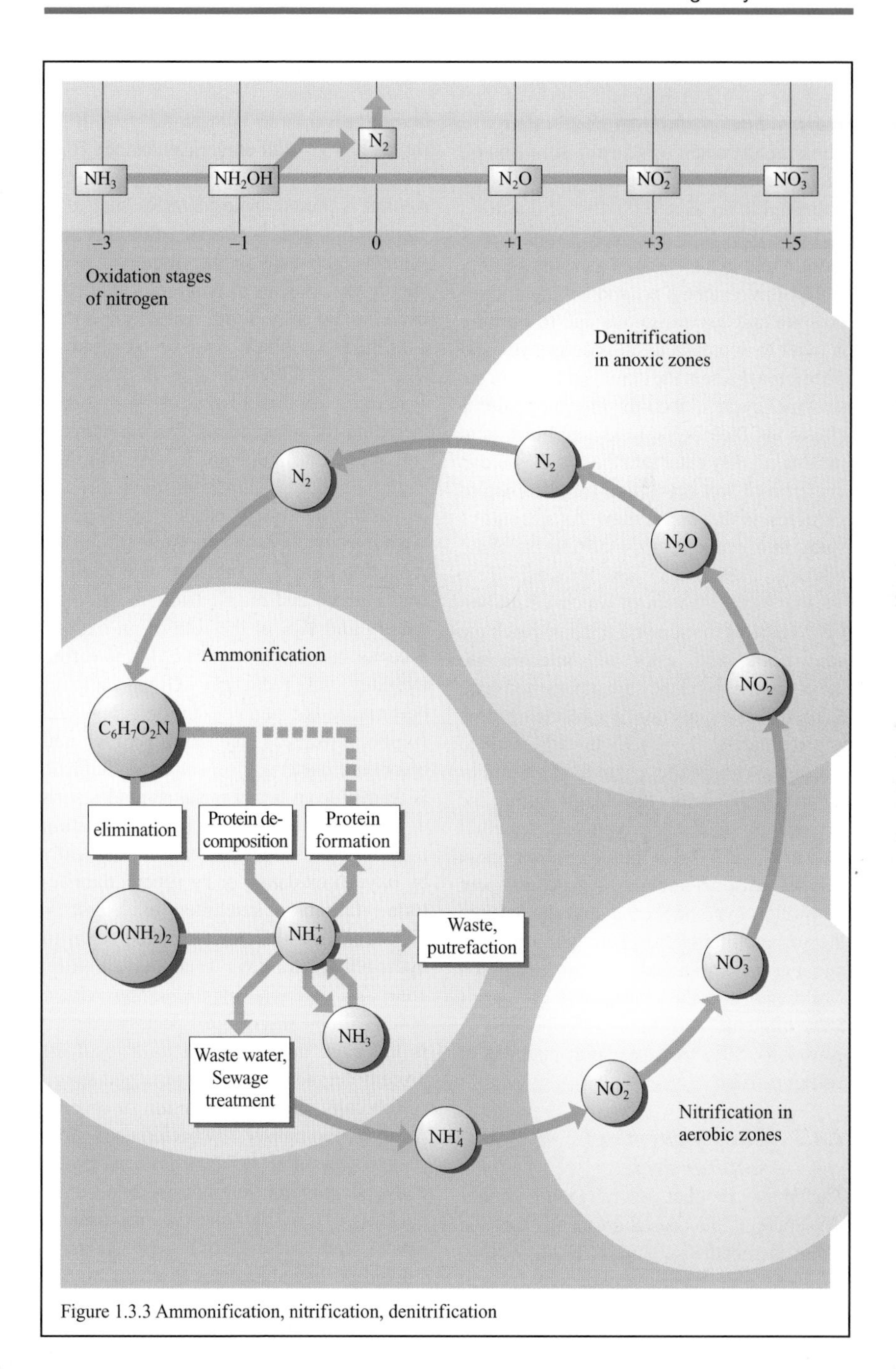

Figure 1.3.3 Ammonification, nitrification, denitrification

1.4 The sulphur cycle

1.4.1 The global sulphur cycle

The sulphur reservoirs in the lithosphere, pedosphere, hydrosphere and biosphere are estimated to be 12 × 10^{15}, 10^{13}, 1.3 × 10^{15}, and 6 × 10^{9} t S, respectively. Some 2 to 3 million tonnes are released annually (as SO_2 or H_2S) by volcanic eruptions. The anthropogenic sulphur emissions due to burning amount to some 75 to 80 million t yr^{-1} and therefore represent the dominant factor in the sulphur cycle, next to the biogenic S emissions (see below). As sulphuric acid (acid rain), they enter into the hydrosphere in the form of wet deposition (or adsorbed to particles as dry deposition). As a result of spray and evaporation, maritime sulphate aerosols (seasprays) are created from oceanic surface waters, of which 1.5 million t yr^{-1} (10% of the sprayed sulphur) reach the continents, and after an intermediate depositing in rivers they are transported once again to the oceans (average residence time approximately 1 year). In addition to hydrogen sulphide, gaseous sulphur compounds from the degradation of biological material include dimethyl sulphide, carbon disulphide, and carbonyl sulphide in small amounts. Altogether about 35 million t yr^{-1} of S each are released into the soil and the hydrosphere. On a percent age basis, the annual sulphur emissions worldwide into the atmosphere consist of 38% biogenic and 38% anthropogenic S emissions, 20% from sea spray, and 4% of volcanic origin.

1.4.2 The biochemical sulphur cycle

The main product of biogenic sulphur conversion in the coastal areas, marshes and moors is dimethyl sulphide $(CH_3)_2S$, with 40 million t S yr^{-1}. On the other hand, vegetation in inland soils produces mainly hydrogen sulphide H_2S, at the rate of 10 million t S yr^{-1} (all numerical data are from Kimmel and Papp, 1988). For example, sulphur is present in amino acids such as l-methionine and l-cysteine. Sulphur and sulphur compounds are very important to the energy metabolism of numerous strains of bacteria and some fungi. Under anaerobic conditions, sulphides can be oxidised to sulphates by thiobacteria. In deep waters, *Beggiatoa* strains oxidise H_2S to form elemental S, whereas *Thiobacillus* strains convert it into sulphate. In this way, the bacteria acquire chemoautotrophically via oxidation the energy needed for forming organic compounds (see Section 1.2 for the reduction of CO_2). In areas with shallow water, green and purple bacteria use light energy and H_2S in the role of an oxygen acceptor in order to convert CO_2 into carbohydrates via reduction. Green sulphur bacteria oxidise sulphides to form elemental sulphur, whereas red and purple and colourless bacteria form sulphate. Sulphate is formed from heavy metal sulphides such as iron sulphides (pyrite, FeS_2 – oxidation number of the sulphur +1) either chemically by oxygen oxidation or by special thiobacteria (aerobic sulphide oxidisers – *Thiobacilli*). Sulphate is reduced again to hydrogen sulphide via desulphurication by anaerobic bacteria of the *Desulfovibrio* group (e.g. by *Sporovibrio* species). Sulphur is fed back into the cycle as a result of protein degradation by bacteria and fungi. Aerobically, the decomposition of organic waste is accomplished by bacteria and fungi such as *Aspergillus* or *Neurospora*; anaerobically, the organic sulphur is reduced by bacteria of the *Escherichia* and *Proteus* genera. In the presence of heavy metal ions, sulphide is bound again at first as a slightly soluble heavy metal sulphide.

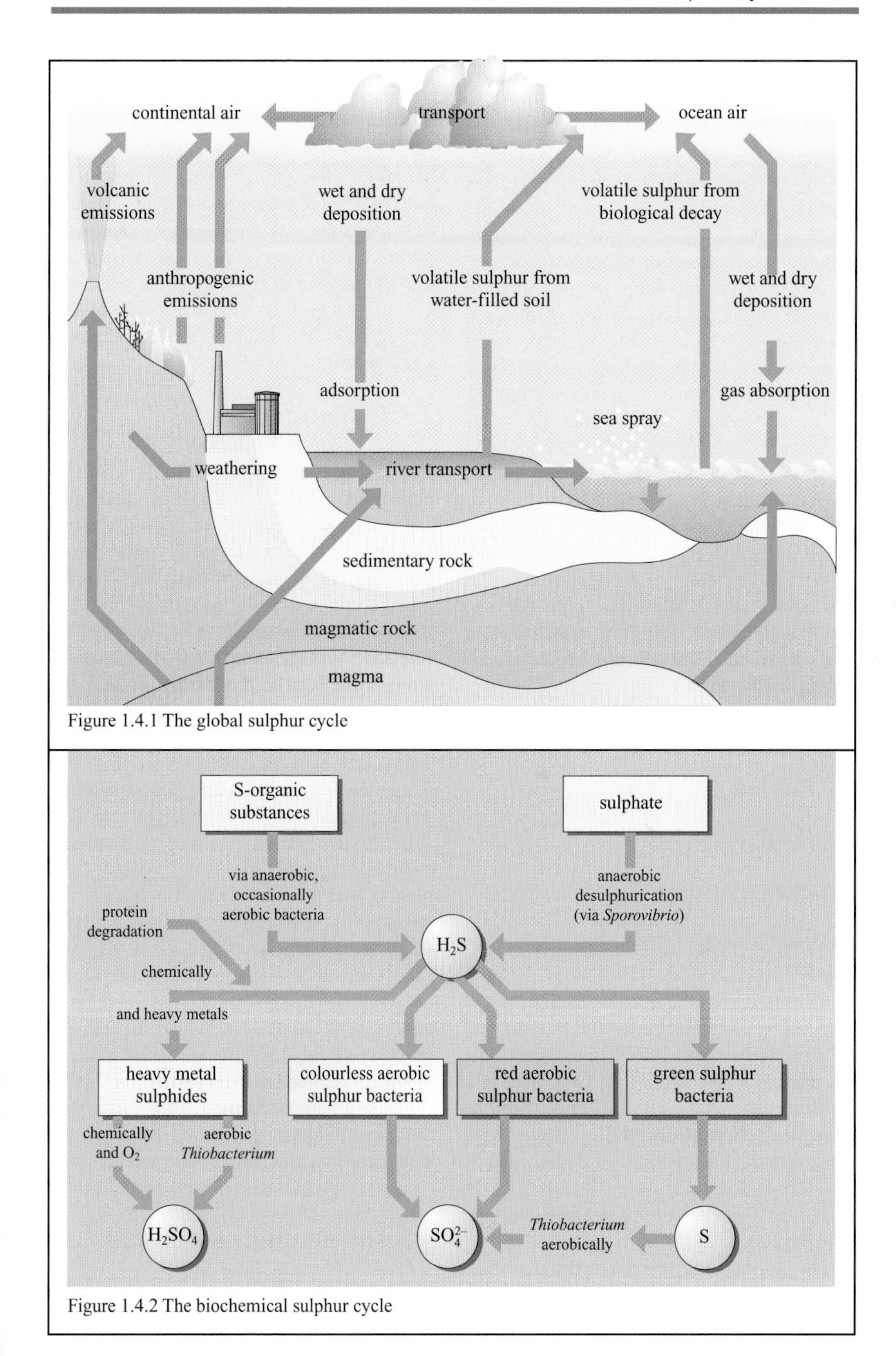

Figure 1.4.1 The global sulphur cycle

Figure 1.4.2 The biochemical sulphur cycle

1.4.3 Environmentally relevant sulphur compounds

Sulphur species in a reduced state are hydrogen sulphide (H_2S), dimethylsulphide ($(CH_3)_2S$) (and methylmercaptan, CH_3SH), carbon disulphide (CS_2) and carbonyl sulphide (COS). They come mostly from natural (geochemical or biological) sources. Sources for the relatively long-lived COS include volcanic activity and biological processes, in which oxidation of the carbon disulphide (CS_2) takes place, e.g. in forest fires. For the sulphur compounds mentioned, the primary sources are biological processes in anaerobic conditions, such as are found in swamps and in tidal lands. For example, dimethyl sulphide is made by algae.

In the atmosphere, sulphur species with an oxidation number of –2 are attacked largely by OH radicals and by oxygen atoms, and are converted via intermediate products to sulphur dioxide. Here are examples for possible reaction mechanisms for carbonoxisulphide (* radicals):

1. $COS + OH^* \rightarrow CO_2 + SH^*$

2. $COS + O \rightarrow CO + SO$

For dimethyl sulphide:

1. $(CH_3)_2S + OH^* \rightarrow CH_3OH + CH_3S^*$

2. $CH_3S^* + 3.5O_2 \rightarrow 2CH_2O + 2SO_2 + H_2O$

Carbon disulphide is oxidised first to CS and SO_2 at a wavelength below 280 nm, and in a second step to COS and atomic O. At wavelengths above 300 nm, first there is a splitting into CO and S, and then an oxidation to form CO and SO_2. The average atmospheric lifetime is 2–4 days for H_2S, 0.1–0.3 days for CH_3SH, 0.8–1.2 days for $(CH_3)_2S$, 10–40 days for CS_2 and more than 100 days for COS.

Due to the long residence times of CS_2 and COS, these sulphur species can diffuse into the stratosphere as source gases and can contribute to the formation of the sulphate layer there via oxidative degradation. The oxidation of sulphur dioxide can occur via oxidation of photoexcited singlet or triplet SO_2 molecules, as a result of oxidation by hydroxy or hydroxyperoxy radicals, or by nitrogen oxides or ozone. SO_2 or sulphuric acid arrives back on the Earth's surface as a wet deposit (dotted line in Figure 1.4.3) or as a dry deposit.

1.4.4 Emissions and transformations

Depending on the conditions, volcanic emissions produce SO_2, H_2S, and also elemental sulfur from the reaction of the two sulphur compounds. All combustion processes of sulphur-bearing fuels create other sources of SO_2 emissions. Sulphur dioxide or sulphurous acid, as well as sulphuric acid (Section 1.4.3), fall from the atmosphere back to the Earth's surface and serve as a contributing factor to deforestation. As a result of anaerobic processes, substances such as hydrogen sulphide are generated from organically bound sulphur. In the presence of sufficient oxygen, sulphates are formed from mineral sulphides such as sulphidic ores, lustre (*'glanz'*) and blende due to weathering processes and microbial processes (Section 1.4.2). These sulphates are largely washed out of soils and finally end up in the oceans. In watt, where reducing conditions predominate (the black colour of the watt comes from iron disulphide, or pyrite), the sulphate can be reduced again to sulphide. Under oxidising conditions, sulphate is formed again; likewise, inorganic sulphur species can be generated anew from organic sulphur compounds as a result of mineralisation. These partial cycles take place largely in soils and in the sediments of streams, rivers and lakes.

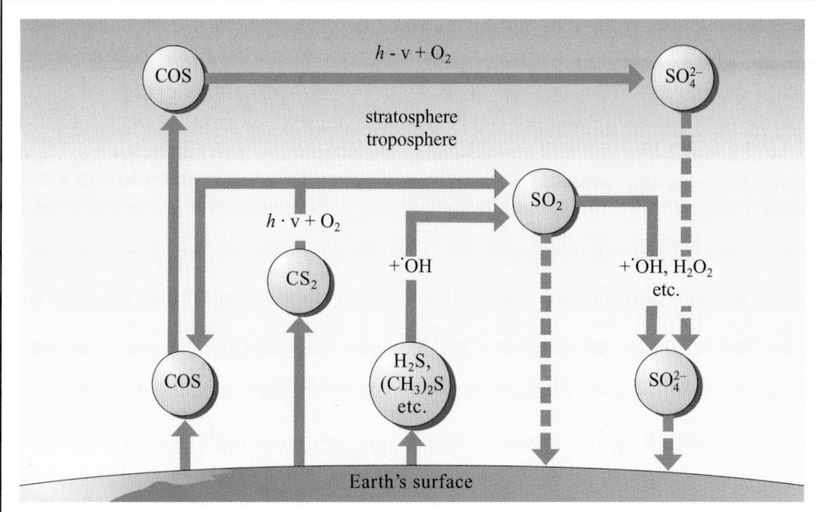

Figure 1.4.3 Environmentally relevant sulphur compounds

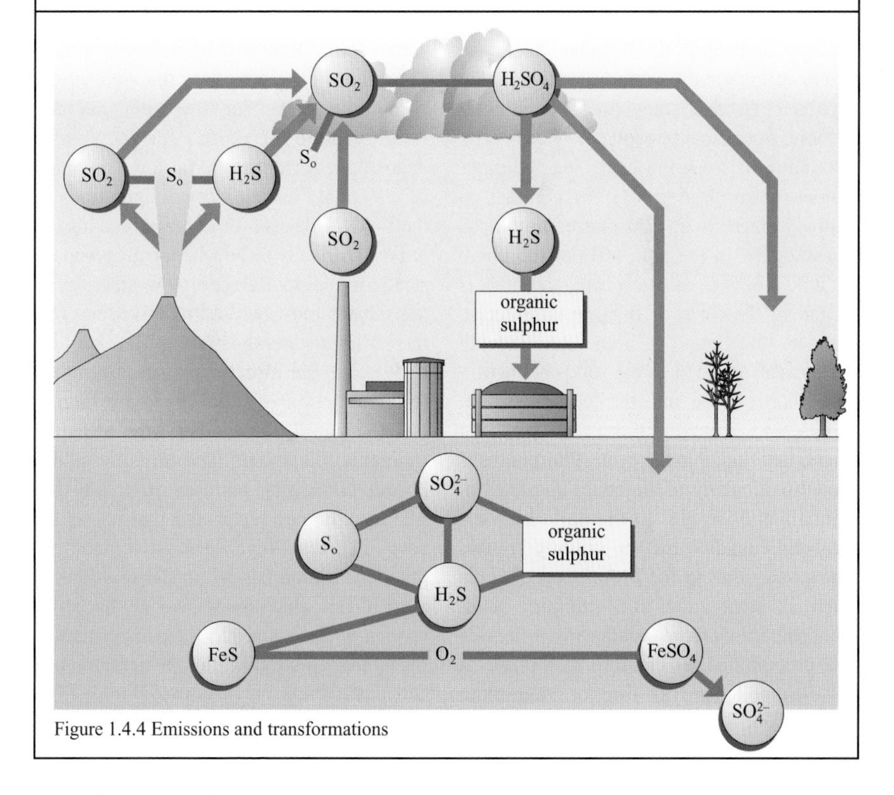

Figure 1.4.4 Emissions and transformations

1.5 The phosphorus cycle

1.5.1 The global phosphorus cycle

With respect to its abundance, phosphorus takes 11th place in the lithosphere (0.7% of the Earth's crust). It occurs with an oxidation number of +5 in phosphate rocks, such as the mineral apatite, $Ca_5[X(PO_4)_3]$, with X = F, Cl, OH. Phosphorus is the most important growth-limiting element in an ecosystem. Phosphate ores in the Earth's surface (as apatite, 32×10^9 t) are mostly of sedimentary, or more rarely volcanic, origin (P in the deep-sea sediments with 10^{12} t, in the hydrosphere with approximately 10^{11} t, and in the land biomass with 2×10^9 t). Phosphate-rich guano deposits can be found on Pacific islands. Guano consists of the excrement of cormorants and other seabirds, and contains calcium phosphate and other nitrogen-organic compounds. It also collects along the coasts of Peru and Chile due to its use as a fertiliser. Because they are only slightly soluble, numerous phosphates (Ca, Fe, Al) are removed from the cycle via sedimentation (13 million t yr^{-1}). In contrast to nitrogen, there is no phosphorus in the gas phase cycle. In the soil, 60% of the phosphorus is present as phosphate, and 40% is stored in the form of organic substances, such as that bound to humus. As soluble phosphate (only 5% of the soil phosphates), phosphorus enters into the food chain – in plants and animals to humans and via excrement back into the cycle. Phosphorus is stored particularly in the bones and teeth of animals and people. In the soil, microorganisms can dissolve the slightly soluble phosphates due to the production of acids such as citric acid and sulphuric acid. Phosphate ions from weathering processes are precipitated as calcium phosphate in alkaline soils and as iron or aluminium phosphate in acidic soils.

1.5.2 The biogeochemical phosphorus cycle

Phosphate-containing rocks reach the surface via diagenesis (Section 1.1), and they can be taken up by plants to a small degree (Section 1.5.1). Phosphorus is required by living organisms as a building block of nucleic acids and of phospholipids (in membranes), and it plays a significant role in energy metabolism as AMP, ADP and ATP (adenosine mono-, di- and triphosphate). The endogenous cycle of phosphorus in phytoplankton occurs very rapidly: phosphates are taken up in 5 minutes and excreted again after 3 days, or they are passed on to the zooplankton, which excretes as much phosphate daily as it takes up. The human organism (70 kg) contains some 700 g P, 600 g of it in the bones. Phosphates also enter into the atmosphere with dust particles due to erosion, and they return back to the Earth's surface via dry deposition. Farmers take up large amounts of phosphate during harvest operations in the fields. However, this phosphorus loss or removal can only be geochemically compensated for to a small degree by weathering, so phosphates must be used in fertilisers. The terrestrial (mineralisation of 0.2 million t yr^{-1}) and the aquatic phosphorus cycle (mineralization of 0.06 t yr^{-1}) are largely independent of one another: both insoluble inorganic phosphate and organic phosphorus-containing residues enter into the sediment. In the aquatic area, phosphates are part of the diet of phytoplankton. Overfertilisation has led to the well-known algal bloom which assimilates the dissolved oxygen with a subsequent decomposition of the biomass (eutrophication). Concentrations of 10 to 100 mg m^{-3} in rainwater come from dust and sea spray.

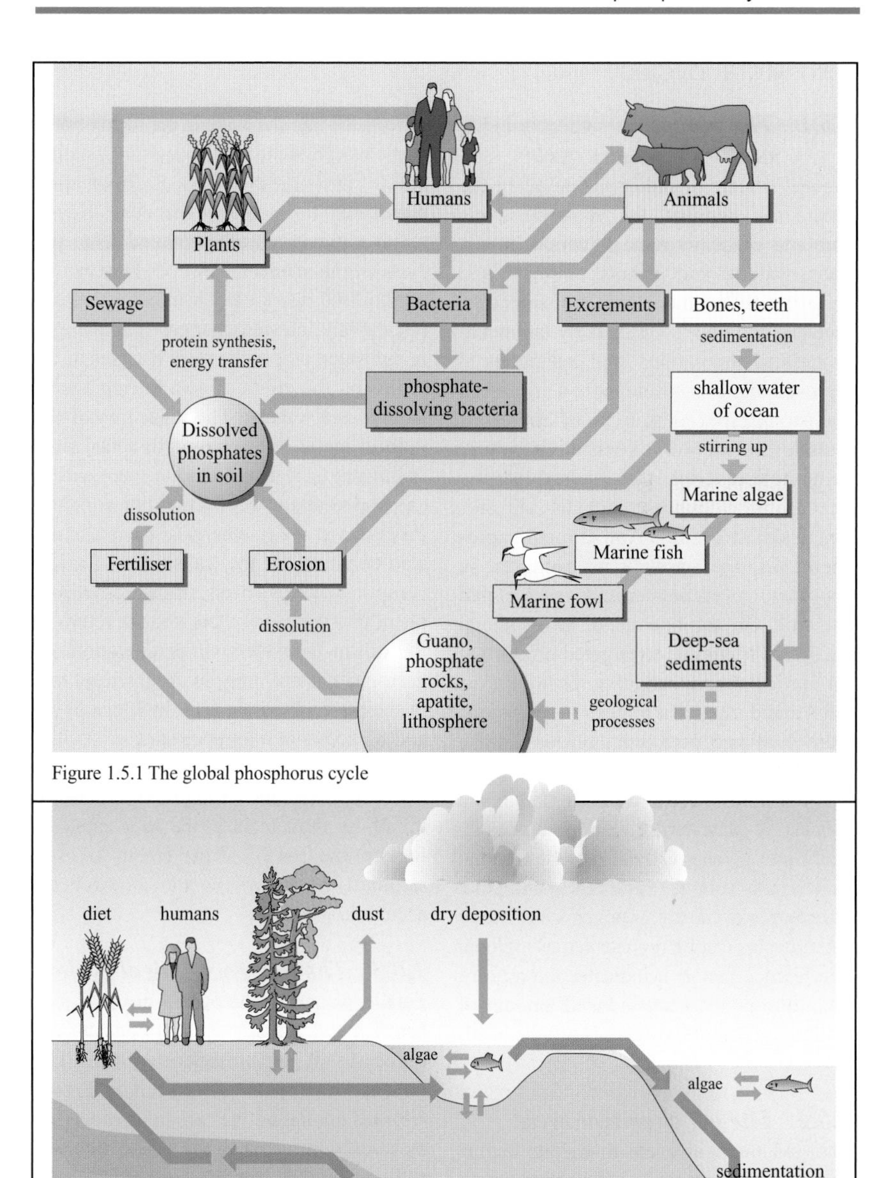

Figure 1.5.1 The global phosphorus cycle

Figure 1.5.2 The biogeochemical phosphorus cycle

1.6 Metal cycles

1.6.1 The global anthropogenous cycle

Since the start of industrialisation in the nineteenth century, the material cycle amounts of heavy metals, which impact upon waters, soil, plants, animals and people, have increased considerably. Industry and trade – particularly the metal-processing operations and refineries – represent significant sources of heavy metal emissions. Special sources of emissions include cement works (thallium), battery manufacturers (lead) and electroplating operations (copper, nickel, chrome, zinc/cadmium, etc.). And through agro-chemicals, metals such as cadmium in phosphate-containing fertilisers can get into the soil. On the one hand, heavy metal emissions occur due to targeted production for particular commercial purposes; on the other hand, they are undesired emissions from coal and petroleum burning, trash burning, and cement and glass production. Through refuse, waste water or sludge and exhaust air (dust particles), they enter especially into or on soils and thereby into the food chain (plants→animals→humans). The degree of the anthropogenous effect of metal cycles can be represented as a global interference factor: it indicates the ratio of the anthropogenously induced amount of material to that of the natural (geochemical) material cycle.

1.6.2 The geochemical cycle

The geochemical cycle of metals begins with the plutonic rock, which then enters into the water and the atmosphere via the Earth's surface or the ocean floor due to volcanic activity. From the weathering of stones, metals are dissolved (chemical weathering) into the water cycle, or they are introduced into the cycle in the form of dust particles (physical weathering as clastation due to mechanical forces). Sedimentation represents the opposite process. Heavy metal compounds are removed from the cycle in this manner; however, they can be remobilised due to geochemical or anthropogenously induced changes (pH changes in the water, biogeochemical changes in the sediment, the effect of complexing agents from waste waters). The logarithm of the ratio of metal concentration in actual river sediments to that in prebiotic sediments is calculated as the geoaccumulation index. Processes in the geochemical cycle that are also common are the transitions of metal compounds in aerosols from the hydrosphere into the atmosphere and the return to the soil or the waters via precipitation. As gases, metals play only an insignificant role in the materials cycle (e.g. hydrides of As and Se or Hg in a gaseous state or volatile metallo-organic compounds such as methyl mercury). The ratio of the relative concentration of an element in the atmosphere to that in the Earth's crust (with Al as a standard) is known as the atmospheric accumulation factor.

1.6.3 The biogeochemical cycle

Finally, as part of the overall discussion of heavy metal cycles, we must consider the processes in the biosphere as well. The geochemical cycle (to the left in Figure 1.6.3) is connected via metabolic processes by microorganisms in the ocean floor, in sediments or sludge, or in water. Microorganisms, plants and small animals serve as food sources for fish. By means of plants and other animals, levels of heavy metals from waters and soils find their ways into our foods.

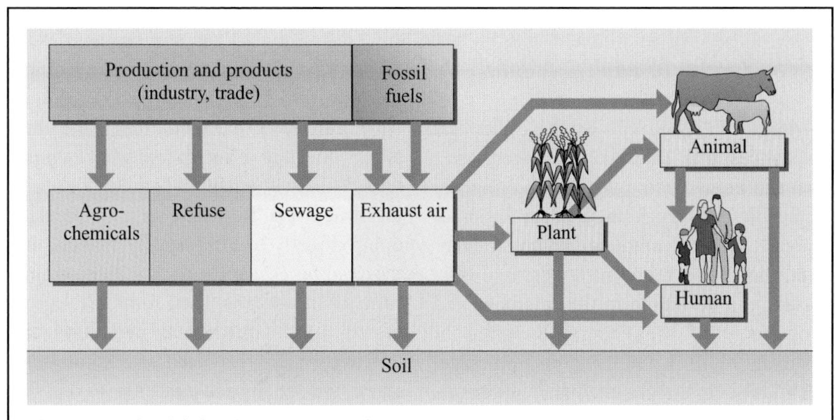

Figure 1.6.1 The global anthropogenous cycle

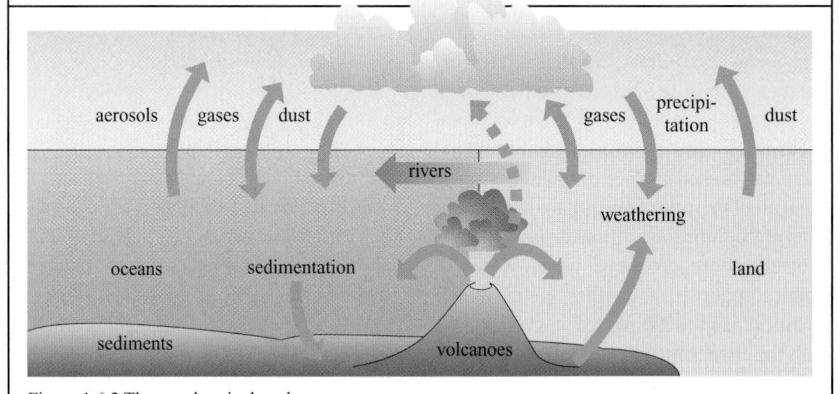

Figure 1.6.2 The geochemical cycle

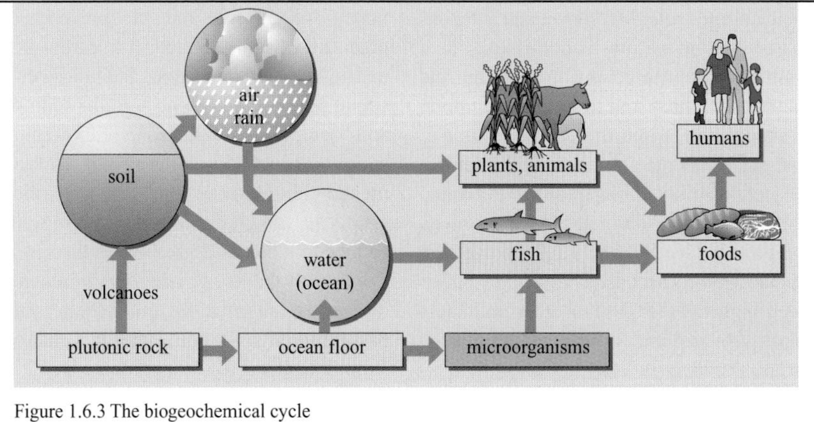

Figure 1.6.3 The biogeochemical cycle

1.7 Special cycles

1.7.1 Cycles of environmental chemicals

Environmental chemicals are those chemical substances that enter the ecosphere as a result of human activity (anthropogenously induced). As a rule, they occur in small amounts or concentrations, but on the other hand they often represent a potential risk. Based on a definition in the environmental programme of the Federal Republic of Germany of 1971, the term 'environmental chemicals' includes chemical elements, compounds of organic and inorganic nature, of synthetic or natural origin. Limiting the term to substances that can endanger living organisms is arbitrary, so that today, any substance introduced into the environment by people is labelled an environmental chemical. The term 'environmental chemical' is used in place of 'pollutant', even if not all of these substances have to be pollutants at the same time. Approximately 60 000 substances are produced world–wide. To protect people and the environment, chemical laws were passed in conjunction with the environmental laws.

Environmental chemicals in the water (Figure 1.7.1, 1) can be subjected to chemical and biological conversion reactions and/or decomposition reactions (e.g. as a result of hydrolysis) following vortexing and convection, which lead to a state of solution or suspension. From the decay of organic material, dead organisms provide the detritus as finely distributed suspended or deposited matter, to which water-insoluble heavy metal compounds (see Section 1.6.2), polycyclic aromatic hydrocarbons (PAHs), polychlorinated biphenyls (PCBs) or other pollutants can be attached and settle. These deposited particles of biogenic origin (→ organic sedimentation) and suspended clay minerals (→ inorganic sedimentation) make possible the processes of adsorption and desorption, which can lead to an equilibrium in the water under constant chemical–physical conditions. The environmental chemicals that enter the waters can be taken up by organisms either directly or after transformation, and they can be excreted either in their original form or in a metabolised form. Biological effects of environmental chemicals can appear immediately or later, after bioaccumulation. We use 'bioaccumulation' to describe the ability of organisms to store substances in their own organism or to concentrate, or enrich, them in their environment (calculated as the bioaccumulation factor).

Environmental chemicals can get on and in soils (Figure 1.7.1, 2; see also Chapter 3) by means of either wet or dry deposition (sedimentation). Transformations on the surface occur due to phenomena such as photoreactions, which can also lead to the volatilisation of directly applied materials. The processes in the soil passage, which can lead to a separation of environmental chemicals after a time, play an important role in the soil itself. On account of the water content, the entrance of oxygen into the upper soil layers, and the presence of organisms, from soil bacteria to earthworms and moles, similar reactions and processes as occur in water can occur here. In addition, soil space filled with roots represents a special absorption pathway for environmental chemicals. Around tree roots, these chemicals can be decomposed and/or transformed to a higher degree due to the special flora and fauna. Finally, environmental chemicals can enter the groundwater after passing through the soil, either in a changed form or unchanged.

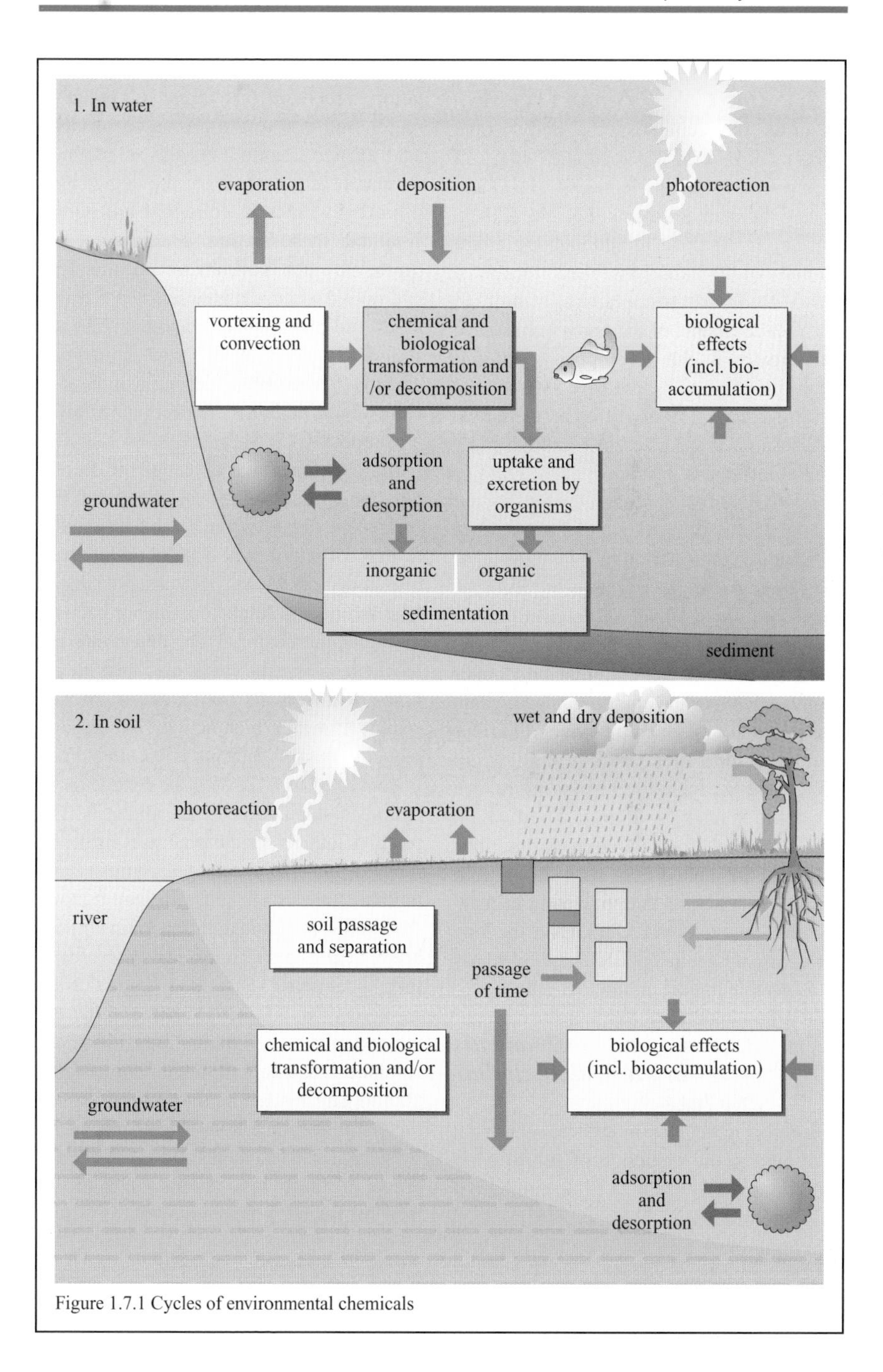

Figure 1.7.1 Cycles of environmental chemicals

1.7.2 *Interrelationships among the C, S, P, N and O cycles*

In order to decompose the body's own substances in large amounts, the living cell requires the principal elements C, H, O and N, and the macroelements P, S, Na, K, Mg, Ca, Fe, Si, Al and Cl. Although the entire biomass of the Earth has a mass fraction of only 0.1% in the crust, practically all of the chemical elements of the Earth's crust have participated in the decomposition during evolution. There are intensive interactions among the cycles of different elements, which can occur due to chemical and especially biochemical reactions.

Photosynthesis is characterised by the uptake of carbon dioxide and the release of oxidation via respiration in a living cell. At the same time as the CO_2 is being incorporated into organic material (shown here in general as CH_2O for carbohydrates), N, S and P are taken up as well for the building of amino acids, nucleic acids, etc. The organic material and oxygen are required in turn for the independent cycles of S, N and C, from which the participation of phosphorus (in the form of ADP) is derived. The C, N, P and S cycles that were previously described in detail are linked here with the oxygen cycle (P stands for the enzymatic energy-yielding processes via ADP). Over the long term, a change in one of the cycles has considerable impact on the four others.

1.7.3 *Bacterial and biochemical cycles in the sedimentation of a lake*

Anaerobic processes in the sediment of a lake lead to the reduction of sulphates to sulphide (Section 1.4.2). The number of sulphate-reducing bacteria reaches a maximum here depending on the depth of the layer of sediment. Biochemical (fermentation) processes take place in the upper region and lead to the formation of lactic acid and acetic acid. Both acids are products of anaerobic glycolysis, or the degradation of carbohydrates in cells that are poorly supplied with oxygen. Macromolecules (proteins, fats, carbohydrates) are first hydrolysed to form monomers in a fermentation and putrefaction process, or they are fermented to form different acids and ethanol (fermentation: decomposition of C compounds; putrefaction: decomposition of N compounds).

In light of the redox potential, the decomposition processes of biomass can be divided into a range from aerobic respiration to denitrification and desulphurication to methanogenesis. It has been observed that in lake sediments, sulphate-reducing bacteria in the uppermost 2-3 cm determine the biochemical events. It has also been determined that methane from a deeper layer in sediments with a high activity of sulphate-reducing microorganisms is consumed and is converted to CO_2 with the formation of hydrogen sulphide from the sulphates. The redox potentials for the processes of desulphurication (sulphate respiration) and methanogenesis are close to one another. Anaerobic methanogenic bacteria require acetic acid as a substrate. The processes in lake sediments described above were observed in Lake Vechten in the Netherlands. The spatial separation of sulphate-reducing and methane-producing microorganisms can be traced back to such factors as the inhibition of methanogenesis by sulphates in sediments. However, coexistence of these two groups of microorganisms has been proven as well.

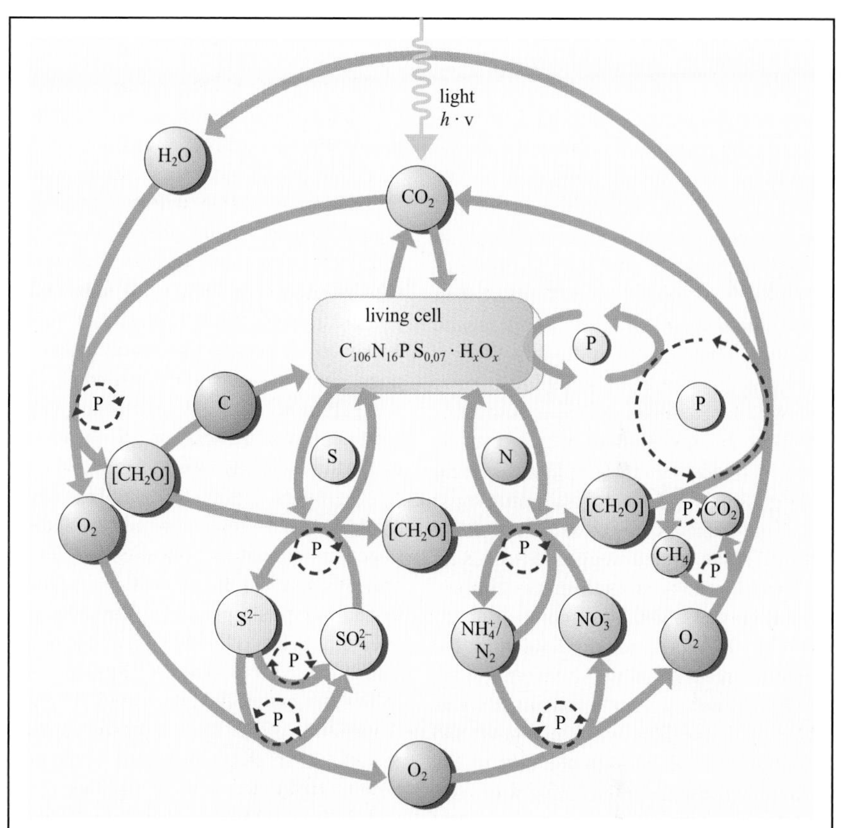

Figure 1.7.2 Interrelationships among the C, S, P, N and O cycles

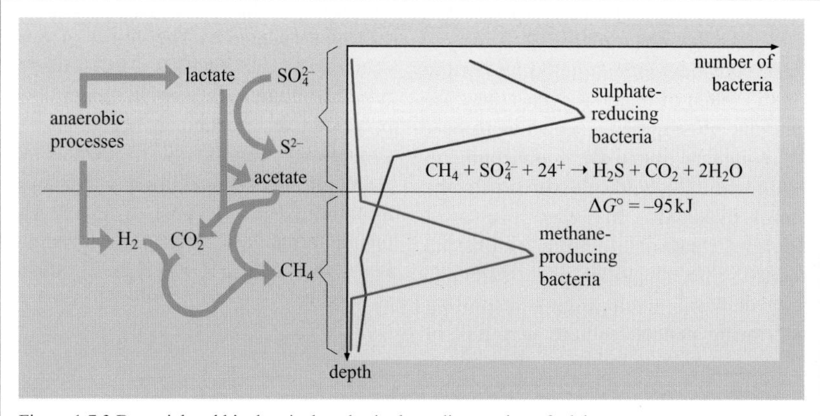

Figure 1.7.3 Bacterial and biochemical cycles in the sedimentation of a lake

1.7.4 Anthropogenously related surface material flow

Figure 1.7.4 shows anthropogenously induced flows of material (with units such as t yr^{-1}), in particular of solid materials that eventually end up on soils. Around 1990, 90% of the population in the old German *Bundesländer* were connected to sewage treatment plants. With a volume of some 8000 million m^3 per year (more than 3000 million m^3 of that as domestic sewage), about 10% of this is sludge, having a dry weight of 4 million tonnes. Sewage sludge has a high water content of 94–97% in raw sludge, which can be reduced to 55–70% by simple mechanical dehydration. Sludge from municipal treatment plants is viewed as part of urban refuse. Of the treated sludge, 45% ends up in landfills, 25% is used in agriculture, 15% is burned, and 0.5% is composted.

Of the 8-fold larger material flow for domestic refuse (about 32 million t), in 1987 some 66% went to a landfill, 2% was composted, and 26% was incinerated and thereby returned back to the soil in the form of exhaust gases (including water and carbon dioxide) and ash.

1.7.5 Material cycles with links to the environment

If one considers human activities, in the overall system of material cycles there are numerous possibilities – shown as paths – for transfer into the environment. Environmental chemicals can enter the natural cycles in this way. The most important pathways from industrial production are exhaust air, waste water and waste disposal. Even when transporting and storing materials there is a risk of transfer into the environment, especially if safety devices fail. Leachate collects in landfills; its path into the environment (especially into groundwater) is determined by the geological nature of the subsoil. Landfill gases are released owing to biological decomposition. Even the regulated use of plant protection agents, fertilisers and other everyday chemicals, which contain readily volatile solvents in particular, leads to material impact on the ecosphere. A portion of residual materials goes into the different environmental compartments following waste water treatment and incineration of waste, and even after the cleaning of exhaust air.

The overall goal of environmental protection technology is to keep the anthropogenously related cycles contained as much as possible, by preventing, reducing or reusing waste products. This requires special cycles of recycling and waste utilisation. Of domestic refuse, the following materials are recycled: glass (some 10% of the total domestic refuse), metals (3%) and paper and cardboard (about 16%). Recyclable plastics (5%) and the vegetable part (biological waste) of about 30% out of a refuse volume of 300–400 kg yr^{-1} per resident are collected separately. The object of waste containment, i.e. landfills in particular for residual materials which cannot be recycled for technical, economic or ecological reasons, is to reduce the pollutant pathways mentioned above to a minimum. It is the goal of all environmental protection methods especially to keep the industrial cycles as closed as possible.

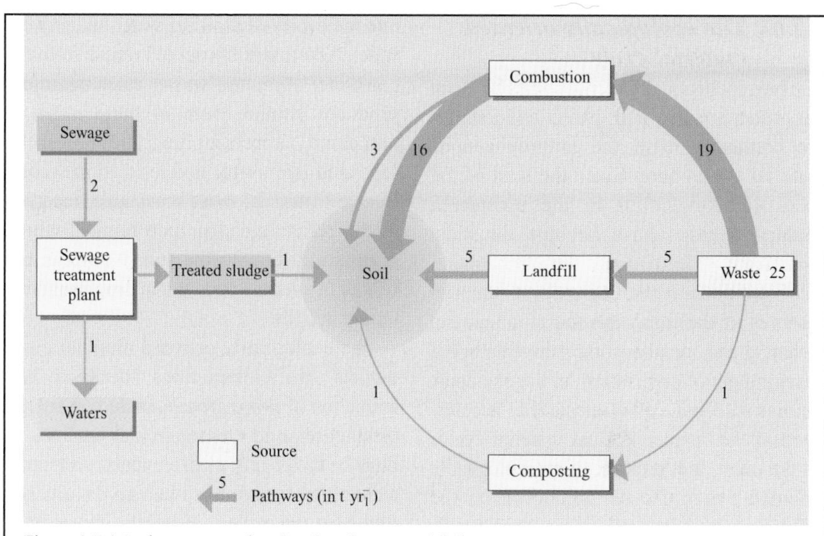

Figure 1.7.4 Anthropogenously related surface material flow

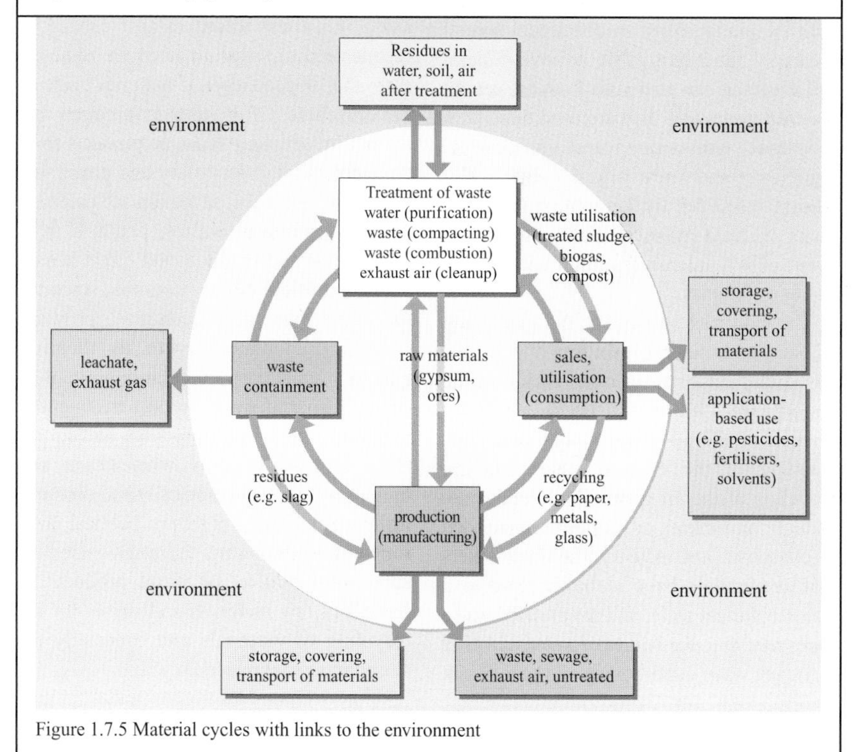

Figure 1.7.5 Material cycles with links to the environment

1.7.6 *The ecologically oriented material cycle*

The biogeochemical material cycles between atmosphere, water, soil, plants and animals are connected with the anthropogenous material cycles here. Since the start of the Industrial Revolution in the nineteenth century, humans have become the most significant factor in the ecosphere. Intervention into the energy and material reserves of the Earth has led to scientific, technical and social accomplishments in the development of civilisation on the one hand, but has also induced a substantial acceleration of the biogeochemical material cycles on the other. It is the task of an ecologically oriented material cycle to coordinate the effects of human activity on the system performance of the biosphere and ecosphere with its relationships, feedback and special means of functioning. We remove mineral and vegetable raw materials from the natural material cycle, and they are used directly in industrial production and processing. Agriculture and forestry avail themselves of natural resources in the area of flora and fauna, wherein animal husbandry represents a particularly intensive portion with a clear ecological impact.

The industrial production and processing of plant and mineral resources and animal husbandry give rise to intervention in the atmosphere. In turn, the ecologically oriented material cycle necessitates the cleaning of the exhaust gases and the recycling of the air if possible. Water pure enough for drinking or service use is needed by consumers and industry. The waste waters that accumulate have to be directed to a water treatment plant, and the purified waste water returns to the bodies of water, and after sufficient water conditioning can be fed back into the cycles of drinking water and process water. Add to this household refuse from the consumer area, and waste water treatment produces sludge. Both of these must be processed as refuse, i.e., they must be separated into usable and residual materials. Usable materials, those recyclable materials in the refuse area, include humus, which returns nutrients to the soil. Non-reusable residual materials from waste processing end up in landfills.

The ecologically oriented material cycle depicts the interactions between the extraction of raw materials, including energy production, and environmental changes. It must be the goal of environmental protection technology to minimise human disturbance of the natural cycles – using the approach of avoid, reduce and utilise. Some 10^5 million $t\ yr^{-1}$ of natural resources that cannot be regenerated in the short term are removed from the biogeochemical material cycle of the ecosphere. Of the total by-products that are not immediately reusable products from the technical transformation and processing of natural resources, at the top are pollution gases, agricultural wastes, residues from incineration and smelting, and faecal matter.

As a subfield of the economic sciences, the environmental economy provides ecological viewpoints with its theories, analyses and cost calculations. It is a strategic goal of waste product treatment to avoid or reuse waste products and to put them in landfills only when those two options are not technologically and economically viable. The progressive political environmental economy already uses a substantial portion of waste products as secondary raw materials and strives for the recycling of materials and especially for closed industrial material cycles.

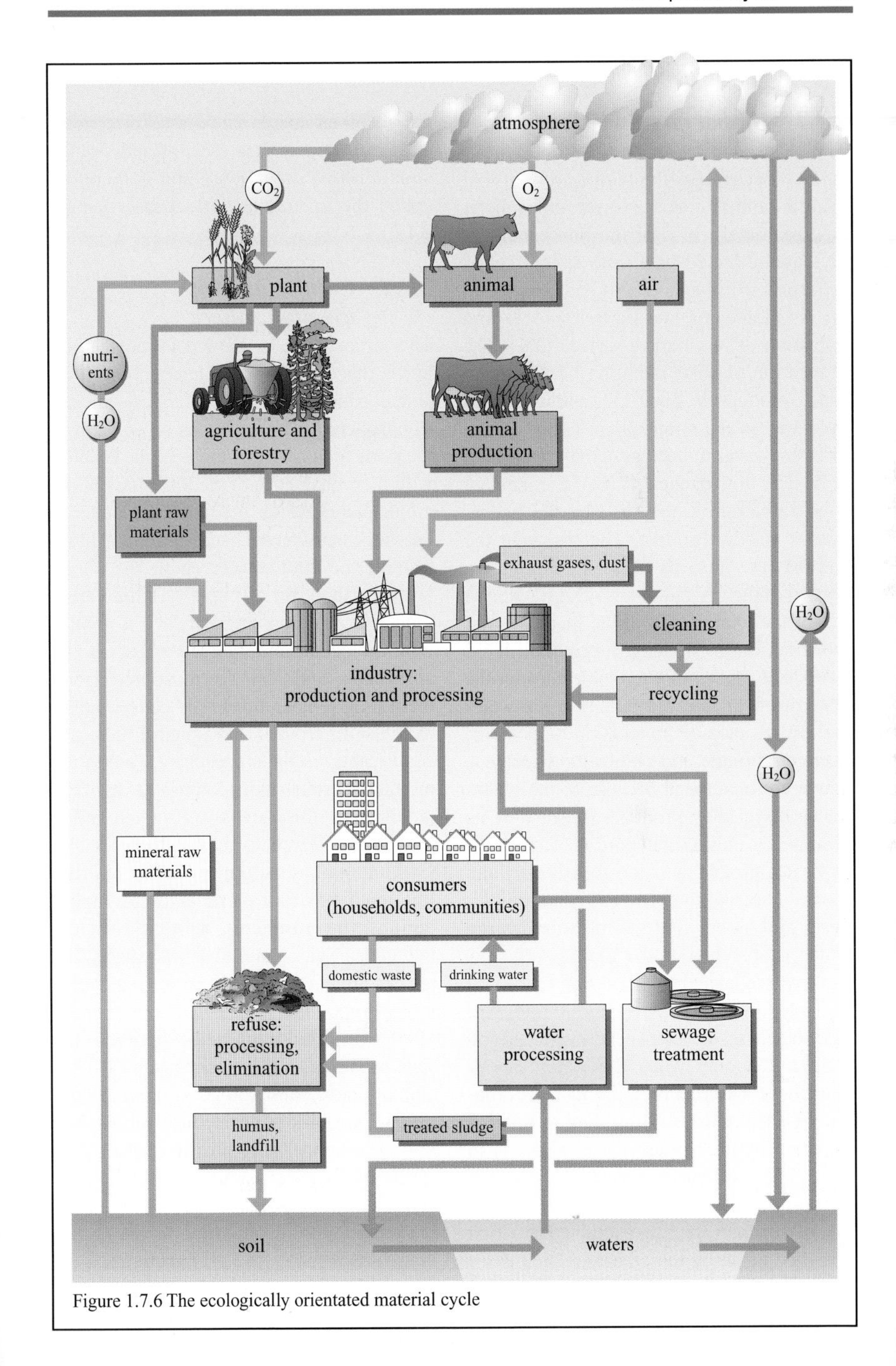

Figure 1.7.6 The ecologically orientated material cycle

2 The Atmosphere

2.1 Physiocochemical processes

2.1.1 Global energy balance

The solar radiation that reaches the Earth's surface determines the temperature at that location and that of the lower atmosphere. Approximately 30% of the Sun's incident rays are reflected back into space due to reflection off clouds (19%), backscattering on air molecules and particles (6%) and reflection off the Earth's surface (3%). The absorption of solar radiation by the atmosphere amounts to some 25% and is induced by ozone in the stratosphere (3%), clouds (5%) and water molecules in the troposphere (17%). In this manner, 47% of the incident solar radiation is absorbed by the hydrosphere and lithosphere. The ozone in the stratosphere absorbs only the short-wave portion of the UV radiation with a fraction of 3% of the total energy of the incident solar radiation, but in so doing it protects life on Earth. The most important molecules for the absorption of infrared radiation are water and carbon dioxide, in addition to methane, nitrogen dioxide and chlorofluorocarbons, which are mentioned because of the greenhouse effect they produce. Because of its absorbency, only a small portion of about 5% of the net infrared radiation goes directly into space. The major portion is absorbed by gases and clouds. Part of the solar energy which is absorbed on the Earth's surface is converted to latent heat (as a stored form). In the transition of water from liquid into gaseous phase, the latent heat is taken up and is liberated only when the water vapour condenses. Convection currents and turbulence (10%), plus the absorption of infrared radiation by greenhouse gases, also add to the release of surface heat. The largest portion of the net infrared radiation is radiated back by the Earth, although it is first absorbed again by the climate-related gases and is stored as thermal energy. They radiate the absorbed energy back in all directions. This build-up of heat corresponds to the situation in a greenhouse and is therefore called the greenhouse effect (see Section 2.1.6).

2.1.2 Composition of the atmosphere

The atmosphere is divided vertically into levels that differ in their temperature level and particle concentration. These are the troposphere (average expansion is 8 km in the polar region and 18 km near the equator, with more than 80% of the total mass of the atmosphere) which mediates the material exchange between the hydrosphere and lithosphere; the stratosphere (with the ozone layer); the mesosphere (temperature reduction from approximately 0 to –90 °C with molecular and atomic oxygen as UV absorber); and the thermosphere (with increasing temperature). The concentration of ionised particles (ionosphere) increases significantly in the stratosphere. Due to the turbulent currents in the lower 100 km of the atmosphere, this part is well mixed and is therefore referred to as the homosphere. Gravitational attraction produces a partial fractionation of air particles in the higher regions (heterosphere), so that only the lightest particles can diffuse into space. The upper limit of the atmosphere as a region that impacts upon human activities lies at about 1000 km. The boundaries between the spheres are called the tropopause (8–18 km altitude), stratopause (50–55 km), mesopause (80–85 km) and thermopause (about 500 km). The temperature profile of the atmosphere can be traced back to the emission and absorption processes (Section 2.1.1), while the temperature drop is due to the partial liberation of energy by heat radiation and the evaporation of water (cloud formation).

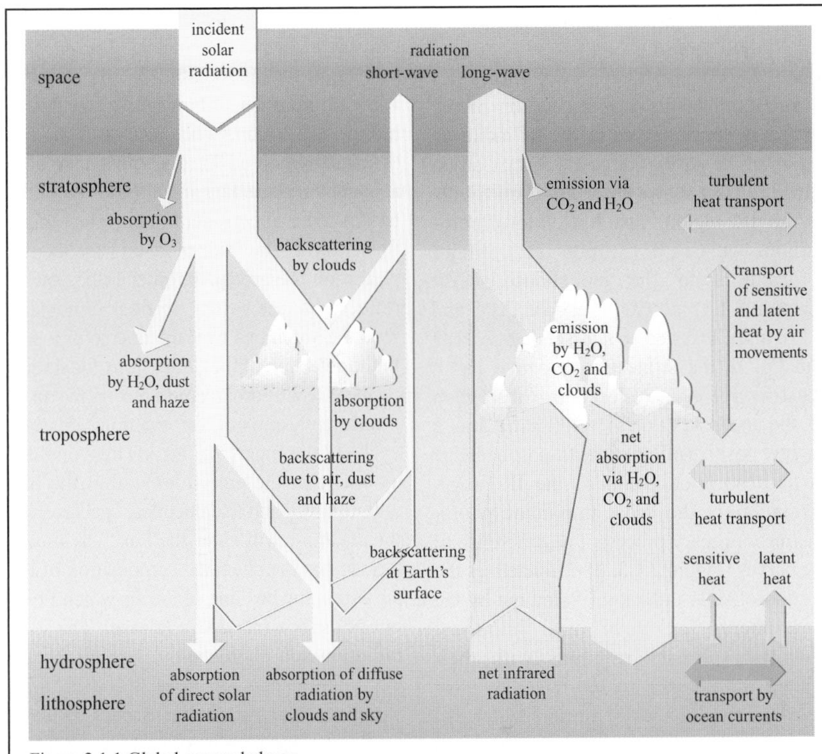

Figure 2.1.1 Global energy balance

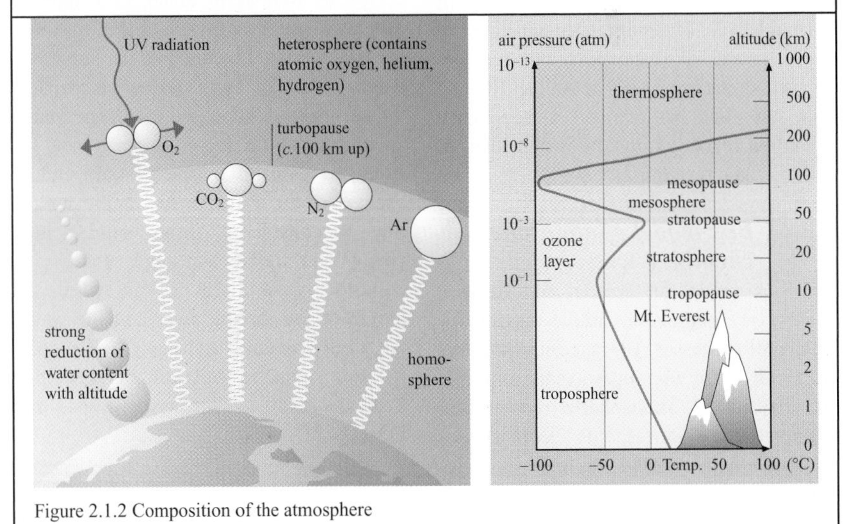

Figure 2.1.2 Composition of the atmosphere

2.1.3 Basic physicochemical processes

Water (H_2O) and ozone (O_3) molecules absorb both in the spectral region of the incoming sunshine and in the IR radiation region of the Earth (Figure 2.1.3, 1). Carbon dioxide (CO_2) shows absorption bands only in the IR region, just like chlorofluorocarbons or CFCs (Section 2.2). Major contributors to the greenhouse effect (Section 2.1.1) are CO_2 (Sections 2.1.5 and 2.1.6) and CFCs and, to a lesser degree, H_2O and O_3. In the stratosphere, ozone has a negative greenhouse effect due to absorption of the sun's UV rays (<300 nm) and a slightly positive effect in the troposphere due to its absorption bands in the IR range. Oxidation is the most important photochemical primary process. The difference of the curves (Figure 2.1.3, 2) characterises the portion of solar radiation filtered out by the atmosphere. The three most important photolytic primary reactions with their critical wavelengths are the relatively slow decay of tropospheric O_3 in molecular and excited oxygen (above 310 nm); the rapid splitting of NO_2 (above 400 nm); and that of formaldehyde. The substances generated in these processes, especially the reactive oxygen atoms or radicals as well as NO and CO, can then be involved in secondary chemical reactions, mostly in the form of radical chain reactions.

2.1.4 Box model of atmospheric chemistry

The box model of Graedel and Crutzen (1994) is used as a simple model for chemical processes. Entrance into the box area can occur via source emissions and as a result of atmospheric movements. Atmospheric movements consist of advection and merging. Advection includes the transport of chemical species by the movement of pockets of air. Merging is a term used for the local inflow via compact vertical movements of pockets of air due to turbulent diffusion. When using box models (see also Chapter 3), one must take into account the placement and dimensions of the box in order to consider the chemical transformations. The box model is frequently placed on the ground in order to be able to reflect changes in the vertical dimension with the variations over the course of a day. In the selection of the dimensions for studies on the effect of urban emissions on the air in a rural environment, for example, this box encloses the municipal area. Thus, one can assume that the emissions within the box will mix well and that chemical species from the clean air will enter the box. The model determines the chemical composition of the air within the box and of the air which flows out of the box (advective release). The solar radiation and photochemical reactions are taken into account (Sections 2.1.1 and 2.1.3). Reactions take place in the gaseous state and on the surface of aerosol particles (Section 2.1.8), as well as in cloud, rain and fog droplets. The average dwell time of the air in the box is derived by dividing the horizontal length of the box by the advection speed. If the air quality is supposed to be determined in the course of a day in an urban area, for example, then all of the time-dependent changes of the characteristics being studied must be considered. These would include changes in traffic and other sources of emission, wind direction and speed, the height of the mixing layer and the wavelength-dependent variations in the solar radiation. Box models can be described mathematically and further developed into 1D up to 3D models.

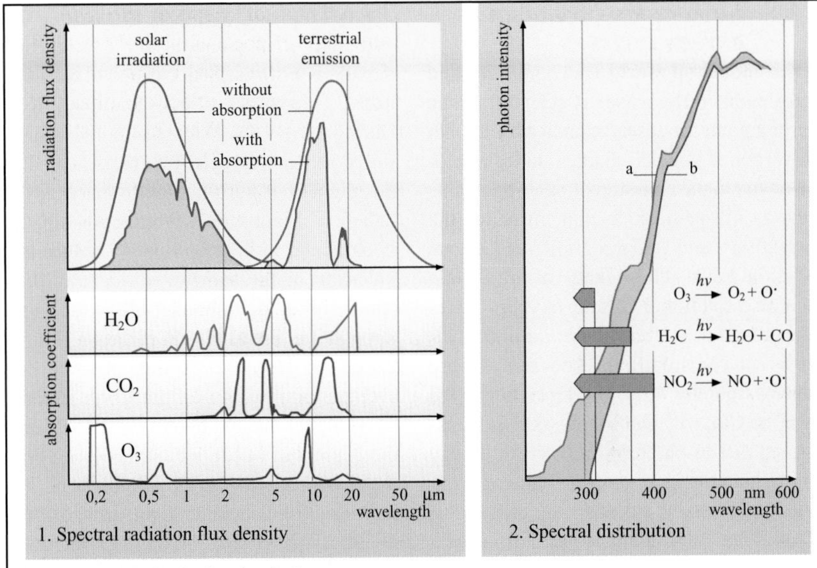

Figure 2.1.3 Basic physicochemical processes

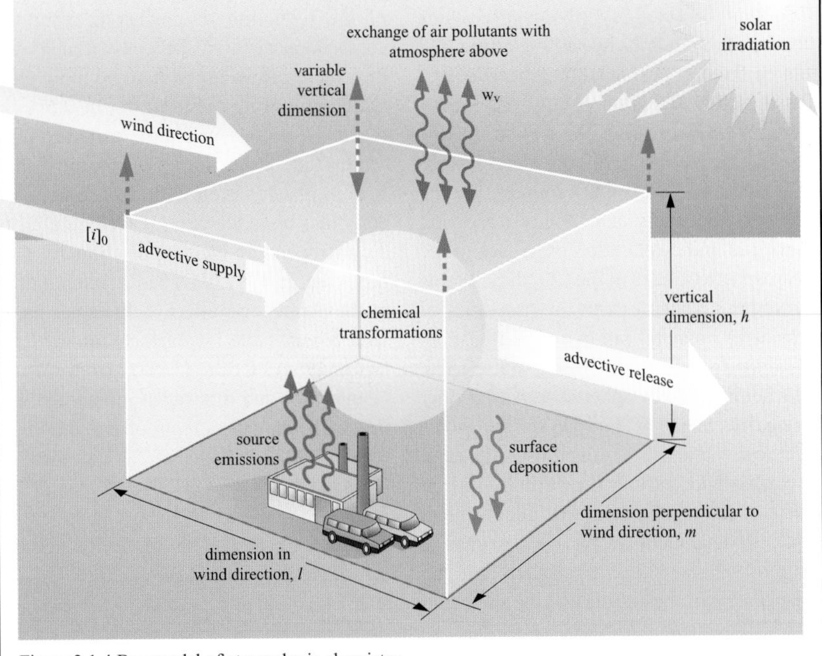

Figure 2.1.4 Box model of atmospheric chemistry

2.1.5 Linking the carbon and oxygen cycles

If the industrial world (burning coal and petroleum, in this case) is selected as the starting point for observation, it adds to both the carbon dioxide discharge into the atmosphere and to the emission of other pollutants such as nitrogen oxide that influence the formation and degradation of ozone (Section 2.2.3). The linkage of the carbon (see also Section 1.2) and oxygen cycles takes place in the biosphere via photosynthesis and respiration. The atmosphere contains roughly 720×10^{12} kg carbon in the form of CO_2, of which one-sixth is transformed due to photosynthesis (with energy uptake and release of oxygen – Section 1.2). One-half of this is stored in the biomass of plants, the other half is used to obtain energy; it is inhaled, whereby CO_2 returns back into the atmosphere. Only 0.1% of the CO_2 that is fixed by photosynthesis is converted into humus by soil organisms and can be fixed in peat, coal or petroleum in a fossil state. Burning energy-rich fossil biomass residues raises the carbon dioxide pool in the atmosphere considerably (see also Section 2.1.6). CO_2 sources for the carbon in the atmosphere include dissolving processes and evaporation processes in the hydrosphere (release of the CO_2 dissolved as hydrogen carbonate) and the processes of respiration, mineralisation and combustion. Photosynthesis is a CO_2 drain. The geochemical and biosphere cycles (the latter being determined by the speed of the photosynthesis and mineralisation of dead biomass) take part in the carbon–oxygen cycle in the atmosphere. The geochemical cycle is dependent on the slow exchange processes between the atmosphere and the deep waters of the oceans and the sediment.

2.1.6 Role of carbon dioxide

Although it is essential for life, carbon dioxide is also the major cause of an additional greenhouse effect (Section 2.1.1). During the day the incident solar radiation is stored as heat, especially in the soil, and in the night it is returned into space as IR radiation. Absorption and reflection of CO_2, in particular in the troposphere, reduce this radiation in the same way that the glass panes in a greenhouse allow sunlight through and retain the IR radiation (Figure 2.1.6, 1). Because of this process, the average temperature on the Earth is some 33 °C higher than with total radiation. Since the beginning of the Industrial Revolution, the CO_2 content in the atmosphere has increased from about 250 ppm around 1760, to 300 around 1920, 330 in 1975 and up to nearly 350 after 1990 with adjustment for seasonal variation (Figure 2.1.6, 2), in spite of the reduction of emissions from the combustion of fossil fuels. Associated with this, there has been an increase in temperature amounting to about 0.3 to 0.6 °C in recent years. In northern latitudes with 'eternal ice' it has increased by more than 5 °C in places. Even with decreasing CO_2 emissions up to the year 2050 (Figure 2.1.6, 3), a temperature increase of up to 7.5 °C is predicted. In 1992, CO_2 had a share of 50% in the greenhouse effect. Other trace gases with considerable involvement are methane (with 19%), CFCs (17%), tropospheric ozone (8%) and dinitrogen monoxide N_2O (6%). The climate activity of the long-lived trace gas N_2O (a factor of 200 compared to CO_2) and that of CFCs (a factor of 3500–17 000) and the short-lived ozone near the Earth (factor of 2000) all add significantly to this situation.

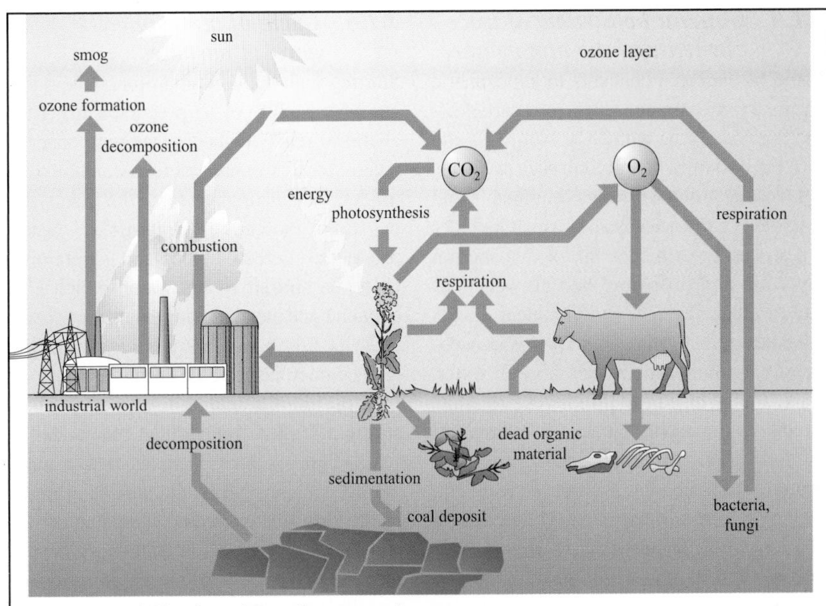

Figure 2.1.5 Linking the carbon and oxygen cycles

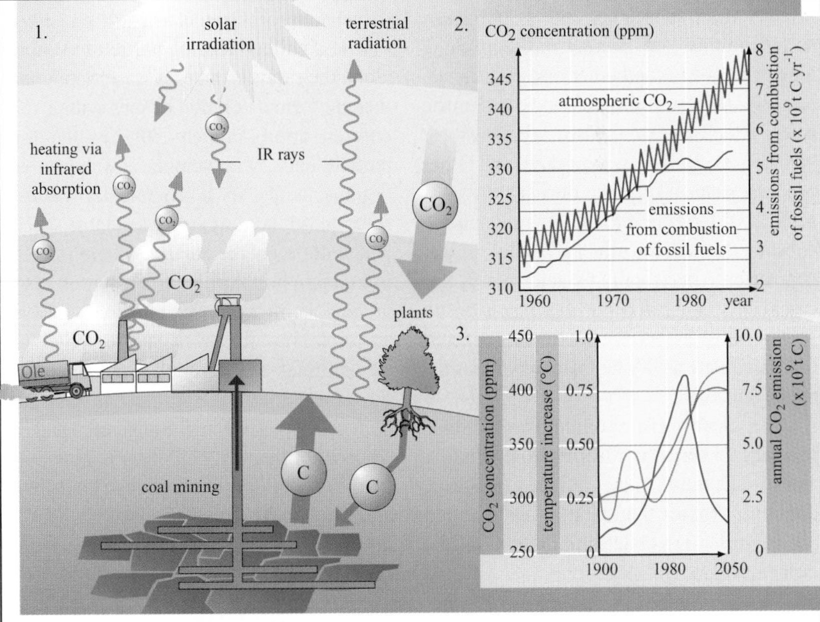

Figure 2.1.6 Role of carbon dioxide

2.1.7 *General behaviour of trace elements in the atmosphere*

Some of the basic concepts in atmospheric chemistry are emission (from dust particles – see Section 2.1.8, with SO_2, NO_2, CO and CO_2 from sources of emission such as plants, volcanoes, combustion engines and chimneys) and transmission, which includes the processes of dispersion or distribution (by wind and diffusion) and also changes due to other physical and chemical effects (chemical transformations). The 'recipient' – usually a compartment such as soil, water, plant, animal or human – receives pollutants directly as atmospheric trace substances and as deposition after a chemical transformation. From the viewpoint of the recipient, they represent an *immission*. Direct effects can be expected particularly with plants, animals and humans. Sulphur dioxide and nitrogen dioxide, for example, are dispatched from a source of emission as primary trace substances, and as pollutants as well in this case. During transport along long routes, these and other materials can be transformed, here into sulphuric and nitric acids, which are then called secondary (or 'acidic' in this case) pollutants. Other processes can take place during transport, such as diffusion into other atmospheric substances such as rain and snow as wet deposition (specifically as acid rain in our case). They can also occur as aerosols or can be bound to aerosols (Section 2.1.8) and can precipitate as dry deposition without the aid of rain.

Other characteristics of atmospheric trace substances concern the origin (natural – anthropogenous), the effect (reactive, climatic, toxic), and the distribution (ubiquitous – as a concentration in supposed clean air regions).

2.1.8 *Cycle of atmospheric aerosol particles*

Starting with the emission of a condensable gas (here, SO_2 or a precursor of H_2S), chemical reactions take place with condensation to a sulphate aerosol as the end product. We distinguish three different types of direct sources for particles. Spray produces aerosol droplets containing chloride, similar to sea water, which also contain trace metals and organic components such as fulvic and humic acids from the surface microlayer of the ocean. As a second source, mechanical abrasion of solid surfaces (fallen leaves, dust blown by the wind) can produce particles that contain almost exclusively organic aerosols (lipids) in the case of leaves, and inorganic oxides and organic material from humus with wind-blown dust – both are organic aerosols. Combustion represents the third particle source, with the formation of soot as a carbon aerosol (Graedel and Crutzen, 1994).

The smallest aerosol particles combine along their travels, take up water vapour and thereby form a solution of ions with a solid core of approximately 30% of the total particle mass, with sulphur as sulphate and organic acids in a chemically reactive solution. In the next stage, insoluble species concentrate on the surface of the particle. Substances that work particularly well in this process are those having a hydrophobic and a hydrophilic component, such as long-chain alcohols (e.g. as *n*-decanol). An organic layer forms, which can take up gaseous molecules such as NO_2 or NH_3 in the next stage and can react with them as well. This is how polycyclic aromatic hydrocarbons (PAHs) and aromatic ketones are formed from arenes. In the final stage of the cycle, the particles reach the soil.

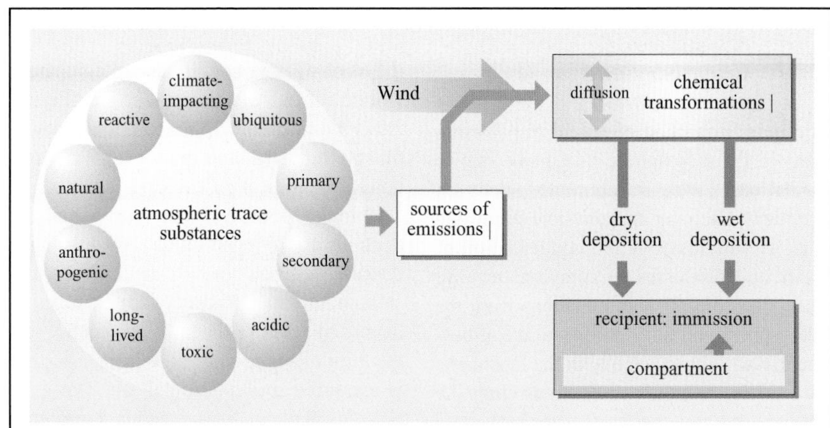

Figure 2.1.7 General behaviour of trace elements in the atmosphere

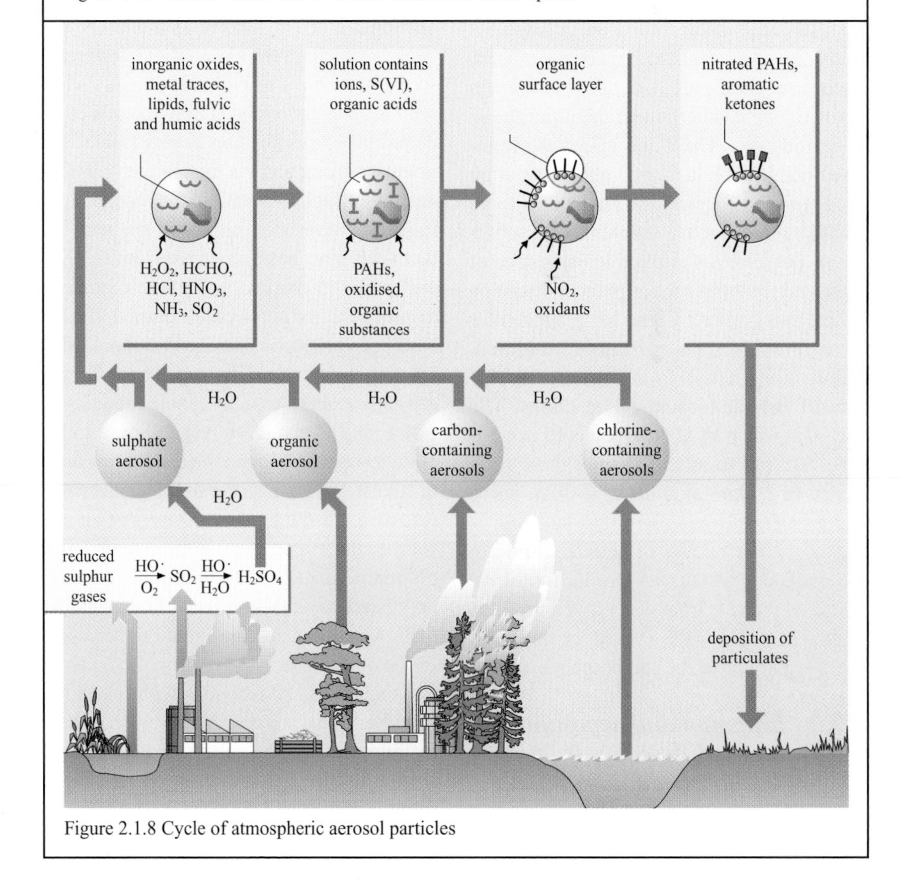

Figure 2.1.8 Cycle of atmospheric aerosol particles

2.2 Ecological photochemistry

2.2.1 Chemistry and photochemistry

The most important chemical transformations in the atmosphere take place due to insolation. We differentiate between chemical reactions and physical processes such as volatilisation and condensation of water and processes of spraying out. We must also consider the transitions from the hydrosphere and lithosphere into the atmosphere, and the exchange with the biosphere. Meteorites, asteroids and comets enter the atmosphere and the Earth's surface from space, and they induce chemical reactions just like the solar radiation. Ecological photochemistry deals with reactions of environmental chemicals with other substances (or with one another) under the influence of solar radiation. The major stages in photochemical reactions are photoactivation (conversion of atoms and molecules to an excited, more reactive state) with the subsequent possibilities of photoionisation or of chemical reactions such as photodissociation (photolysis), whereby ions are generated for other processes (isomerisation, addition, elimination, decay). Photosynthesis is also one of the photochemical reactions. The region in which all of these reactions occur is also referred to as the chemosphere (at a height of 10 km up to about 150 km above the Earth's surface). An important role is played by the release of activated oxygen from ozone, which can produce reactive hydroxyl radicals from water (Sections 2.2.4 and 2.2.6). The OH radical is the most reactive particle in the chemosphere.

2.2.2 Emission and deposition under different weather conditions

The significance of transport processes of air pollutants becomes especially clear when different weather conditions are compared. Under normal weather conditions (Figure 2.2.2, 1), warm air from households, power plants, industry and motor traffic goes upward; a vertical air exchange takes place. The major pollutants that the heated air contains are nitrogen oxide (NO_x), sulphur dioxide, carbon monoxide and hydrocarbons. A small portion of the pollutant is deposited again in the emission centre – in the form of a deposition. The greater portion is captured by horizontal air flows at altitudes of about 1000 m and is transported further (transmission process – see also Section 2.2.7). The transmission of air pollutants can occur over thousands of kilometres, so that the effects of pollutant emissions are seen in areas far removed from the source of emission. The deposition occurs finally via gradual sinking and via raining out – preferentially at higher elevations. Photochemical reactions have occurred along the way, so that acids have been generated from sulphur and nitrogen oxides, and photo-oxidising agents such as peroxyacetylnitrate (PAN) have been produced from hydrocarbons after reacting with ozone. The deposition of such substances leads to acidification of bodies of water and damage to vegetation in the pollution phenomenon called *'Waldsterben'* or 'forest death'. In an atmospheric inversion (Figure 2.2.2, 2), the higher layers of air are warmer than those close to the ground. The pollutant-filled exhaust gas from densely populated areas cannot reach higher elevations. The pollutant concentration increases. In the summer in large metropolitan areas, the photochemical reactions with ozone (to form PAN) that are possible here also lead to the formation of photochemical smog (from smoke and fog), sometimes referred to as summer smog or Los Angeles smog.

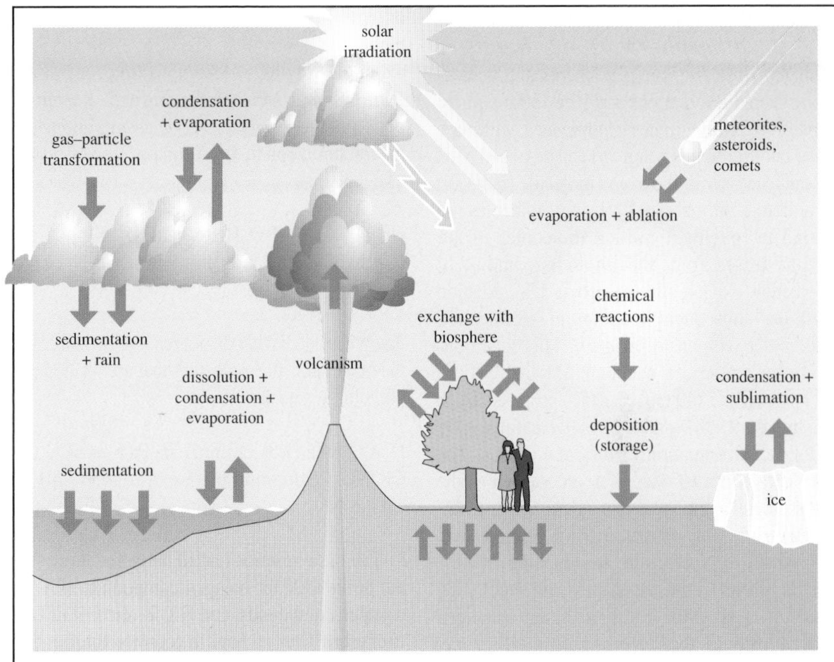

Figure 2.2.1 Chemistry and photochemistry

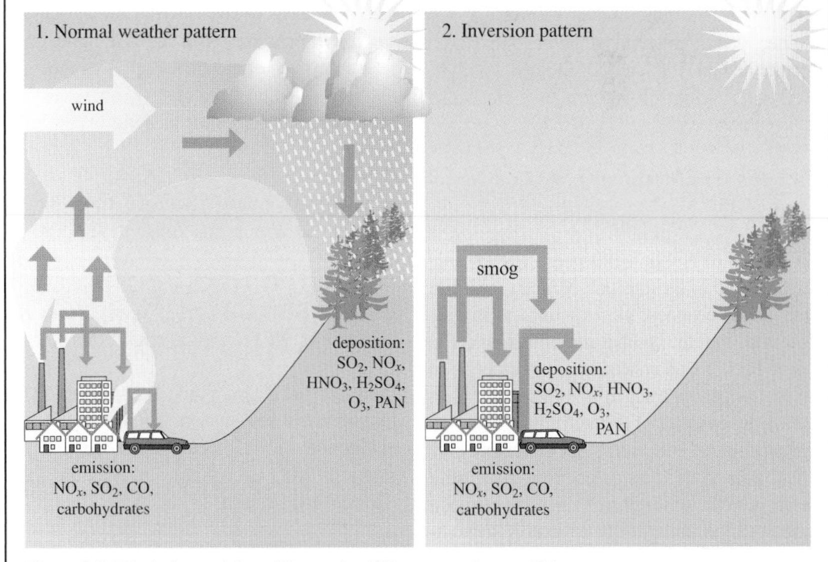

Figure 2.2.2 Emission and deposition under different weather conditions

2.2.3 *Catalytic cycles of atmospheric ozone chemistry*

The extraordinarily complex chemical processes and their associations in the stratosphere (including those in the ionosphere, although their contribution is not great) can be traced back to the reactivity of the ozone molecule (box OX) – although only about 10 ozone molecules are found in roughly 1 million molecules of air. Ozone arises from the photodissociation of molecular oxygen after absorbing UV radiation and the subsequent reaction of atomic and molecular oxygen to form O_3. The decomposition of ozone can occur photolytically (net: $2O_3 + h\nu$ ($\lambda \leq 1140$ nm) → $3O_2$ – proposed by S. Chapman, 1930), and especially catalytically with the participation of nitrogen oxides. The nitrogen oxides NO and NO_2 are viewed as the most significant catalysts (Paul J. Crutzen, 1970); they are generated from ground-level emissions of N_2O in the stratosphere: N_2O + $O(^1D)$ → NO (^1D: electronic excited state). NO_x molecules start the catalytic chain reactions which lead to the decomposition of ozone. Another catalytic reaction is started by OH and HO_2 radicals that are formed from water: H_2O + $O(^1D)$ → 2HO. Chlorine is responsible for another type of catalytic ozone decomposition (formulated in 1974 by R. Stolarski and R. Cicerone); chlorofluorocarbons are also involved here:

$CFCl_3 + h\nu$ ($\lambda < 260$ nm) $+ nO_2$ → CO_2 + HF + 3 (Cl· or ClO·)

Chlorine enters the atmosphere via volcanic eruptions and by the transport of methyl chloride from the troposphere as a product of marine algae and from the combustion of biomass. The natural background concentration of chlorine in the stratosphere is smaller by a factor of 5 than that of the emitted CFCs and other industrial organic chlorine compounds. CFCs require only a few years to be transmitted from the Earth's surface into the stratosphere. The solar radiation there (at an altitude of 20–25 km) is so high-energy that chlorine atoms and chlorine monoxide molecules that are even stronger ozone-destroying decomposition catalysts than NO and NO_2 can be formed. Reactions between the catalysts serve as an important natural limitation to the decomposition of ozone. The reactions:

HO· + NO_2 + M → HNO_3 + M
(where M is a reaction partner such as N_2 or O_2)
and ClO· + NO_2 + M → $ClONO_2$ + M

involve products that do not react with ozone. On the other hand, the molecules that are formed can photodissociate again:

$HNO_3 + h\nu$ ($\lambda \leq 330$ nm) → HO· + NO_2 and
$ClONO_2$ (chloronitrate) + $h\nu$ ($\lambda \leq 450$ nm) → ClO· + NO_2

The basic reactions listed here are repeated in the boxes ClX for halogen compounds, NX for nitrogen compounds, and HX for the OH radical chemistry. They all have in common the fact that they lead to the destruction of ozone. Next to the ClX box, the chemical equations are shown for the reaction of hydrogen chloride and chlorinated hydrocarbons with the OH radical and after absorption of radiant energy as a process of photodissociation (photolysis). These are the most important photodissociations:

O_3 → O_2 + O
($\lambda \leq 360$ nm, partial residence time 2000 s according to Wagner and Zellner, 1979),

NO_2 → NO + O ($\lambda \leq 420$ nm, 125 s)
HNO_2 → NO + OH ($\lambda \leq 390$ nm, 360 s) and
H_2CO → H_2 + CO ($\lambda \leq 360$ nm, 20 000 s)

More detailed diagrams and explanations on the formation and decomposition of ozone are found in Section 2.2.4; others for ozone and the catalytic NO_x cycles are in Section 2.2.5 (Graedel and Crutzen, 1994).

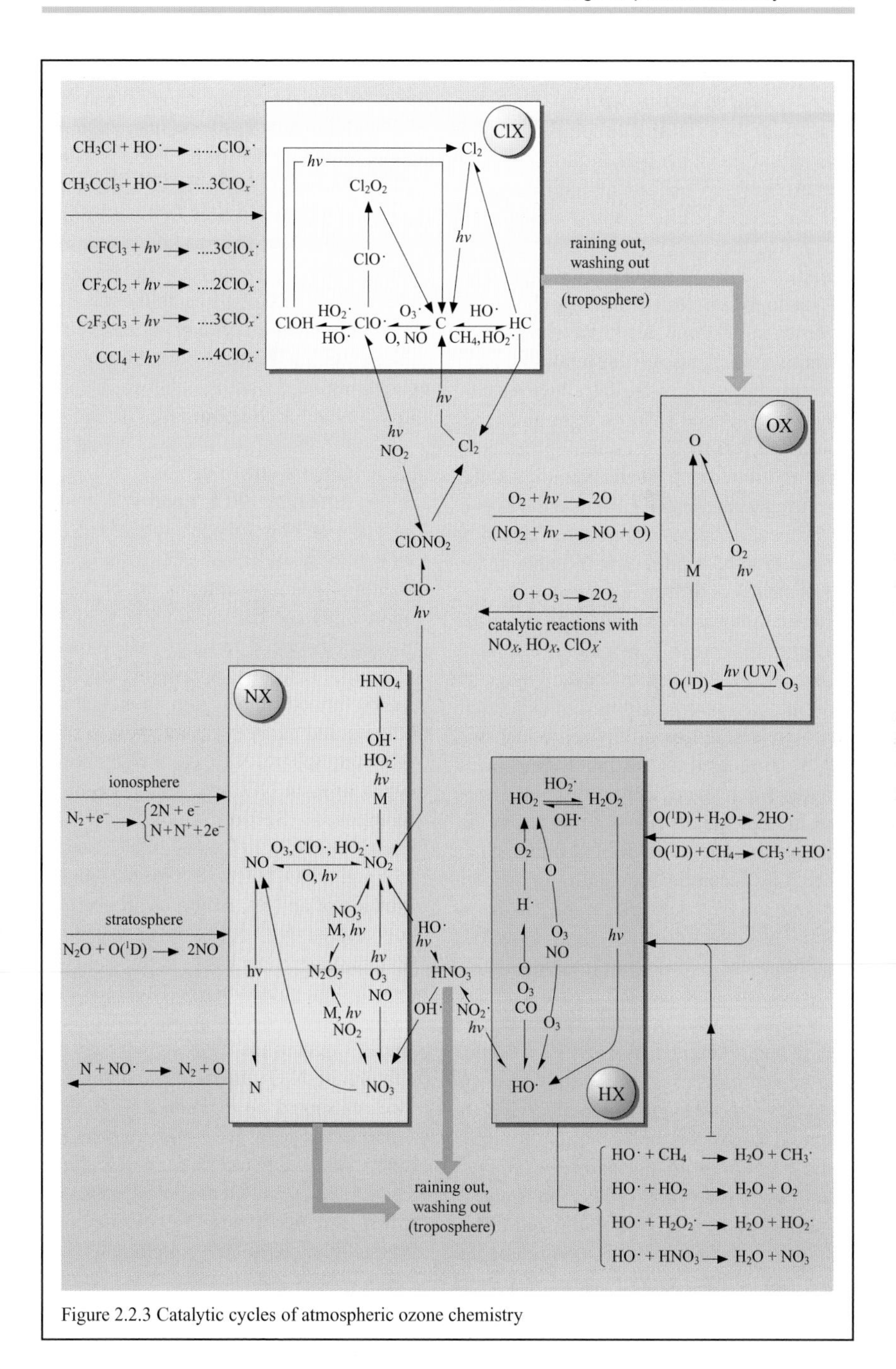

Figure 2.2.3 Catalytic cycles of atmospheric ozone chemistry

2.2.4 *Ozone: formation and decomposition*

The ozone layer at an altitude between 15 and 50 km above the Earth's surface (in the stratosphere) is of existential significance for life on Earth because it absorbs the short-wave, high-energy UV portion of the solar radiation. Even the public have become increasingly interested in research on the destruction of ozone due to the emission of certain trace gases, especially CFCs. Spectroscopic studies and, later on, measurements with altitude balloons determined and confirmed that the Earth's atmosphere does have an ozone layer, whereupon in 1930 the English geophysicist S. Chapman developed a chemical mechanism for ozone formation and a kinetic model which explain the phenomenon of the stable ozone layer.

In the stratosphere (Section 2.1), under the influence of 'hard' UV radiation with wavelengths below 240 nm, molecular oxygen is split into two oxygen atoms. Depending on the wavelength, two different mechanisms can be considered for this process. At wavelengths where $\lambda < 243$ nm, there is first a transition of the oxygen molecule from the base triplet state to an excited triplet state. Then the decomposition into two triplet oxygen atoms takes place: $O(^3P)$. At a continuous absorption of light at wavelengths where $\lambda < 176$ nm (maximum $\lambda = 147$ nm), after the excited triplet state has been generated, the oxygen atoms decay into a base and an excited state: $O(^3P) + O(^1D)$. Ozone is formed as a triple-atom oxygen modification due to the action of an oxygen atom on an oxygen molecule, in a trimolecular reaction with the cooperation of a reaction partner M. The tropospheric concentration of $O(^3P)$ is estimated to be 10^{-8} ppm. In contrast to the stratosphere, in the troposphere NO_2 is the primary source of $O(^3P)$ at wavelengths shorter than 420 nm. Once formed, ozone can also be decomposed by photolysis, whereby either $O(^3P)$ is formed at wavelengths below 1180 nm (by photons) or excited $O(^1D)$ is formed due to the action of UV radiation, depending on the radiant energy. An additional cleavage takes place as the result of an impact reaction between ozone and the excited oxygen that has been formed. An equilibrium develops between the formation and decomposition of ozone.

2.2.5 *Ozone and the catalytic NO_x cycles*

The cycle among ozone, oxygen and the nitrogen oxides is called a photostationary condition. Primary sources of N_2O are tropical soils and the surface waters of the northern Atlantic. The long-lived nitrogen compound (lifespan approximately 150 years) is formed largely due to the activity of nitrifying and denitrifying microorganisms. In the stratosphere, N_2O reacts with excited oxygen to form NO. In Cycle 1, ozone is decomposed (Section 2.2.4) with the formation of NO_2. In Cycle 2, which occurs in the night, NO_2 reacts with ozone with the formation of an NO_3 radical as an intermediate and of N_2O_5. The NO_3 radical is extremely unstable in light and can therefore be easily split photolytically. Thus, N_2O has an effect on the decomposition of ozone. Sample calculations have shown that the doubling of N_2O emissions (combustion processes should be mentioned as another source) leads to a 10% reduction in the ozone level. In the cycle we are describing, NO, NO_2, O, O_2 and O_3 are being constantly formed and decomposed; they are in a dynamic (photochemical) equilibrium.

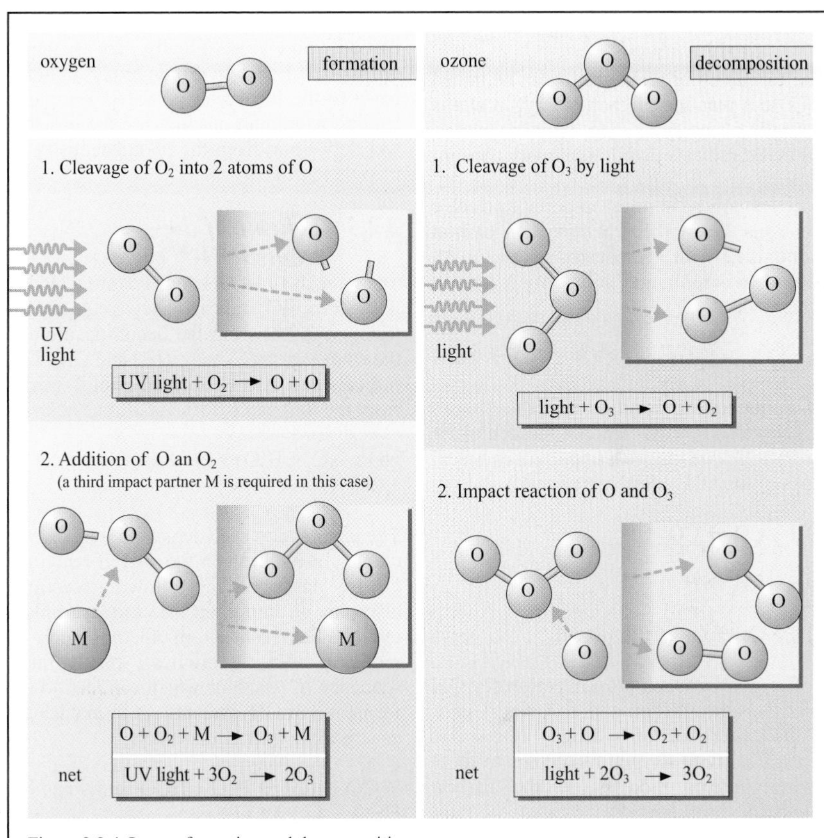

Figure 2.2.4 Ozone: formation and decomposition

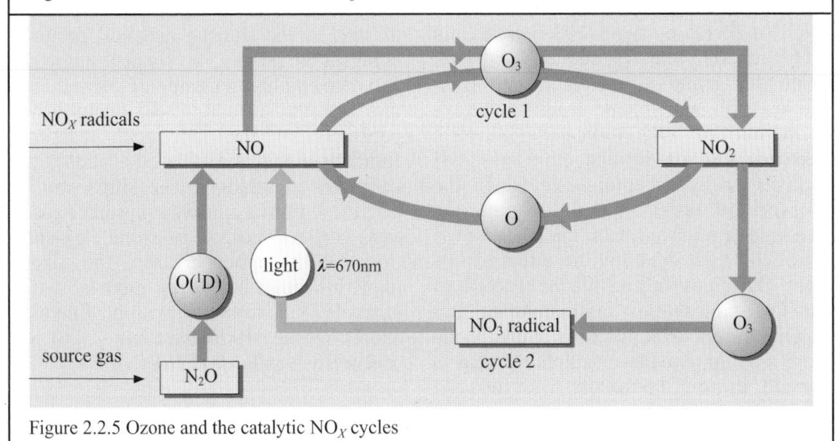

Figure 2.2.5 Ozone and the catalytic NO_X cycles

2.2.6 *Photochemistry of the OH radical*

The formation of OH radicals can be traced back to a reaction of excited oxygen atoms from the ozone cycle with water molecules. Hydroxyl radicals exhibit stationary concentrations in the troposphere of 10^5 to 10^6 particles cm^{-3}. The most important possible reactions lie in the reaction with carbon monoxide, with hydrocarbons and with ozone, wherein the following reaction scheme occurs:

$O_3 \rightarrow O_2 + O(^1D)$ and $O(^1D) + H_2O \rightarrow 2{\cdot}OH$

as a quenching reaction for the singlet oxygen $O(^1D)$ + M (impact molecule) $\rightarrow$ $O(^3P)$ + M, and the most important chemical sinks of the OH radical are

$\cdot OH + CO \rightarrow CO_2 + H$

$\cdot OH + RH \rightarrow H_2O + R$

$\cdot OH + O_3 \rightarrow O_2 + HO_2$

The main consumption reaction for OH radicals is the reaction with CO, which leads to the generation of hydrogen atoms that recombine with oxygen molecules to form hydroxyperoxy radicals. If the carbon monoxide concentration increases, the ratio of HO_2/OH radicals increases as well. However, in the presence of nitrogen oxides (NO_x) it decreases again. NO radicals and HO_2 radicals yield OH radicals and NO_2 again after giving up an oxygen atom. Even the reaction with ozone leads to the re-formation of OH radicals. Hydrogen peroxide can be formed from two OH radicals or by accepting one H. In the presence of water (by raining out), the peroxide is removed from the cycle and is returned to the cycle by the effect of UV light. Other possibilities for the reaction of the OH radical exist with chlorine and sulphur compounds, whereby hydrochloric acid, sulphur dioxide or sulphuric acid is formed. In turn, the reaction with methane (Section 2.2.7) leads to the generation of carbon monoxide. Finally, the reaction of hydroxyl and hydroxyperoxy radicals leads to the formation of water. The paths leading to raining out indicate the possible wet deposition from the photochemistry of the OH radical.

2.2.7 *Scheme of the catalytic HO_x cycles*

The radicals of H, OH and HO_2 are included in the term HO_x radicals. They are generated (Section 2.2.8) from the decomposition of the source gases CO, H_2, H_2O and CH_4. OH radicals can also be formed photolytically from the nitrogen oxides via nitrous acid:

$NO + NO_2 + H_2O \rightarrow 2HNO_2\ (+ h\nu) \rightarrow 2{\cdot}OH + 2NO$

The inner cycles each depict the influence of ozone. In addition to the partial reactions already described, the following reactions also play an important role in these linked cycles. The reaction of alkanes such as methane with OH radicals results in a sequence of reactions which can lead to the formation of OH_2 radicals via formaldehyde as a reactive intermediate step:

$H_2CO + \cdot OH \rightarrow H_2O + HCO\cdot$ and
$HCO\cdot + O_2 \rightarrow CO + HO_2\cdot$

In Figure 2.2.7 all of the reactants are shown in circles, the starting and end products being in the boxes. The concentrations for dynamic equilibrium states are derived from the known velocity and equilibrium constants of the individual reactions. Building on this, and using calculations from computer simulations, the shifts can be calculated from the increase in source gases, especially with respect to ozone concentrations. With higher alkanes, the alkane oxidation described using methane as an example leads to the formation of peroxyacetyl nitrate, PAN (Section 2.4.6) via acetaldehyde radicals (CH_3CO–).

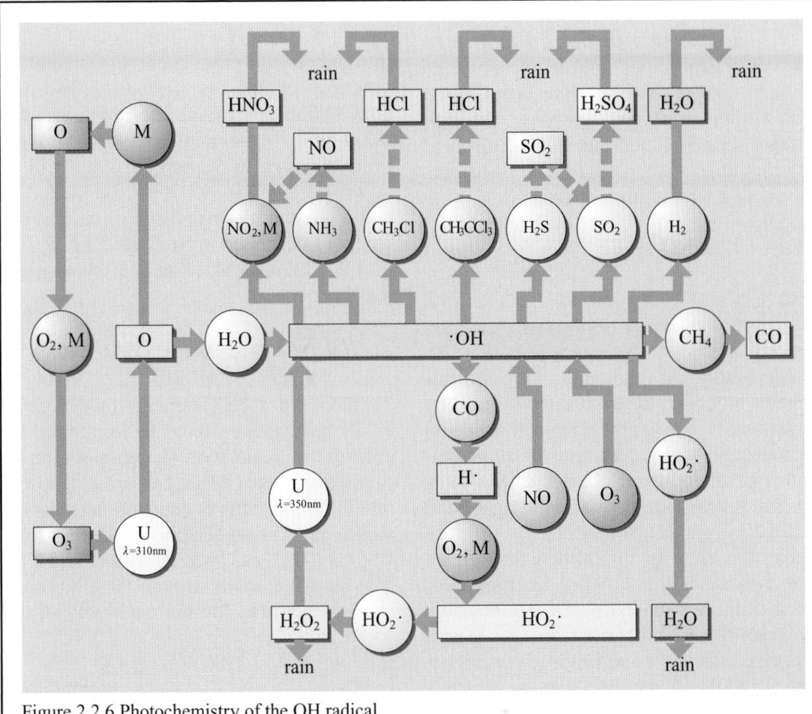

Figure 2.2.6 Photochemistry of the OH radical

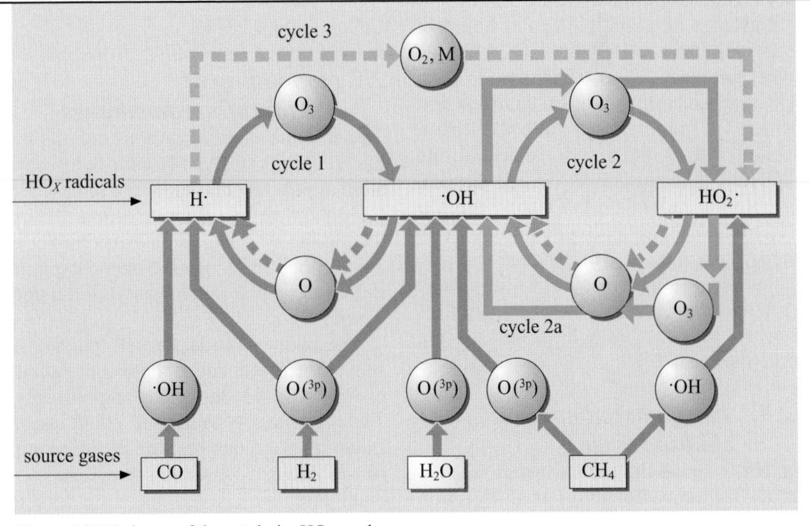

Figure 2.2.7 Scheme of the catalytic HO_X cycles

2.2.8 *Scheme of halogen photochemistry in the atmosphere*

The most important source for chlorine atoms is methyl chloride (chlorofan, CH_3Cl) with a lifetime of about one year. Although another source of chlorine, NaCl aerosols at altitudes up to 3 km, provides chlorine atoms by reacting with nitrogen trioxide:

$$NO_3 + Cl^- \rightarrow NO_3^- + Cl\cdot$$

both Cl· and NO_3^- are so unstable that they do not enter the troposphere from the atmosphere.

An additional and important anthropogenic source of chlorine is represented by all of the halogen hydrocarbons produced on an industrial scale as solvents or as propellants (as fluoro/bromo/chlorohydrocarbons, several of which are shown on the right-hand side of Figure 2.2.8). The CFCs are very stable; they enter the troposphere through the stratosphere without being decomposed. Their stability at higher elevations increases with additional numbers of fluorine atoms. Chlorine radicals can be produced photolytically from these compounds in the presence of solar radiation ($\lambda < 220$ nm), whereby one chlorine radical can catalytically destroy tens of thousands of ozone molecules before it disappears into the HCl sink. The chlorine radical can react further with ozone to form a ClO radical. Depending on their reaction partner, ClO radicals can follow different reaction paths. With NO_2, chloronitrate ($ClONO_2$) is formed; with excited oxygen, the chlorine radical forms again; and the OH_2 radical leads to hypochlorous acid. Finally, dimerisation can produce unstable dichlorodioxide. The reaction with methane is the major source for the deactivation of chlorine radicals, which leads to the formation of HCl.

2.2.9 *The global atmospheric chlorine cycle*

The NaCl above the sea salt spray has the largest fraction, although it has no effect on the troposphere or stratosphere (Section 2.2.8). An almost equal amount of HCl is introduced into the cycle by means of volcanic explosions and especially by combustion processes. The processes that occur with biogenic CH_3Cl have already been discussed in Section 2.2.6 and will be described again in detail in Section 2.2.10 below. Accumulations of CFCs and of the reactions described in Sections 2.2.6, 2.2.10 and 2.2.11 take place in the troposphere and stratosphere.

2.2.10 *Scheme of the catalytic ClO_x cycle*

As is true for HO_x radicals (see Section 2.2.7), catalytic cycles can be formulated for chlorine radicals with the cooperation of ozone and of the OH radical, excited oxygen and UV light. This is depicted here for the source gases methyl chloride (biogenic, from kelp) and CFCs as propellants R11 and R12. The most important steps of the ClO_x cycle (first described by Molina and Rowland) are:

$$O_3 + h\nu \rightarrow (\lambda < 1180\ \text{nm})\ O_2 + O\cdot$$

$$Cl\cdot + O_3 \rightarrow ClO\cdot + O_2$$

$$ClO\cdot + O\cdot \rightarrow Cl\cdot + O_2$$

2.2.11 *Splitting of chlorofluoromethanes*

Two chlorine atoms split off as radicals from the CFC R11 (CFM-11) with one fluorine atom at wavelengths below 230 nm and from CFM-12 with two fluorine atoms at wavelengths below 2 nm. As described in Section 2.2.10, they influence the ozone cycle and reduce the ozone concentration of the troposphere.

The ozone hole is formed, particularly above Antarctica, due to the accumulation of reactive particles during the long polar night. These particles get their full effect catalytically via solar radiation, with the rising of the sun in August, which is springtime at the South Pole.

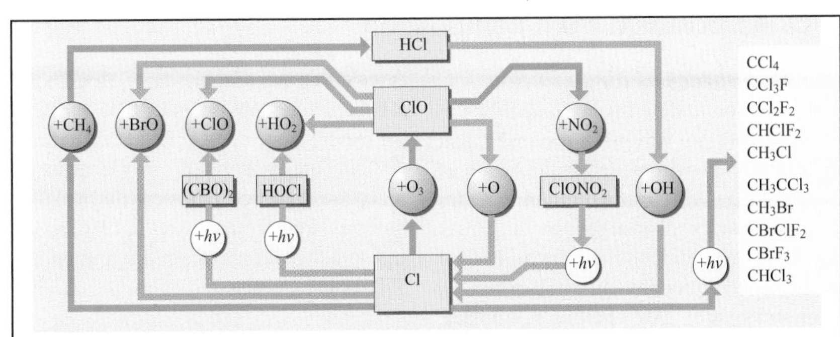

Figure 2.2.8 Scheme of halogen photochemistry in the atmosphere

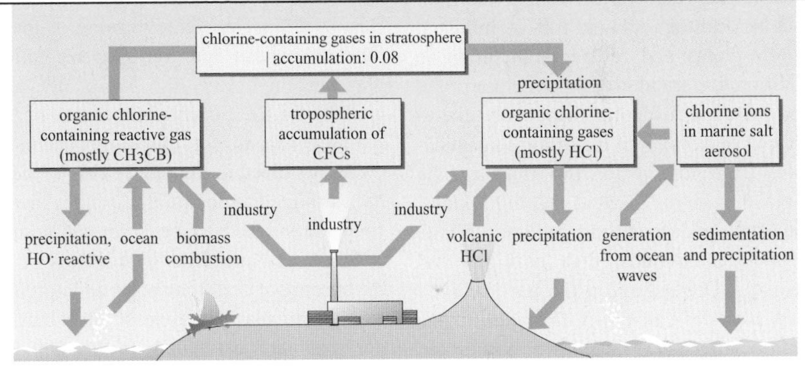

Figure 2.2.9 The global atmospheric chlorine cycle

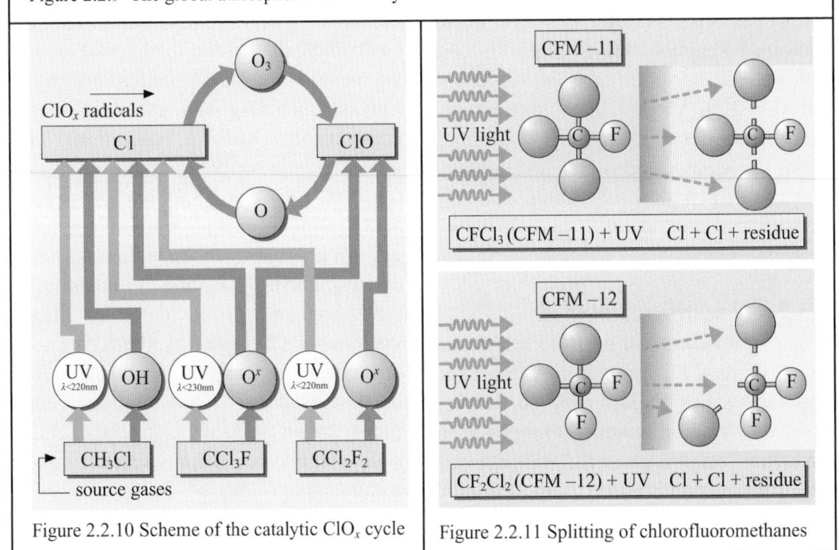

Figure 2.2.10 Scheme of the catalytic ClO_x cycle

Figure 2.2.11 Splitting of chlorofluoromethanes

2.3 Air pollution from combustion

2.3.1 *Sources of air pollution*

Air pollution is described as impurities in the air due to various substances that endanger the environment or the health of living things. The term emission includes the substances discharged by a causer (emitter). According to the German Federal Immission Protection Law (BImSchG), emissions are 'Air pollution, noises, shaking, light, heat, radiation and similar phenomena' that come from a plant or factory. Emitters include traffic, industry, power plants and refuse incineration in addition to agriculture, service enterprises and private households (especially due to heating devices). The major air contaminations from combustion processes can be divided into six groups: sulphur oxides, nitrogen oxides, carbon oxides, volatile organic compounds, dust particles and aerosols. Depending on the particle size, dust particles can either fall back to the Earth's surface near the emitter, or they can reach a recipient (human, animal, vegetation, building) as immissions via transmission. Reactions take place in the atmosphere because of solar radiation (Section 2.2). Air pollutants contribute to the phenomena of the greenhouse effect, ozone depletion, deforestation and the acidification of bodies of water and the impacting of soils, and they can also induce certain diseases such as allergies.

2.3.2 *Quantification of air pollutants according to source*

Based on the Federal Immission Protection Law, the states in Germany in densely populated areas are required to monitor air pollution. Today an entire EU-wide network of air measuring stations is in operation (measuring SO_2, NO_x, CO, and in some cases ozone and dust particles). Households are the main contributors of SO_2; traffic is the major source of NO_x. Estimates indicate that in the year 2000 in Germany, a reduction of particle emissions down to 0.47 million t (from 0.56 million t in 1986) and of CO to 6.2 million t (from 8.9 million t in 1986) can be expected.

2.3.3 *Emissions from burning vegetation*

Besides the technogenic burning of fossil fuels, emissions from burning vegetation play an important role regionally and globally for the current composition of the atmosphere and for the biogeochemical cycles described in Chapter 1. This includes the burning of agricultural waste, the use of firewood and the burning of vegetation such as forests and heathland. In south-east Asia, the burning of rice straw plays an important role in particular. The areas of forest burned per year are estimated to be 600 000 hectares (ha) in the Mediterranean area, 4 million ha in North America, and 10 million ha (comparable to the total forest area in Germany) in the boreal coniferous forest of Eurasia including the subarctic tundra (lands of the Russian Federation). This generates not only classical emissions (Sections 2.3.1 and 2.3.2), but also organic acids and halogen hydrocarbons up to the polychlorinated dibenzodioxins and dibentofurans. Flying laboratories (including one from the Max Planck Institute for Chemistry in Mainz) are used for studies of extensive vegetation fires. The initial evaluations of these analyses have shown that an especially high level of chemically and photochemically reactive carbon compounds is released during a Siberian forest fire compared to a savanna fire.

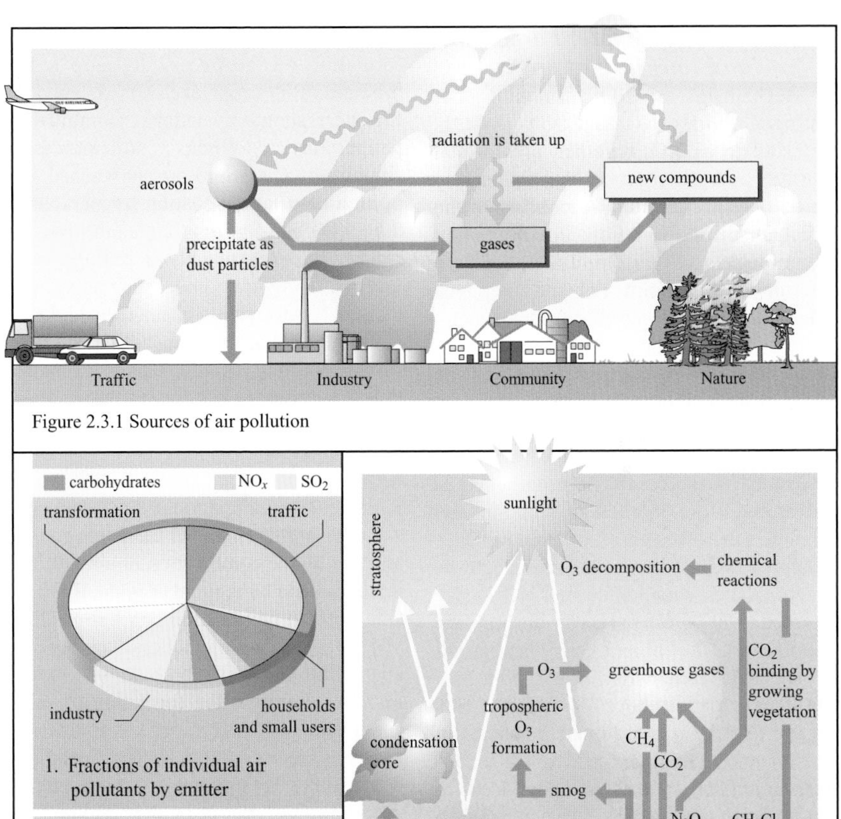

Figure 2.3.1 Sources of air pollution

Figure 2.3.2 Quantification of air pollutants according to source

Figure 2.3.3 Emissions from burning vegetation

2.3.4 Nitrogen oxide: Formation and lowering of levels

1. Mechanisms of NO formation. The nitrogen oxides NO and NO_2, usually referred to as NO_x together, are formed during combustion processes, whereby up to 95% NO is emitted. According to the mechanism of formation, in combustion technology we differentiate between thermal NO, prompt NO and fuel NO. Thermal NO is formed from atmospheric nitrogen by splitting the N_2 molecule with the interaction of oxygen radicals in the zone of the combustion products. Only at temperatures above 1570 K in the combustion chamber does the content of thermal NO increase. The prompt NO is generated from molecular nitrogen and organic substances via short-lived radicals, whereby NCH and CN radicals occur as intermediates. Finally, the fuel NO is also formed via CN radicals during the combustion of coal and heavy heating oil in the hydrocarbon-rich flame front as a result of oxidation above 870 K. Some 80% of the total NO_x emissions are formed in this manner. The high reactivity of hydrocarbon radicals in the flame front leads to the splitting of molecular nitrogen and then via oxidation to form prompt NO.

2. Dependencies of thermal NO formation. The calculated dependencies of the formation of thermal NO on the temperature and on the excess air ratio λ (ratio of air to fuel) at a constant dwell time of 1 s (assuming ideal mixing conditions) show that NO formation occurs only above 1600 K and that it increases considerably with increasing temperature above this point (hence the name). In addition, as one would expect, NO formation is inhibited by low excess air ratios (Figure 2.3.4, 2a). Figure 2.3.4, 2b shows the dependence of NO formation on the excess air ratio λ for different dwell times; each λ is assigned by the corresponding caloric combustion temperature. When stoichiometric combustion conditions are approached, i.e. at high caloric combustion temperatures, the NO equilibrium concentration is reached relatively quickly. In addition, with increasing dwell time, the maximum thermal NO formation is moved to higher λ numbers. Thus, to reduce thermal NO formation, one should strive for small excess air ratios at low combustion temperatures and short dwell times in the zones of high temperature (1600 K).

3. Procedures for lowering NO. Technically a significant thermal NO reduction can be attained by the combination of air and fuel gradation and by recirculating the flue gas. One optimal combustion concept consists of a good mixing of fuel and combustion air, i.e. an adiabatic combustion with no loss and therefore the formation of 'hot' combustion chamber walls and an internal flue gas recirculation. In the example depicted, by the appropriate dividing up of the addition of fuel or by the addition of a secondary fuel (natural gas during the combustion of coal), one obtains locally substoichiometric combustion zones in which the nitrogen monoxide is reduced again via hydrocarbon radicals. As a result of air gradation (with thermal decoupling between a substoichiometric and a superstoichiometric stage), a temperature drop is achieved (Figure 2.3.4, 2a and 2b). The central point in all fuel engineering measures is the lowering of the combustion chamber temperatures, via the recirculation of cold flue gases, etc.

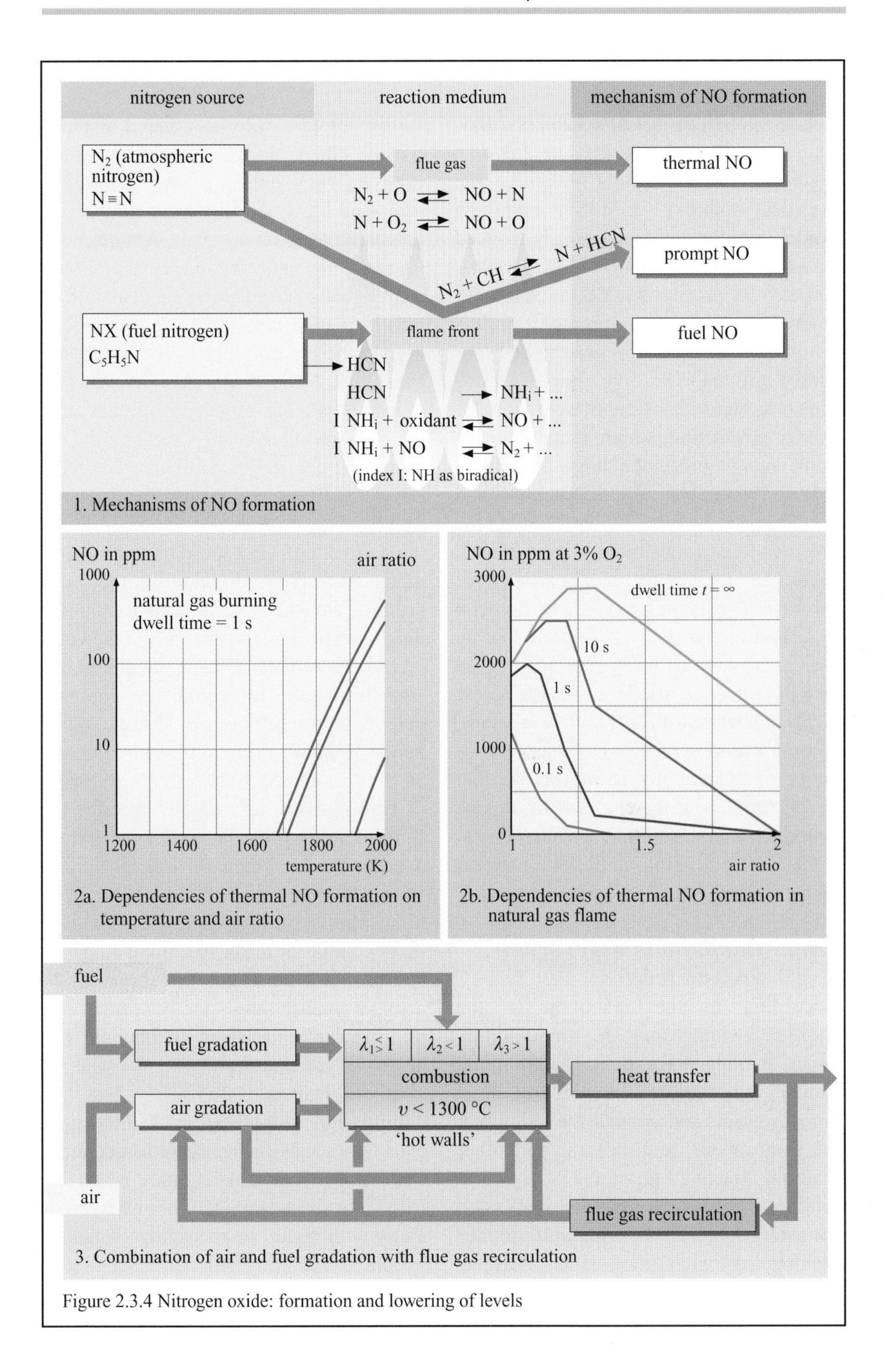

Figure 2.3.4 Nitrogen oxide: formation and lowering of levels

2.4 Anthropogenous pollutants and their effects

2.4.1 Anthropogenous emissions in West Germany in 1984

Figure 2.4.1 (based on findings of the chemical industry in 1987) gives an overview of the anthropogenously induced emission of pollutants in 1984 in Germany (see also the prognosis in Section 2.3.1). In environmental technology, anthropogenously induced emissions are divided into source groups. The five source groups shown in Figure 2.4.1 are power plants for energy production, which as a stationary point source present a large dispersion range of more than 50 m; households and small businesses as a stationary area source (average dispersion range 10 to 20 m); industry as a stationary point or area source (average dispersion range <50 m); traffic as a mobile source (with reference to an individual vehicle) or as a line source on heavily travelled roads and with a low dispersion range (<0.5 m) of the emissions, and the use of solvents. The main contribution of CO and NO_x is from traffic, that of SO_2 from power plants, that of organic materials from the use of solvents (in paint shops, dry cleaning and the chemical industry), and that of dust from industry.

2.4.2 Immissions near ground level

According to the definition in the BImSchG, immissions are 'air pollution, noise, shaking, light, heat, radiation and similar environmental activities that affect humans, plants and animals, the soil, water, the atmosphere, and cultural and other material assets'. Figure 2.4.2 depicts a differentiation of the expected immission concentrations for the pollutants sulphur dioxide, nitrogen monoxide, nitrogen dioxide and ozone, the concentrations of which are as a rule less than 1 mg m^{-3} outside peak loads. We differentiate between rural areas, densely populated areas and highly impacted inner city areas compared to clean air areas. Although the latter are not subjected to special emission sources, the global distribution of air pollutants has led to measurable concentrations even in these areas (such as Antarctica and the Alps). Airborne particles and carbon monoxide are not included here; they play an important role in the quality of municipal climates in particular. The most important factors influencing the diffusion of air pollutants (see also Section 2.1) are wind direction, wind speed and turbulence of the atmosphere. In addition, the properties of the source itself, such as chimney height, waste gas temperature and the volumetric rate of flow of the waste gas, play a significant role. The volumetric flow rate determines the lift, i.e. the effective source height from which the diffusion proceeds. Models developed to estimate the transport of air pollutants; they usually describe the diffusion of the air pollutants starting from a point source based on transport with wind and on mixing with air as a result of atmospheric turbulence (e.g. the Gauss model, valid for areas 10–20 km from the source). Tall chimneys are supposed to achieve smaller immission concentrations as a result of widespread distribution. The acidity of lakes in Scandinavia showed the false conclusions of this policy. The VDI commission 'Air Quality Maintenance' establishes maximum immission concentrations (MIC values), as either long-term (MIC_L) or short-term (MIC_S) impact, especially with regard to protecting groups of people at risk.

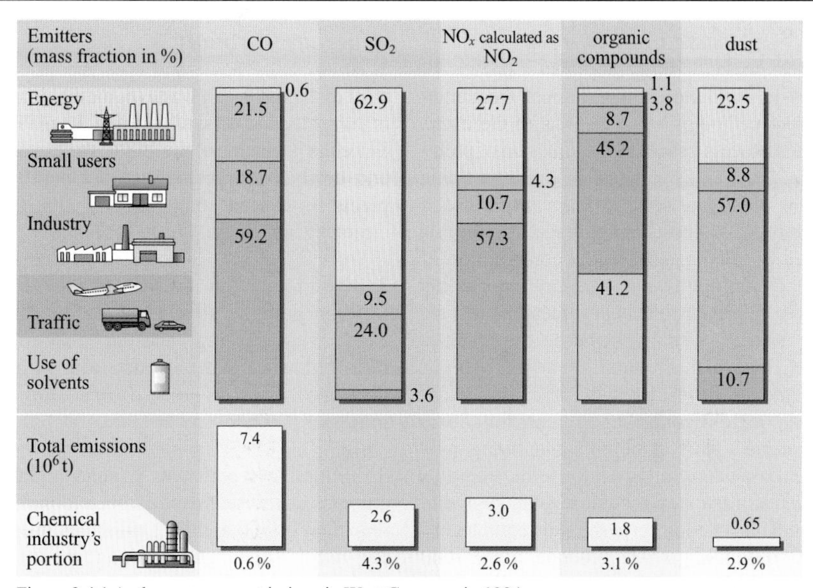

Figure 2.4.1 Anthropogenous emissions in West Germany in 1984

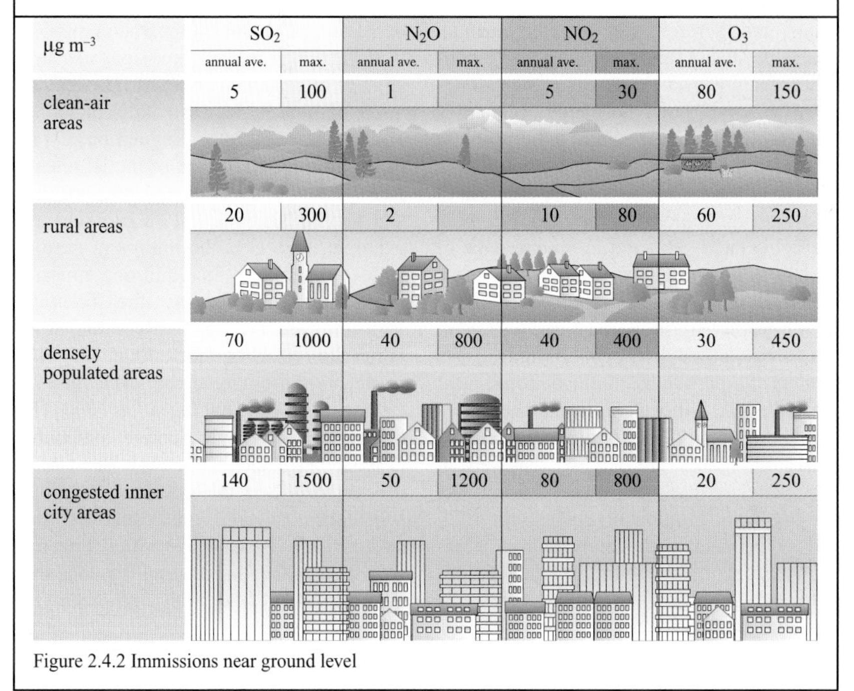

Figure 2.4.2 Immissions near ground level

2.4.3 *Pollutant flows during waste incineration*

Next to landfilling (see also Chapter 4), waste incineration is the most frequently used method of waste disposal in Germany. This process generates the emission of pollutants as well as combustion residues. Some 300–400 kg of solid residues are produced from 1 t of household garbage. Part of this can be used as slag in the construction of roads and pathways, whereby the proportions of leachable heavy metals must be monitored. Usually, however, the residues are deposited in domestic garbage dumps. Residues from fly ash precipitation or cleaning (Section 2.5) contain toxic materials such as dioxins and furans plus heavy metals, so they have to be stored as special waste in places such as underground landfills. In addition to the solid residues, 1 t of household garbage generates 5000–7000 m^3 of flue gases during combustion. The latter contain inorganic and organic pollutants, and therefore they must be cleaned. Polychlorinated dibenzodioxins and dibenzofurans, polycyclic aromatic carbohydrates (PACs) and polychlorinated biphenyls (PCBs) represent especially toxic problem substances. If they are not already present in refuse, they are generated in flue gases during incomplete combustion (PACs, carbohydrate C_nH_m) and in the presence of halogen-organic compounds in the ashes (CBs, dioxins/furans) in the course of the combustion, e.g. in the cool-down stage at gas temperatures between 470 and 270 °C and in the particulate removal zone. Dust particles are removed from the flue gases using electronic filters (or tissue filters). The inorganic gases are largely removed in downstream flue gas washers using wet or dry neutralisation processes (for HCl, HF, SO_2) and with catalysts and high-temperature methods for nitrogen oxides (Section 2.5).

2.4.4 *Waste incineration in Germany*

1. Amounts of waste and slag. About one third of the municipal waste (domestic waste) is currently incinerated in some 50 plants. Since the beginning of the 1980s, the existing and newly constructed waste incineration plants have been equipped with flue gas cleaning technology.

2. Residues from waste incineration. Correspondingly, after 1980 the residues from wet washing and dry sorption increased, whereas the amounts of filter ash decreased due to the use of dry sorption processes (Section 2.5).

3. Emission of pollutants. In spite of larger amounts of refuse, there has been a marked reduction in SO_2 and HCl emissions due to the use of flue gas cleaning facilities. Technological improvements in the combustion process also resulted in a decrease in CO and hydrocarbon emissions (CH), although the NO_x emissions increased slightly or remained constant. The incineration of household waste falls under the Waste Incineration Regulation (17th Regulation for Execution of the BImSchG: Regulation for Incineration Plants for Refuse and Similar Materials – 17th BImSch). For example, 850 °C is the minimum permissible temperature. In addition, emission limits are specified by the Technical Instructions (TI) for Air or the 17th BImSchV (1990), including for dust particles (10 mg m^{-3}), sulphur oxides and CO (50 mg m^{-3}), NO_x (200 mg m^{-3} as NO_2), HCl (10 mg m^{-3}), inorganic fluorine compounds (1 mg m^{-3}) and dioxins/furans with a total of 0.1 ng m^{-3}. Special powdery inorganic substances (metals) are combined into classes: e.g., Tl + Cd + Hg with 0.2 mg m^{-3}, Cd + Tl with 0.05 mg m^{-3} and Hg alone with 0.05 mg m^{-3} in Class 1.

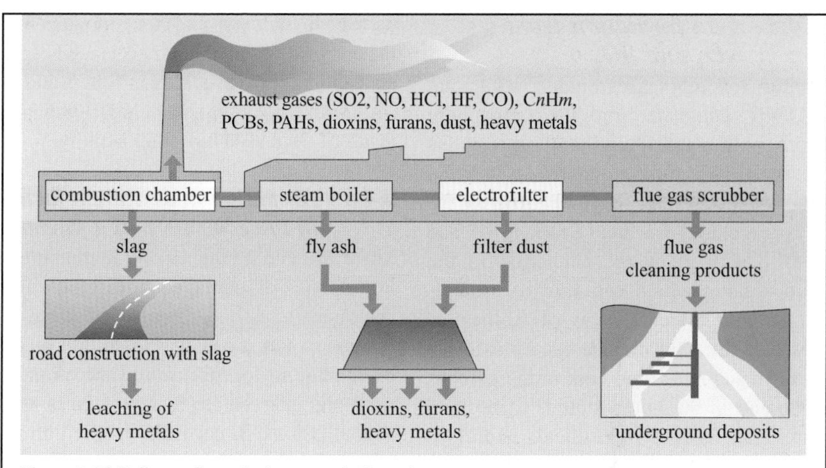

Figure 2.4.3 Pollutant flows during waste incineration

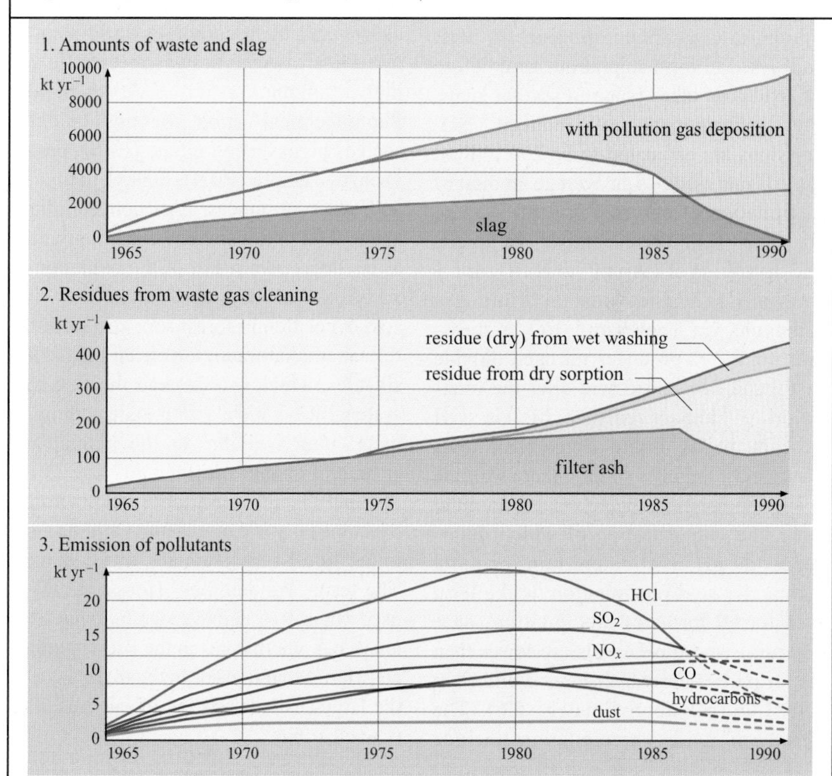

Figure 2.4.4 Waste incineration in Germany

2.4.5 *Acid formation from NO_x and SO_2*

Acidic reacting gases such as SO_2 and NO/NO_2 can enter into the atmosphere from both natural and anthropogenous sources. The anthropogenously induced fraction is >50% for SO_2 and >30% for the nitrogen oxides (see also Sections 1.3 and 1.4). These gases are converted in the troposphere, which can lead to precipitation with extremely low pH values, as low as 2. Sulphur dioxide can be oxidised to form sulphuric acid both in the gaseous state and in the liquid state (i.e. in cloud droplets). Ozone, OH radicals and H_2O_2 (see also Sections 2.2.5 – 2.2.7) play an important role in these processes. In the liquid state, sulphur dioxide is first hydrolysed to form sulphurous acid before the oxidation takes place in several steps. The anthropogenously induced SO_2 emissions are estimated to be 200 million t yr^{-1}. If one assumes an average amount of precipitation of 600 mm yr^{-1}, an average pH value of 4.2 corresponds to an annual acid deposition of 1.8 g m^{-2} of H_2SO_4 or 2.3 g m^{-2} of HNO_3. The oxidation of nitrogen monoxide via ozone and HO_2 radicals (Section 2.2.5) produces nitrogen dioxide firstly and then nitric acid after the interaction of impact partner M via OH radicals. In the liquid state, both acids disassociate largely into their anions nitrate or sulphate and hydrogen ions (hydronium ions), the concentration of which determines the pH level. 'Acid rain' or more accurately 'acid precipitation' is the term used for all forms of precipitation (dew, fog, rain) whose pH levels are lower than that of pure water which is in equilibrium with CO_2 (pH approximately 5.6). The effects of acidic precipitation include damage to buildings and stone monuments, the corrosion of metallic objects, and especially an increase in acidity in waters and in soils with the subsequent acidification as a result of decreased buffering activity.

2.4.6 *Air pollution as a stress factor in the ecosystem of a forest*

Since the end of the 1970s, increasingly large areas of damaged coniferous and deciduous trees have been investigated in relation to forest damage research. Causes of damage to forests include the climate, pests and intervention by foresters, as well as immissions, in particular acidic immissions, photo-oxidants, heavy metals in dust particles, and so on. Ozone, nitrogen oxides and hydrocarbons should also be listed in areas with heavy street traffic and high amounts of UV radiation. Photochemical smog contains oxidants such as peroxyacetyl nitrate (PAN), peroxybenzoyl nitrate (PBN), dialkyl peroxides and other substances. PAN concentrations above 0.02 ppm damage vegetation within a few hours as a result of the oxidation or acetylation of SH groups in enzymes. The acid deposition in forest soils results in the release of aluminium ions from the natural silicates, which can damage the fine root system, heavy metals with toxic activity as metal ions available to the plants, and structural changes in the soils due to the destruction of important soil colloids. Symbiotically important fungi and bacteria in the forest ecosystem are also damaged. The term 'Waldsterben' (forest death) is used when tree damage occurs over large areas, e.g. via flue gas in the Riesengebirge and Harz Mountains or in the tropics due to the large-scale harvesting of raw materials (fire clearing).

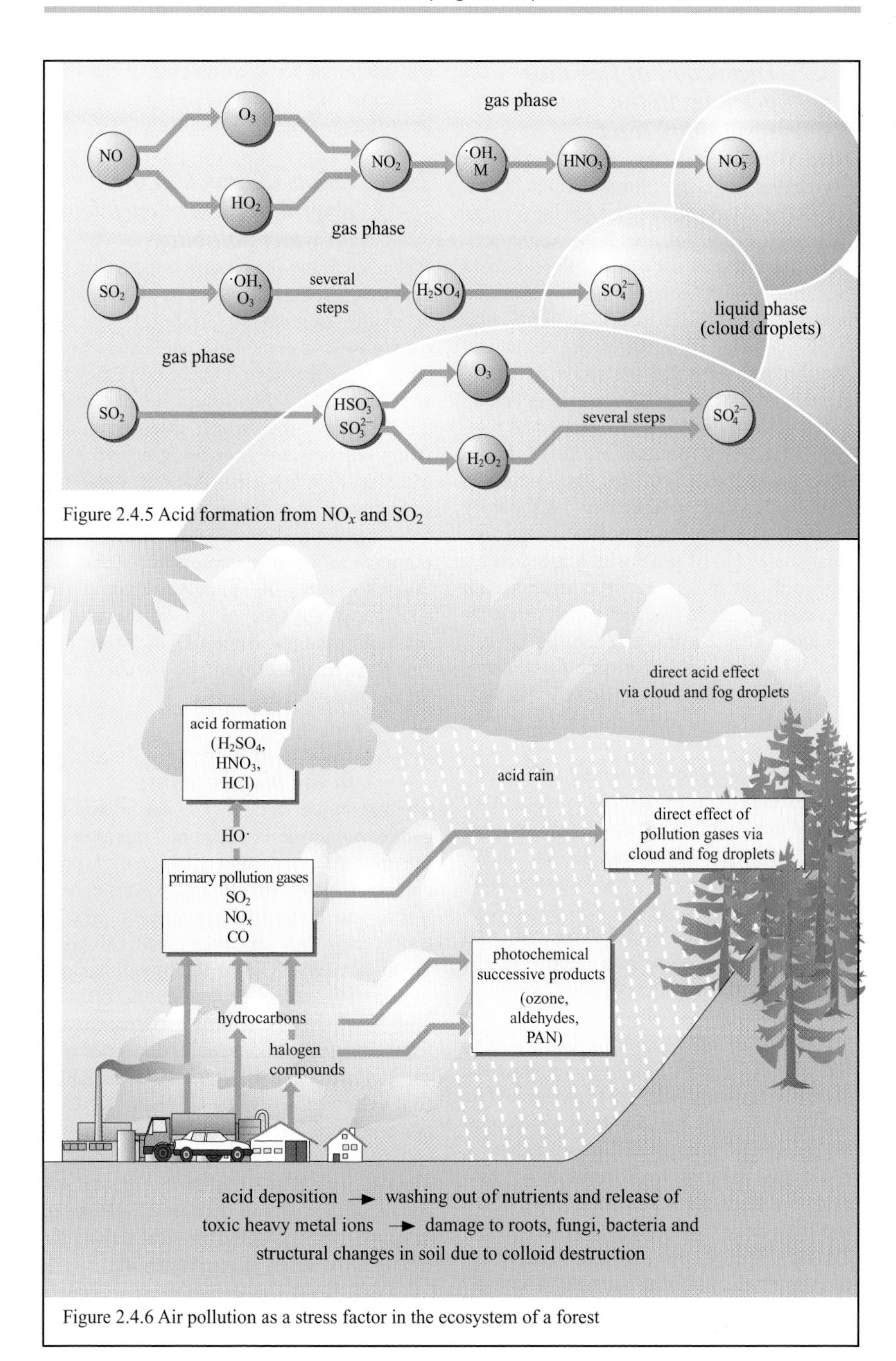

Figure 2.4.5 Acid formation from NO_x and SO_2

Figure 2.4.6 Air pollution as a stress factor in the ecosystem of a forest

2.4.7 *Deposition of fine dust particles in the human respiratory tract*

The 'MWC' lists (Maximum Workplace Concentration) have limits which for a number of chemicals are given in the form of particulates (e.g. for metals or pesticides). The definition used by the Senate Commission for Testing Unhealthy Working Substances of the German Research Centre (DFG) reads: 'Particulates are dispersed distributions of solid materials in gases, generated by whirling. Dust particles belong to the aerosols, along with smoke and fog.' Particulate air pollutants are divided into three groups: grit (10–200 μm), which is technically easy to remove and hardly relevant with respect to health issues; fine particulates (1–10 μm), which are increasingly difficult and expensive to separate with decreasing particle size and which can enter the lungs after 10 μm; and aerosols (= 10^{-4} cm), which are also called fine particulates. Together, fine particulates and aerosols form airborne particles, which are pathologically relevant.

From an occupational safety standpoint, dust particles that can enter the respiratory tract ('total deposited dust') are divided again into 'nose–pharynx–larynx dust', 'tracheo-bronchial dust', and 'alveolar dust'. Larger particles are deposited in the areas of the nose, pharynx and larynx, whereas smaller particles penetrate deep into the lungs. The first two kinds of particles can be removed through the cleaning mechanism of the respiratory tract (movement of the cilia, formation of phlegm), or they can enter the digestive system with the phlegm. The particles with the greatest toxic effects are from soot (from unfiltered diesel exhaust), PAHs (as unburnt hydrocarbons in auto exhaust), from acids adsorbed to particles, and from heavy metal particles (such as Pb, Zn, Cd). The 'TI for Air' also establishes thresholds for airborne particles (with no consideration for the contents of the dust particles) of 0.15 and 0.30 mg m^{-3} (long-term and short-term, respectively).

2.4.8 *Points of attack in the respiratory tract dependent on water solubility*

Water solubility is a decisive factor for the activity of pollutants in the respiratory tract or in the ambient air. Materials that are readily water-soluble have damaging effects on the eyes, larynx and trachea, but they are also captured by the mucous membranes in the nose and throat, which produces an irritating reaction with coughing or sneezing. Materials that are less soluble in water can penetrate as far as the bronchial tree and are less irritating because of the smaller numbers of nerve receptors there and the thinner layer of mucus. Hydrophilic substances advance as far as the alveoli (including the phosgene $COCl_2$, ozone and fine particles of CdO) and lead to changes in the pulmonary tissue.

2.4.9 *Effects of CO in the bloodstream*

The maximum workplace concentration for carbon monoxide is 30 ml m^{-3} (ppm) or 33 mg m^{-3}. At 20 °C and 1.013 bar, 1 ppm corresponds to 1.165 mg m^{-3}. Levels in road traffic can be as high as 50 ppm, and in traffic jams they can even reach 140 ppm. Upon uptake of CO in the blood, haemoglobin (Hb), or oxyhaemoglobin (HbO_2) with inhaled oxygen, which is responsible for the transport of oxygen, is converted into carboxyhaemoglobin, COHb (CO has a 300-fold greater affinity for Hb than O_2). At a concentration of 74 ppm CO in the atmosphere, 10% of the Hb in the blood is blocked for oxygen transport. Figure 2.4.9 shows the correlation between COHb level, effective time of CO and physical activity for different CO levels in respiratory air.

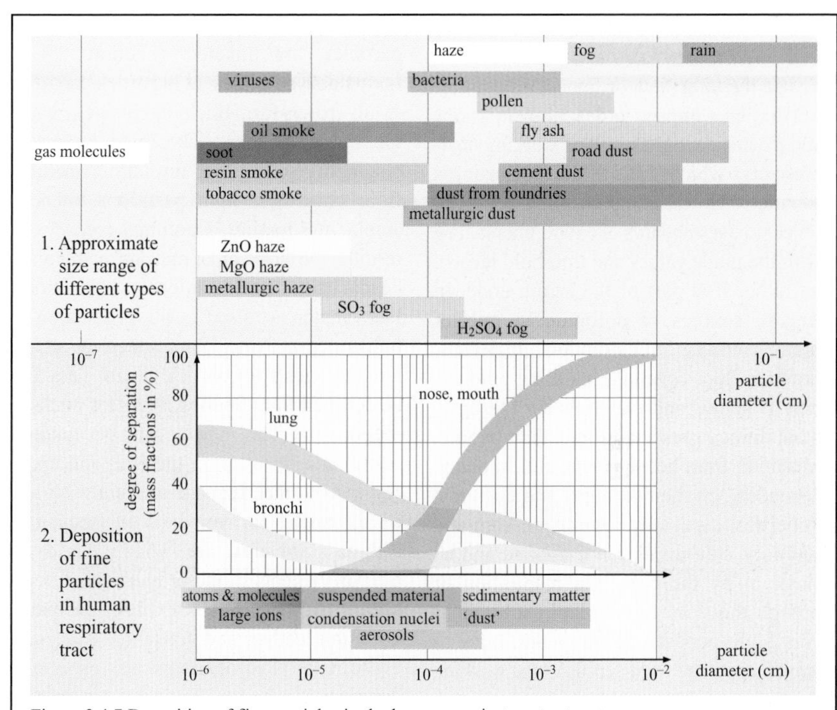

Figure 2.4.7 Deposition of fine particles in the human respiratory tract

Sites of attack	Water solubility	Substances
eye, larynx, trachea	high	NH_3, HCl, $HCHO$, S_2Cl_2, $CH_2 = CH\text{–}CHO$
bronchi, bronchioles	average	SO_2, Cl_2, Br_2, $RCOCl$, $R(NCO)_2$
bronchioles, alveoli, capillaries	slight	O_3, O_2, NO_2, $COCl_2$, CdO,

Figure 2.4.8 Points of attack in the respiratory tract dependent on water solubility

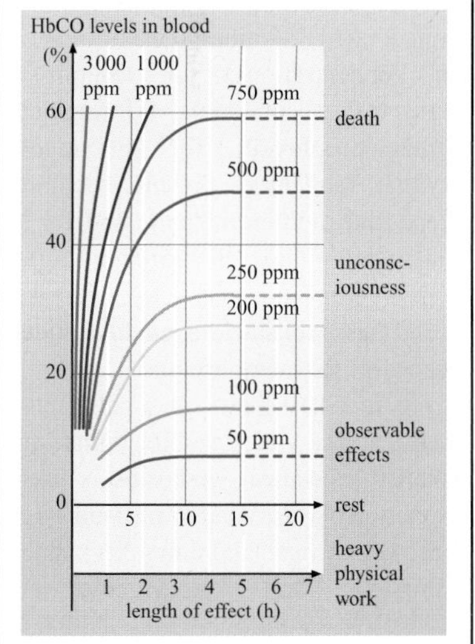

Figure 2.4.9 Effects of CO in the bloodstream

2.4.10 Sources of pollution in living and working environments

We use the term 'indoor impact' in contrast to workplace impact to describe emissions indoors (residential as well as working environments) which are not due to business operations. However, indoor air quality is not subject to the standards of environmental law (with the guide values and threshold levels), but rather it is part of the health code. In general, sources of pollution in interior rooms include construction materials, furniture, floor coverings, cleaning products such as sprays and materials for cleaning, maintaining, preserving and disinfecting, exhalation from house plants, and smoking. Depending on their physical and chemical properties, especially vapour pressure and solubility, substances can evaporate and be taken up by the body via respiration or through the skin. Less volatile substances with high persistence can contribute to chronic exposure to materials in the working and living environments. For example, the German Federal Board of Health has recommended as quality standards for interior impact MRC values (maximum room air concentrations) for several wood preservative materials (formaldehyde, pentachlorophenol), for styrene/toluene /xylene, tetrachloroethylene, trichloroethylene and dichloromethane and for some fungicides like dichlofluanid (as a pentachlorophenol substitute). Likewise, the WHO has published air quality guidelines applicable for interior rooms.

Figure 2.4.10 shows other pollutants with their sources, which are not limited just to volatile substances. Asbestos is used for support in plastics and for insulation (against fire, noise, heat and humidity), and has fibrogenic and carcinogenic effects (asbestosis and lung cancer) if the fine particles are inhaled (Section 2.4.7). Formaldehyde is present in particle board, in resins (urea-formaldehyde resins), wood glues and lacquers. The most important respiratory sources are furniture, panelling, floor coverings, light partitions, tobacco smoke and textiles. Continual contact can result in the irritation of mucous membranes, headaches and allergic reactions. Isocyanates are used in the production of polyurethane foams, floor adhesives and as binding agents for particle board. They can be detected from flooring layers for weeks in indoor air. The result can be mucous membrane irritation in the eyes and upper respiratory tract, damage to the alveoli and on up to the formation of an isocyanate asthma. Solvents are widely used as auxiliary agents in the production of items such as adhesives, textiles, lacquers, imitation leather, and household materials. Chlorinated hydrocarbons are especially dangerous to health. Pesticides are used as fungicides (see above), for example. Ozone, with its typical slightly pungent odour as an irritating gas, comes from photocopying machines and laser printers; nitrogen oxides from combustion plants (Section 2.4.4); phenols from artificial resins (mucous membrane irritation, damage to liver and kidneys, plus changes in the blood); polycyclic aromatic hydrocarbons from tobacco smoke, automobile exhaust, incomplete combustion, and grilling meat products. All of these are carcinogenic. Polychlorinated biphenyls from sealants and PVC floor coverings (storage in fatty tissue) can induce liver damage. As a radioactive inert gas, radon comes from rocks in the ground and from construction materials.

Figure 2.4.10 Sources of pollution in living and working environments

2.5 Principles of air quality control

2.5.1 Techniques for cleaning exhaust air

1. Dust and aerosol separator. The separation principle in the particulate removal procedure lies in the action of external forces, whereby a relative movement of the dispersed particles with respect to the carrier gas is achieved. Thus, by taking advantage of gravity and force of inertia, particles with a diameter >10 μm enter zones in which they are no longer caught up or transported by the gas flow. Inertia force separators utilise techniques in which a separation of the particles from the gas stream takes place via mass-proportional field forces (gravitational, inertia and centrifugal force). The best separation in this field is achieved by a centrifugal force separator, or cyclone. The gas is fed tangentially into the cylindrical container (with a conical base). A vortex is generated, and the particles are spun against the wall of the container and from there flow downward in a spiral motion in the form of strands of solid matter. The purified gas flows radially from the outside to the inside and exits through the top. Independently of the fineness of the particles, higher grades of separation can be achieved with filtering separators, wherein the filter medium can consist of layers of fibres (fabric filters) or of grainy layers. In electrostatic filters (electrical filters), the separation process occurs in four intermediate steps: (i) The particles are charged in the electrical field. (ii) The charged particles are transported to the precipitating electrode. (iii) Adhesion takes place and layers form. (iv) Finally, the layer of particles is removed. These filters are used mainly in fly ash precipitation in power plants, in iron metallurgy, in the chemical industry and in the cement industry. The separators which use water (wet washers) bring the particles dispersed in the gas stream in contact with a washing fluid and separate the resultant particle–fluid mixture from the gas. They are used especially when sticky or readily flammable particles need to be removed. Scrubbing towers are the oldest type of construction. Jet washers work using the principle of a water pump. In a rotating washer, the gas that needs to be cleaned flows from below into the cylindrical container. A dense veil of droplets is formed in the washing zone via rapidly rotating spray disks. In the process, the dust particles attach to the drops of liquid and are spun with the drops out to the container wall. The most frequently used high-powered washer, the Venturi washer with a Venturi tube, from which the finest droplets escape due to the high shearing effect of the gas flow, is suitable for separating particles having a diameter of 0.05 to 0.2 μm. The high separation capability is due to the high gas acceleration in the most narrow cross-section (the throat) and to the resultant high relative speed between the dust particles and droplets (Bank, 1994).

2. Gas separators. The separation of vaporous and gaseous pollutants can be performed using the processes of absorption or adsorption. A $Ca(OH)_2$ solution (Sections 2.5.2–2.5.4) is used for removing sulphur from flue gases. Activated charcoal can be used as an adsorbent for the removal of odours, or a biological sorption procedure (bioscrubbing) can be applied, in which odoriferous substances are broken down using microorganisms (Sections 2.5.9). Thermal afterburning procedures are used for the elimination of organic substances (hydrocarbons, amine, etc.), whereas catalytic afterburning is used in areas such as cleaning automobile exhaust (Sections 2.5.6).

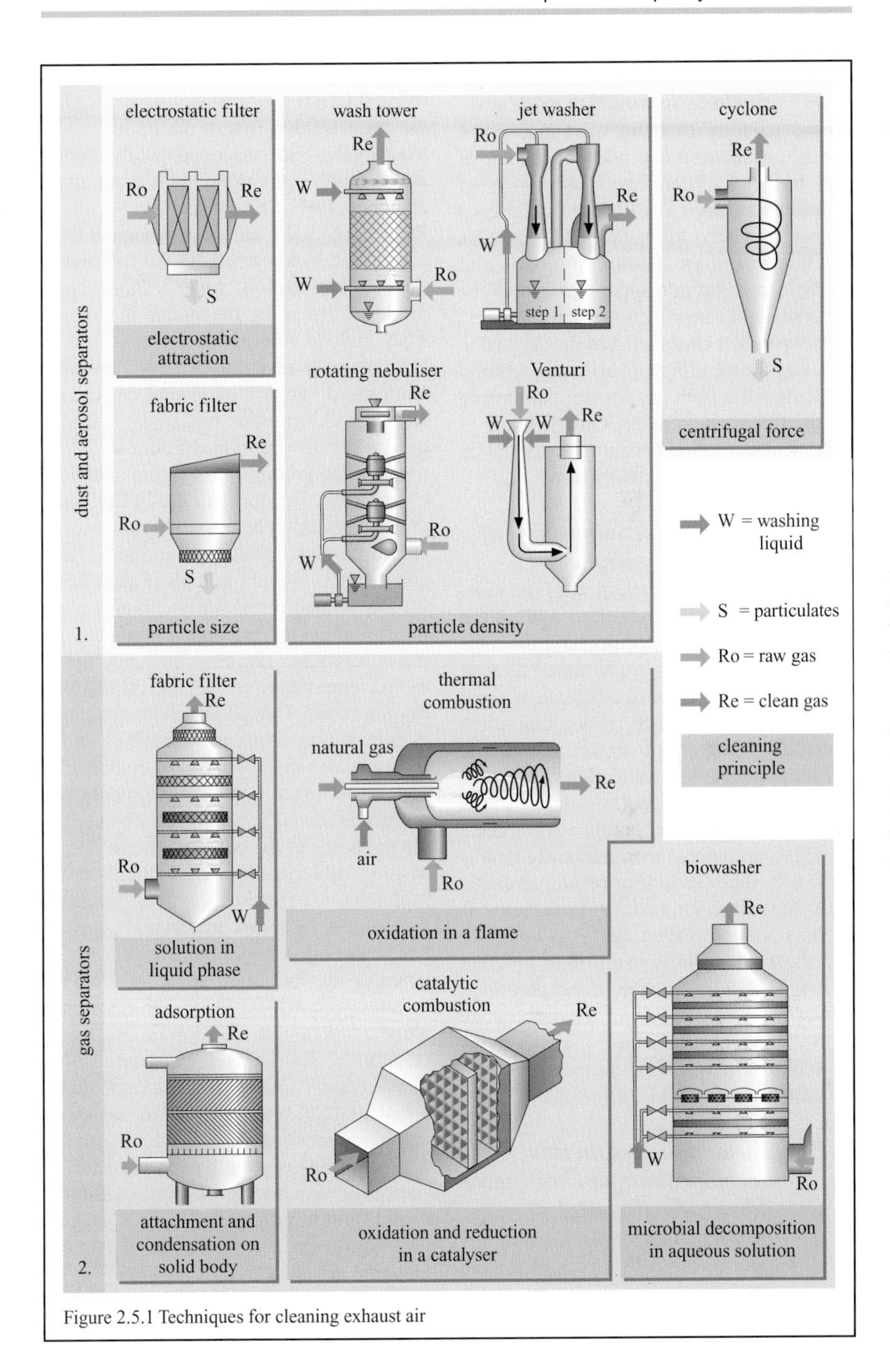

Figure 2.5.1 Techniques for cleaning exhaust air

2.5.2 *Comparison of different dust particle separation systems*

The separating capacity of a given dust particle separator is dependent on the size of the particles. Therefore, the degree of fraction separation $T(x_P)$ is used in order to depict the relative efficiencies. This factor is the quotient of the amount of a particular particle size fraction separated and of the amount of the same fraction in the feedstock. The separation characteristics determined in this way for the different particulate removal systems differ from one another particularly in the fine particle range. Only filtering or electrical separators have a high separation capacity for particles smaller than 5 μm.

2.5.3 *Thermal decomposition of organic materials*

In an ideal tubular reactor, in which no intermixing takes place along the length of the tube and all of the molecules have the same residence time due to an assumed plunger flow profile, one can make statements about the decomposition (thermal decomposition, including oxidation) of organic materials in the air in comparison with CO at 750 °C ($CO + \cdot OH \rightarrow CO_2 + H\cdot$) and as a factor of the dwell time. A large number of organic materials decompose thermally more rapidly than CO; these include aliphatic, aromatic and chlorinated aliphatic hydrocarbons, as well as polychlorinated biphenyls or PCBs. On the other hand, several other problem materials have considerably slower decomposition kinetics, including chlorobenzene (CB), tetrachlorobenzene (TCB), hexachlorobenzene (HCB) and especially tetrachlorodibenzodioxin (see also Section 2.5.4).

2.5.4 *Formation and thermal decomposition of chlorinated dibenzodioxins and dibenzofurans*

With respect to the path of origin for the occurrence of polychlorinated dibenzodioxins (PCDD) and dibenzofurans (PCDF), we must mention first of all the fuel itself, whereby these substances are hardly decomposed during combustion in the combustion chamber (see Section 2.5.3) Secondly, PCDD and PCDF can be regenerated from chlorinated hydrocarbons in the fuel and/or in the combustion gases. These 'pre-products' are called precursors in general. They include chlorinated aromatic hydrocarbons such as chlorobenzene, multiply chlorinated benzene or phenol derivatives and polychlorinated biphenyls. Thirdly, dioxins can also be formed from non-chlorinated hydrocarbons and inorganic chlorine compounds. The formation of PCDD and PCDF from chlorobenzene, chlorophenol or chlorophenoxy acids starts at 500 °C; it is about 100 °C higher for PCBs (Figure 2.5.4, 1). The second formation mentioned plays the most important role in homogeneous gas phase reactions. The experimentally determined temperature ranges are valid for an oxygen excess. They show that the formation of pollutants takes place preferentially in the temperature range of the post-combustion zone. Therefore, the above-mentioned precursors should be completely destroyed in the hot zones of the combustion chamber if at all possible (Section 2.5.3). With an air ratio of $\lambda = 1$ (Section 2.4.4) and a dwell time of 2 s, PCBs are increasingly decomposed above 800 °C (Figure 2.5.4, 2), whereby the formation of PCDF shows a maximum at 900 °C here. The decomposition curves for 1,2,3,4-TCDD show (Figure 2.5.4, 3) that the decomposition kinetics for TCDD are not accelerated until above 800 °C. They are dependent mostly on the temperature and, to a much smaller degree, on the oxygen partial pressure. In order to prevent the formation or emission of dioxins and furans, the 'TI for Air' require minimum temperatures of 1200 °C and 6% O_2 in the exhaust gas.

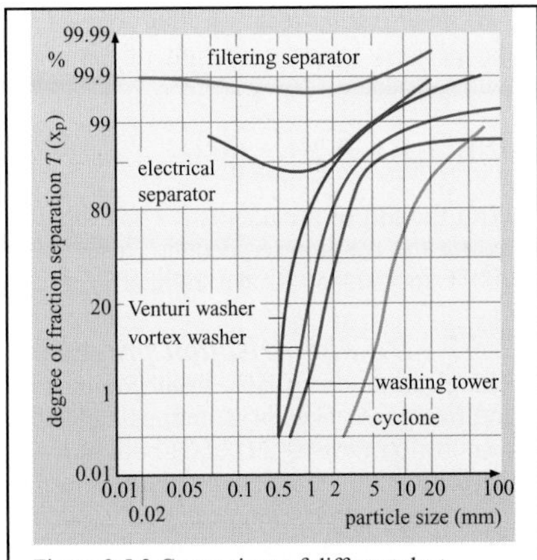

Figure 2.5.2 Comparison of different dust particle separation systems

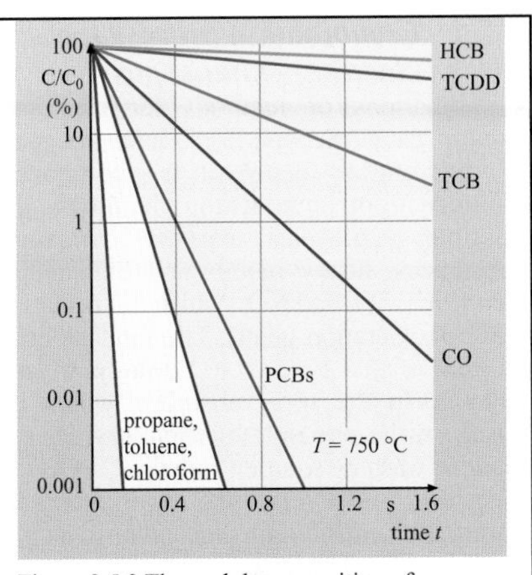

Figure 2.5.3 Thermal decomposition of organic materials

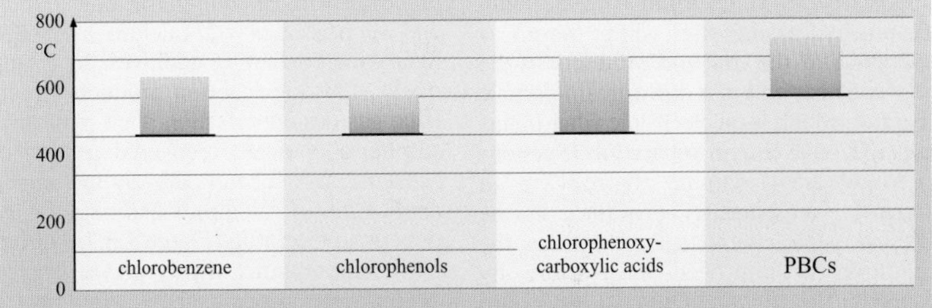

1. Temperature ranges of dioxin and furan formation for various precursors

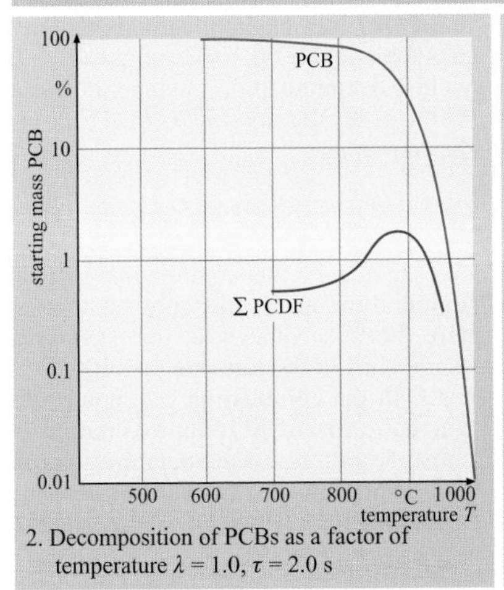

2. Decomposition of PCBs as a factor of temperature $\lambda = 1.0$, $\tau = 2.0$ s

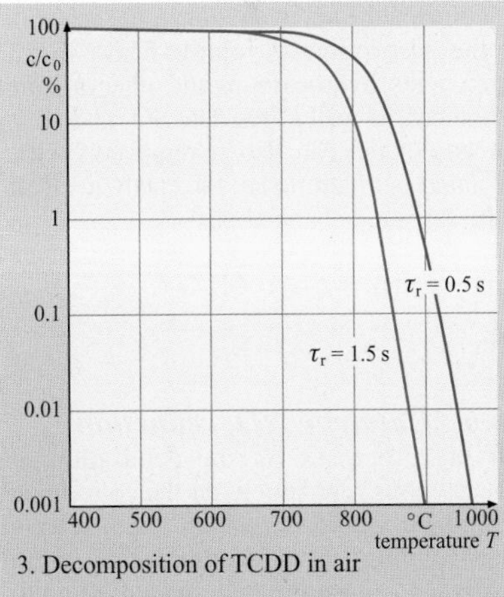

3. Decomposition of TCDD in air

Figure 2.5.4 Formation and thermal decomposition of chlorinated dibenzodioxins and dibenzofurans

2.5.5 *Combination process for cleaning exhaust gas*

Combinations of methods for cleaning exhaust gases (Section 2.5.1) have the goal of retaining as many of the pollutants emitted from combustion facilities as possible, e.g. dust particles, organic substances, heavy metals and the acidic gases HCl, SO_2 and NO_x, which can occur in waste incineration plants. The fundamental aspects of this type of technology are as follows. In the spray absorber, which is based on the principle of spray drying, an alkaline washing solution (usually $Ca(OH)_2$) is sprayed into a scrubbing tower, where it comes in contact with hot exhaust gases (120–160 °C). The heat energy is transferred to the droplets, such that drying takes place there due to the evaporation of the water in the gas flow. SO_2, HCl and HF react with calcium hydroxide, after which there is a separation of the dry end products in the spray absorber and in a downstream particle separator, which is an electronic filter in this case. Effective dioxin separation (Sections 2.5.3 and 2.5.4) can be attained via an extensive fine-particle separation at the lowest possible operating temperatures; the separation is improved even more by adsorption to activated charcoal. In a third step, spray absorbers and electronic filters are connected to activated charcoal reactors for the adsorption of volatile heavy metal components, hydrocarbons and other organic substances, as well as residual SO_2 and HCl. Nitrogen oxides can also be removed via the introduction of ammonia; the catalytic effect of the activated charcoal is:

$$6NO + 4NH_3 \rightarrow 5N_2 + 6H_2O$$

$$2NO_2 + 2C \rightarrow 2CO_2 + N_2)$$

2.5.6 *Catalytic NO_x reduction*

Drying processes for the denitration of exhaust gases rely largely on the conversion of nitrogen monoxide, which is not readily water-soluble, with ammonia to form molecular nitrogen and water vapour. An undesirable reaction that occurs in the process is the oxidation of ammonia with oxygen:

$$4NH_3 + 3O_2 \rightarrow 2N_2 + 6H_2O$$

With the aid of a catalyst, it is possible to reduce the reaction temperature from about 900 °C to 180–450 °C at a slight NH_3 excess.

2.5.7 *Desulphurisation processes*

The elimination of SO_2 from combustion and flue gases takes place immediately after combustion (at 800–1100 °C) both at lower temperatures (e.g. in power plants after the boiler) and at higher temperatures. For desulphurisation in flue gas (Figure 2.5.7, 2), we use drying processes (additives are $Ca(OH)_2$, $CaCO_3$ or activated charcoal), semi-drying processes with calcium-containing additives, and wet processes with calcium- and sodium hydroxide-containing additives. In a subsequent step, the reaction products gypsum (flue gas desulphurisation plant gypsum) or sulphuric acid are recovered as valuable substances. A second pathway includes the application of calcium-containing additives directly in the firing (Figure 2.5.7, 1). At the high temperatures which prevail there, a decomposition occurs first, yielding CaO and H_2O (from calcium hydroxide) or CaO and CO_2 (from calcium carbonate). Partial pressures of $p(H_2O) = 0.2$ and $p(CO_2) = 0.1$ yield decomposition temperatures of 450 °C ($Ca(OH)_2$) or 750 °C ($CaCO_3$). Then the conversion

$$CaO + SO_2 + \tfrac{1}{2}O_2 \rightarrow CaSO_4$$

takes place in the combustion gas. The temperature level must not be too high here, because otherwise the equilibrium would shift to the left side (at 1100 °C and 6% O in the combustion gas: equilibrium concentration of SO_2 approximately 100 ppm). Kinetics and temperature windows as a function of the various calcium-containing additives play an important role in these processes.

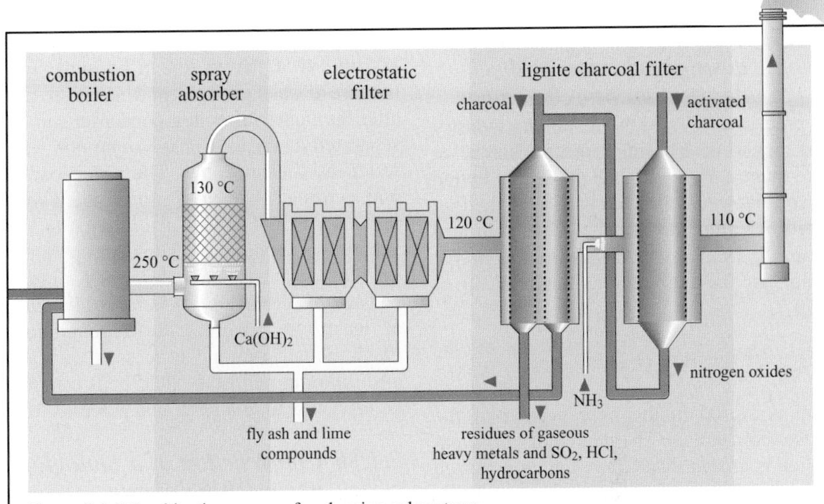

Figure 2.5.5 Combination process for cleaning exhaust gas

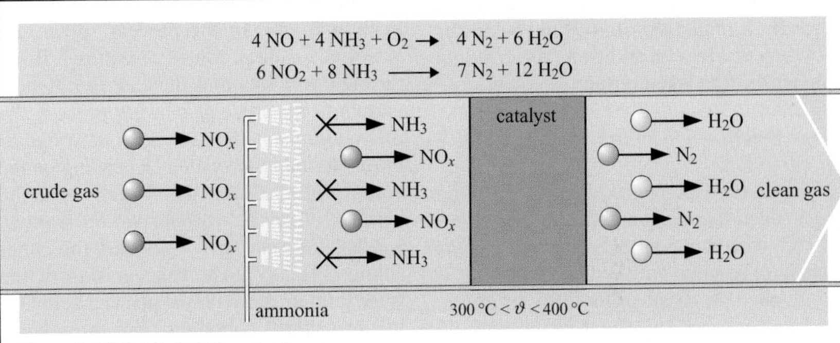

Figure 2.5.6 Catalytic NO_x reduction

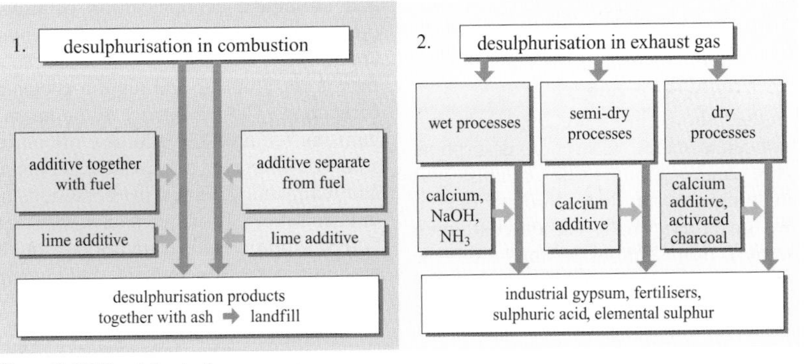

Figure 2.5.7 Desulphurisation processes

2.5.8 *Simultaneous processes for desulphurisation and denitration of exhaust gas*

A process developed by the mining industry in Essen works by using activated charcoal in two stages. First, sulphur dioxide is bound adsorptively to activated charcoal in the first adsorber after the dust has been removed from the exhaust gas, then it is converted to sulphuric acid at temperatures of 80 to 150 °C:

$$SO_2 + \frac{1}{2}O_2 + H_2O \rightarrow H_2SO_4$$

and is stored in the pore system of the activated charcoal. The degree of SO_2 separation is greater than 85%. Ammonia is then mixed with the exhaust gas and is directed to the second phase. The catalytic processes described in Section 2.5.5 take place here, whereby a reduction of NO_x to molecular nitrogen occurs. In addition, the residual sulphur dioxide is converted:

$$SO_2 + 2NH_3 + H_2O + \frac{1}{2}O_2 \rightarrow (NH_4)_2SO_4$$

The loaded activated charcoal can be regenerated thermally at temperatures above 300 °C, whereupon an SO_2-enriched gas (as a result of reaction with carbon) is generated for further processing (to form sulphuric acid or liquid SO_2). The first large-scale plant of this type began operating in the Arzberg lignite power plant near Bayreuth in 1987.

2.5.9 *Adsorption plant for solvent recovery*

The plant shown in Figure 2.5.9 uses four intermediate stages: adsorption, desorption, drying and cooling. The exhaust gas containing solvent flows through one or more adsorbers (with carbon-containing material) from bottom to top, until the vaporous solvent breaks through. The desorption (and therefore the regeneration of the adsorbents) usually occurs with steam at 120 to 140 °C (from top to bottom). The mixture of water and solvent which is present after the liquefaction in a condenser can be separated again in a phase separator with water-insoluble solvents. The water loaded with activated charcoal is usually dried with process gas and is thus also cooled to the adsorption temperature. Thus, an operation of the kind we have described always consists of several adsorbers for loading and at least one adsorber for regeneration. Continuously running processes with the adsorption in multistage fluid beds can also be used instead of a fixed-bed process.

2.5.10 *Construction of a biological gas-washing plant*

Bioscrubbers can be operated in an activated sludge or a trickling filter process (see Section 3.4.2). In the process shown, an activated sludge–water mixture flows through two washing devices in counter-current to the crude gas. In this process, the microorganisms (Section 2.5.1) are free to move around in the water. Depending on the composition of the gas which is to be cleaned, it may be necessary to operate one washer alkaline (pH 7–9) and the other acidic (pH 4–7). The regeneration of the wash fluid loaded with pollutants takes place in the settling container. The three basic processes of biological gas washing are transfer of material, chemical neutralisation, and biological decomposition of usually odoriferous substances such as aldehydes and ketones, amines, low-molecular-weight fatty acids, aromatic and suphfur-containing compounds. Optical activity of the microorganisms requires the addition of nutrient solutions (nitrogen–phosphate compounds), and maintaining certain pH conditions and oxygen levels. Biofilter plants with earth filters as carriers for microorganisms are used in water treatment plants and in animal recycling plants.

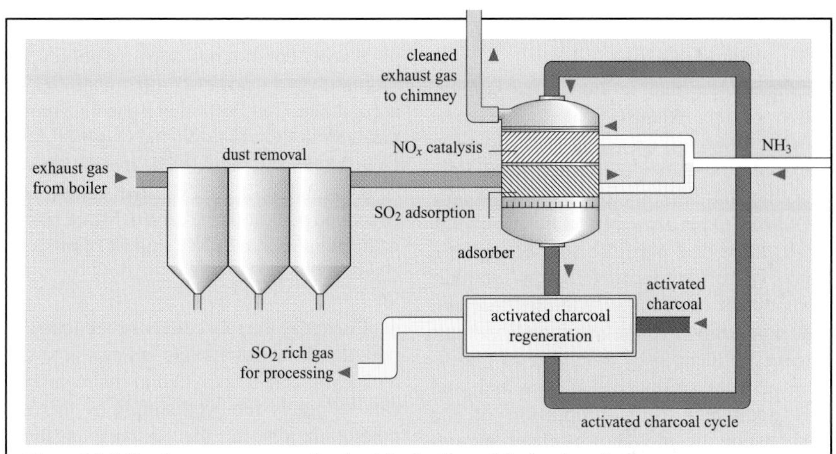

Figure 2.5.8 Simultaneous processes for desulphurisation and denitration of exhaust gas

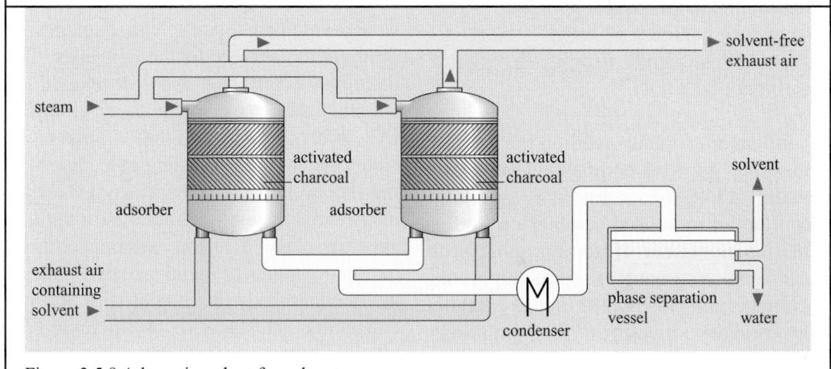

Figure 2.5.9 Adsorption plant for solvent recovery

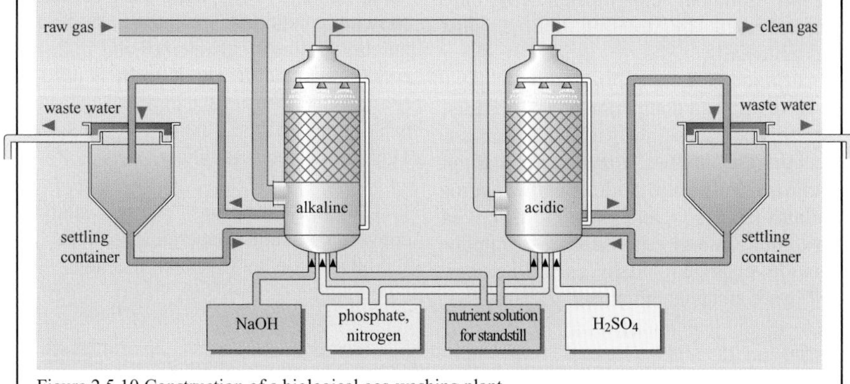

Figure 2.5.10 Construction of a biological gas-washing plant

2.5.11 *Automobile exhaust and its cleaning*

1. Parameters that influence the composition of the exhaust. Due to the short reaction times (10 ms idling up to <1 ms at maximum revolutions), exhaust gases from gasoline and diesel motors contain not only CO_2 and CO but also unburnt hydrocarbons, nitrogen oxides, particles (diesel motors), lead (gasoline motors), and sulphur compounds (especially in diesel motors). The composition of the automobile exhaust gases is influenced by a number of parameters. Primary influences are the fuel (and any additives) itself, proportion of the combustion air, plus motor-related parameters, from the ignition system to the construction and operation of the motor. 'Secondary parameters' include the possibility of subsequently treating the exhaust gas (Figure 2.5.11, 4).

2. Influence of the air ratio on the exhaust emissions. As in all combustion processes (Section 2.5.4), the air ratio λ is especially important in motorised combustion. In the emission minimum of certain substances (such as hydrocarbons), a maximum occurs for another component (NO_x in this case), so that primary measures for emissions reduction can only represent compromises. Therefore, a general reduction of all pollutant components can only be achieved by treating the exhaust gas afterwards.

3. Exhaust gas components. The exhaust gas from both diesel and gasoline motors can contain the limited components carbon monoxide, aliphatic and aromatic hydrocarbons (HC), NO_x and particles, as well as unlimited substances such as sulphur dioxide, hydrogen, aldehydes and ketones, plus trace amounts of amines, phenols, NH_3 and HCN. The particles do not consist of pure soot: compounds such as polycyclic aromatic hydrocarbons and metal oxides accumulate. Combustion in a gasoline motor yields 99% CO_2, H_2O, N_2 and O_2 plus 0.85% CO, 0.8% NO_x and 0.5% hydrocarbons. Approximately 2.5 g NO_x per vehicle are generated per kilometre driven. Motor traffic is the main source of NO_x and CO emissions (Section 2.4.1).

4. Currently, the most effective reduction of pollution from automobile engines is to use a catalytic converter. CO and hydrocarbons can be largely removed simply by thermal post-oxidation in the exhaust system. However, catalytic cleaning is necessary to eliminate the NO_x emissions. Unleaded fuel is a prerequisite, since lead damages or destroys the surface of the converter. The three-way converter was developed to eliminate the three components or groups of CO, hydrocarbons and NO_x. This is an embedded system consisting of a ceramic or metal converter. The actual catalyst is a layer of precious metals (platinum and rhodium, approximately 5:1) on a honeycombed carrier (ceramic or metal), coated with an intermediate layer (wash coat with activity-enhancing substances) to increase the effective surface area. Oxygen is necessary to convert hydrocarbons and CO, whereas the elimination of NO_x must take place under reducing conditions. The most favourable λ number for this is between 0.8 and 1.02 (the converter window). The λ probe is used to continuously measure the oxygen level in the exhaust gas and the mixture (Figure 2.5.11, 1) is adjusted accordingly.

5. Chemical reactions in the catalytic converter include conversions of HC, CO and NO_x plus other (residual) reactions.

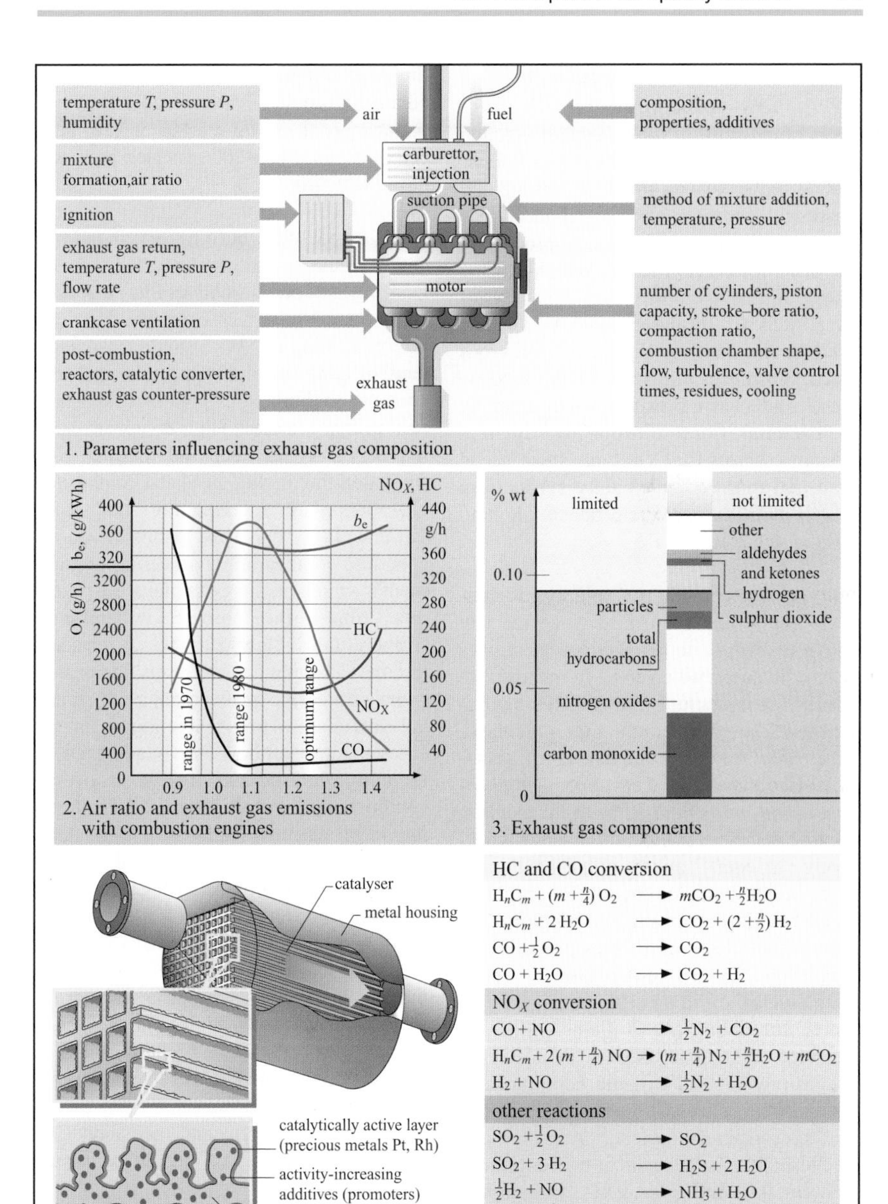

2. Air ratio and exhaust gas emissions with combustion engines

3. Exhaust gas components

4. Construction of catalyser

5. Chemical reactions in catalyser

Figure 2.5.11 Automobile exhaust and its cleaning

2.5.12 *Landfill gases and low-pollution combustion*

1. Composition of landfill gases (Section 4.6.6). Residential landfills act as uncontrolled biofermenters in which aerobic and anaerobic processes take place. Landfill gas biogas or landfill gas is the end product of the decomposition processes which provide energy for the microorganisms. It consists of roughly 55% CH_4 and up to 45% CO_2. The mixture also contains about 2% odoriferous trace substances. In stage II the organic waste contents are broken down by a variety of bacterial flora to form fermentation products such as H_2, CO_2, formic acid, acetic acid and more complex fatty acids, from which methane bacteria produce CH_4 and CO_2 in stages III and IV.

2. Gas chromatograms. Capillary gas chromatography has been used since the late 1970s to analyse numerous trace substances in landfill gases, which come from the refuse itself. These include halogenated hydrocarbons, sulphur compounds, alcohols, aldehydes, ketones, and esters of short-chained fatty acids. The upper gas chromatogram was obtained with an electron capture detector (ECD), and it shows selectively halogen- and sulphur-containing compounds such as hydrogen sulphide, dichlorodifluoromethane and trichlorofluoromethane (R12 and R11 as propellants), dichloromethane, chloroform, 1,1,1-trichloroethane and tetrachloroethane. In the lower chromatogram, a flame ionisation detector (FID) was used as the detector, which registers all organic compounds – among them benzene, toluene, ethylbenzene, the three isomeric xylenes, *n*-hexane and cyclohexane. They occur in concentrations between 0.02 and 500 mg m^{-3}. The particularly odoriferous compounds methanethiol, dimethylsulphide, 2-propanethiol, and 2-butanethiol could be determined. Altogether, so far more than 400 substances have been detected, some 150–200 of which have also been identified.

3. Low-pollutant burning off of landfill gas. Because of the high methane content, landfill gas is combustible and has a heating value for 2.3 m^3 which corresponds to one litre of heating oil. With small amounts of landfill gas, an environmentally friendly combustion (burning off) without using heat is of primary interest. High excess air ratios are necessary to keep the thermal NO_x formation low in well-insulated combustion chambers with subsequently 'hot' walls (around 1000 °C). This results in corresponding combustion engineering requirements, which make possible a sufficient dwell time in the combustion chamber at temperatures below 1300 °C. A targeted thermal decoupling is necessary in order to reach a state with smaller amounts of air and yet sufficiently low temperatures in the combustion chamber. This is achieved by reducing the thermal insulation of the combustion chamber. Figure 2.5.12 shows the resulting flow pathway for the low-pollutant combustion of landfill gas. To utilise the energy from landfill gases, we use electricity conversion with gas engines (in small and medium-sized landfills with low trace substance burden) and with steam turbines or steam motors (in large landfills, including those with high trace substance burden). Procedures for preparing landfill gas include the removal of H_2S and chlorofluorocarbons (using washing and activated charcoal processes), and possibly CH_4 enrichment via CO_2 depletion (using molecular sieves).

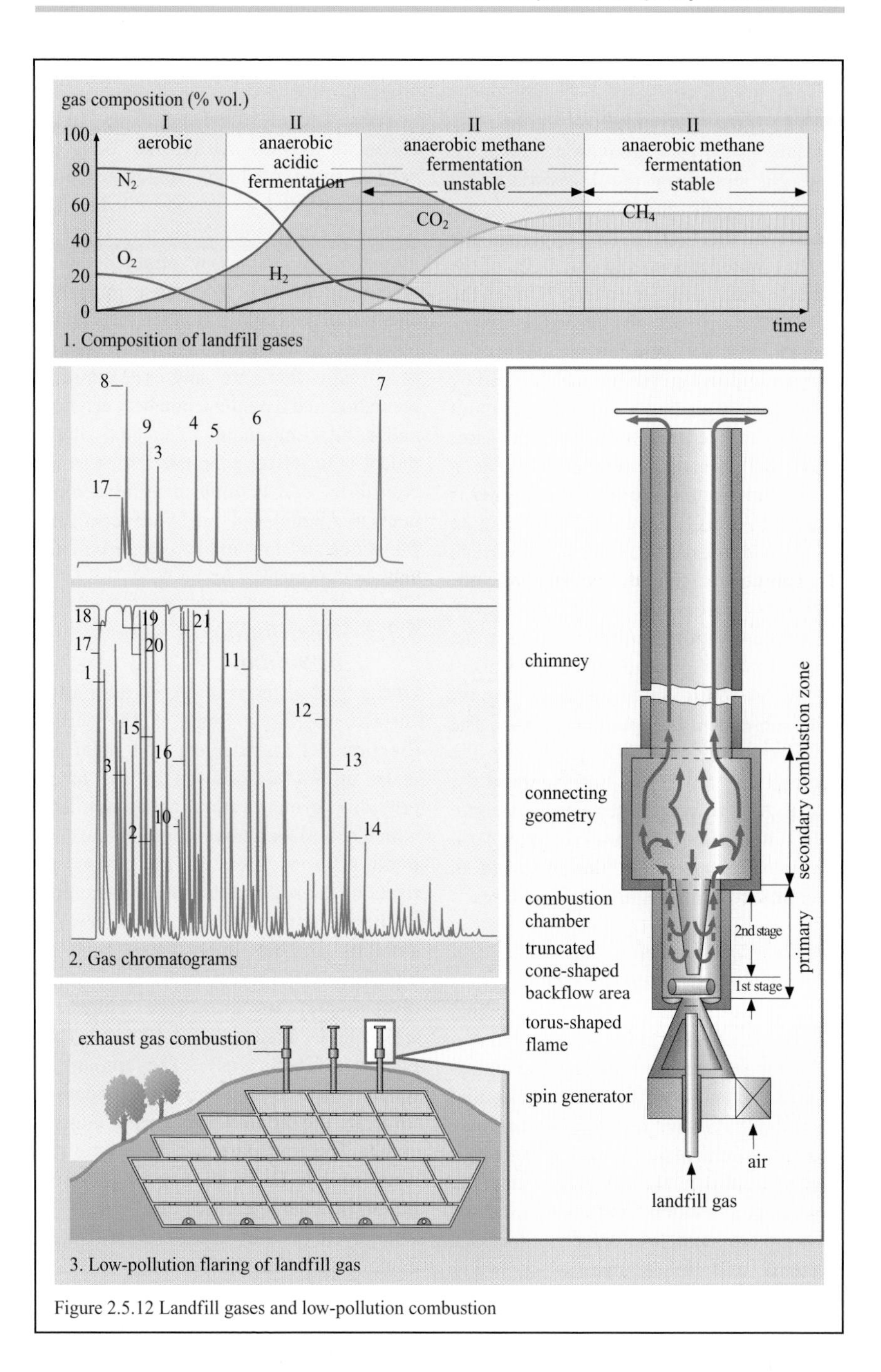

Figure 2.5.12 Landfill gases and low-pollution combustion

3 Water

3.1 The Earth's water cycle

3.1.1 The hydrological cycle

Oceans, seas, rivers, groundwater, polar ice and glaciers form the hydrosphere in liquid, gaseous and solid states. Some 97.3% of the total water supply on the Earth – which amounts to only 0.3% of the Earth's mass and yet covers 71% of the Earth's surface – is contained in the oceans. The global water cycle can be thought of as a gigantic distillation plant which is fed by solar energy. It has a capacity of about 420 000 km^3 per year, 85% of which comes from the ocean. Evaporation is at a maximum when the surface of the water is warm, the air is dry, and a high distribution of the water vapour can occur via vertical distribution (high wind speed). Evaporation from surfaces of plant growth is called evapotranspiration. The most important sink for water in the atmosphere is the condensation of the water vapour with subsequent precipitation via rain. The dwell time of water is 8–10 days in the atmosphere, but 1700–3000 years in the ocean. As a solvent, water plays a decisive role in weathering processes (Chapter 4), in transport processes in the soil, and in chemical conversions in rivers and lakes.

3.1.2 Box model for water chemistry

As a simple diagram, the zero-dimensional box model of water chemistry (see Section 2.1.2 for information regarding the atmosphere) depicts several characteristics that also play a decisive role in more complex models. It demonstrates two ways in which chemical substances (species) can enter the box: as precipitation from above from the atmosphere and via rivers, drainage systems and other streams of water (supply). The model assumes that all of the chemical species within the box, the defined space in which chemical reactions are to be viewed, are mixed well. The loss of chemical species from this space is indicated by the arrow directed downstream (delivery). The most important changes in the composition are induced via reactions between ions and molecules, in the streams that enter and exit, via sedimentation and the interaction between fluid and solid components. The goal of the model is to reflect parameters such as ion equilibria, conversions of solid components to a liquid state (and vice versa), and the water and ion uptake capacity of the soils in the catching basin.

3.1.3 Distribution of water by volume

Of the total water available on the Earth's surface, 97.3% is found in the oceans (Section 3.1.1). Glaciers and polar ice make up 2.1%, followed by the layers providing groundwaters (called aquifers) with 0.6%. With a total water volume (in a liquid state) of 400×10^6 km^3, lakes and rivers or the soil moisture each make up a total of only 0.1×10^6 km^3 in volume. The amounts in the atmosphere and the biosphere (0.013 and 0.001×10^6 km^3, respectively) are even less. Thus, the atmosphere contains only 1/100 000 of the Earth's available water. The amount of water present as vapour and in condensed form in the atmosphere varies considerably. The transport and distribution of water on the Earth are two of the most important characteristics of climate. The water cycle (as a hydrologic cycle) is kept in motion by solar energy.

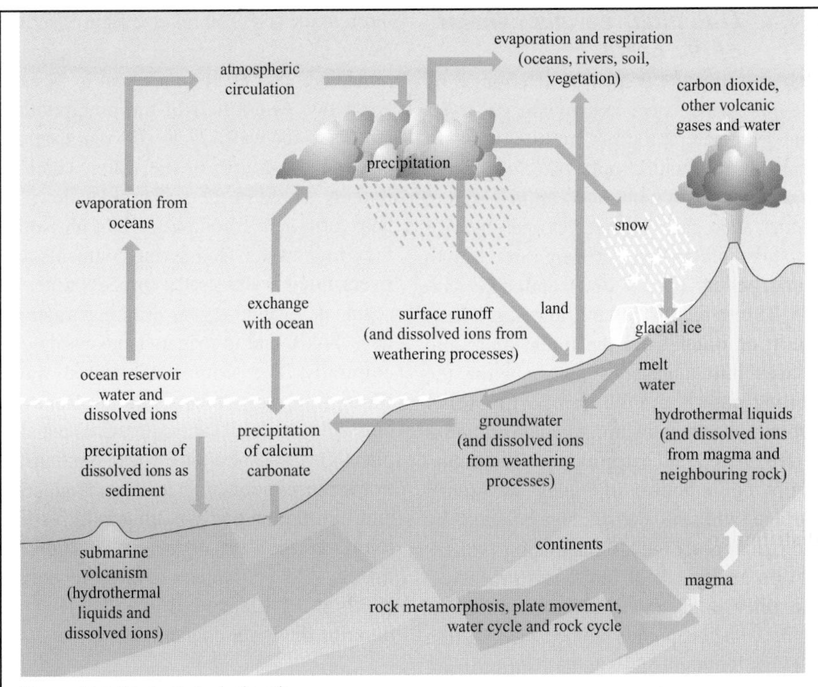

Figure 3.1.1 The hydrological cycle

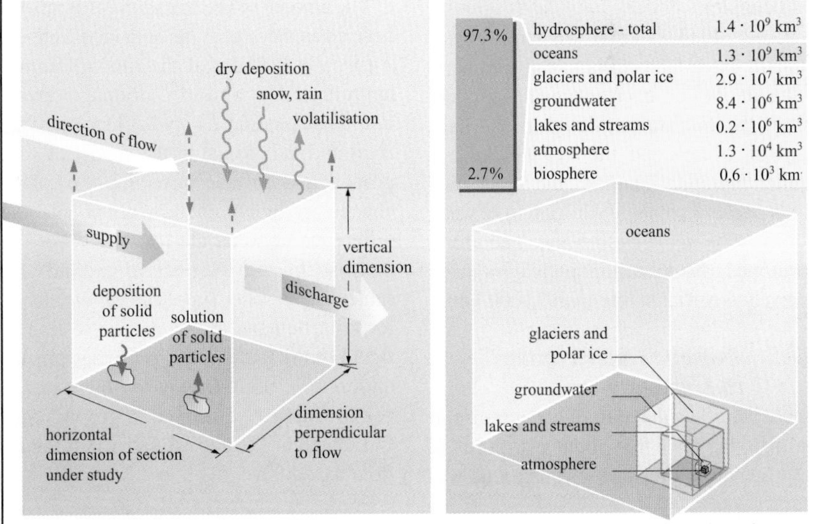

Figure 3.1.2 Box model for water chemistry

Figure 3.1.3 Distribution of water by volume

3.1.4 *Quantitative water content of the Earth*

The overall water content is determined by the following ways and means of water transport: precipitation (in the form of drizzling rain with droplets 0.05–0.25 mm in diameter, rain with drops up to 3 mm in radius, snow from ice needles and other ice crystals, ice pellets (frozen rain drops), snow pellets (1–2.5 mm) and hail with 2.5–25 mm radius); volatilisation (as a transition of water from the liquid and solid phases into a gaseous phase below the boiling point); and the transport of water vapour. The phenomenon of volatilisation is differentiated into categories of evaporation (from open bodies of water, moistened surfaces and soils without any influence by biological processes), transpiration (volatilisation as a result of biological processes; the release of water from the pores of plants), and evapotranspiration (see Section 3.11 as the sum of all volatilisation from areas covered with vegetation, which plays a vital role in the tropical rainforest, for example).

Some 40 000 km^3 are transported annually from the oceans to the continents in the air. Part of this is retained as groundwater; the rest flows through rivers back into the oceans. Of the 111 000 km^3 in annual precipitation onto the land, two-thirds comes from the volatilisation of water from reservoirs of fresh water. The aforementioned evapotranspiration plays a quantitatively significant role with 71 000 km^3.

3.1.5 *Average water content in Germany*

Figure 3.1.5 is a diagram of water transport and use. It shows that about two-thirds of the water from precipitation makes its way through the soil. The relative figures shown in the diagram are valid for an average precipitation of 803 mm per year in western Germany. About half of the precipitation returns to the water cycle via volatilisation. Around one-fourth of the water which is initially stored in the soil trickles down to help form new groundwater. Not only does the groundwater feed springs, streams and rivers, but it is also used as process water by segments of industry, as drinking water by households, and to irrigate lands used agriculturally. The temporarily stored water follows the path of volatilisation from not only cultivated soil (as evaporation) but also plants (via transpiration). These fractions of evapotranspiration, referred to as productive and non-productive volatilisation, respectively, are present in a ratio of approximately 1:3. A water content equation (Section 3.1.10) is used in soil science to describe the water content:

$$N = A + V_{ET} + (R - B)$$

The amount of precipitation N (in mm) is defined as the sum of the amount of run–off A (usually the sum of the run–off from a hydrologically uniform drainage area), evapotranspiration V_{ET} (volatilisation of an area) and the change in the supply of soil, ground and surface water (R – B). This change in supply in turn represents the difference between the reserves R (or the increase of the total above ground and subterranean water supply of the watershed for a particular time) and the use B. The figure shows these values as average precipitation N, drainage into the sea A, volatilization V_{ET}, storage in groundwater and temporary storage R and volatilisation from the soil B.

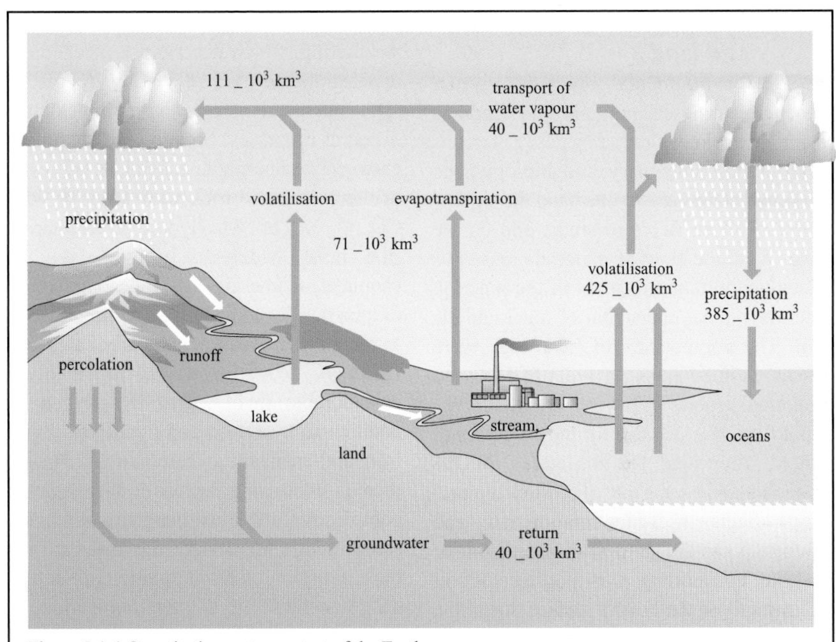

Figure 3.1.4 Quantitative water content of the Earth

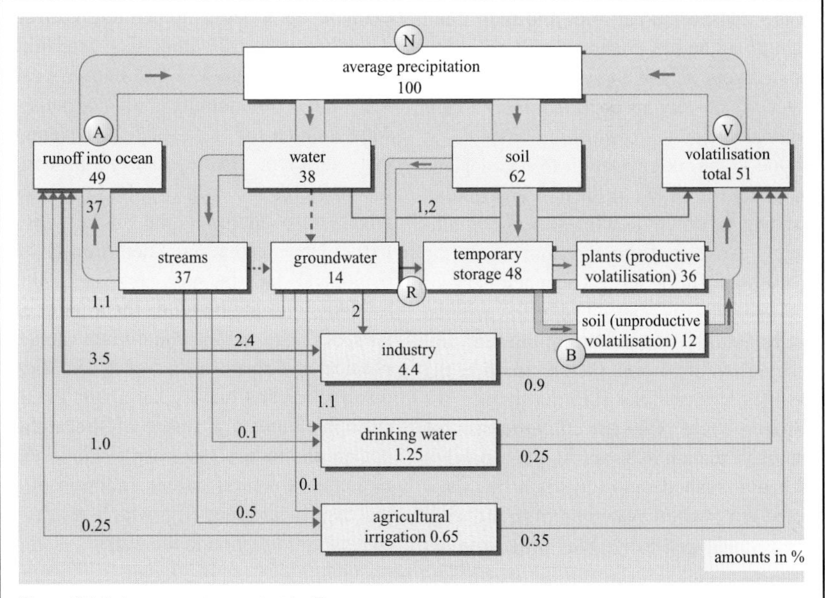

Figure 3.1.5 Average water content in Germany

3.1.6 *Interactions between water and land*

In the environmental study of water, ecochemistry uses methods and findings from hydrology and physics (Sections 3.1.1–3.1.5), chemistry and biology. The flowchart shows on the one hand the significance of these fields of study, and on the other hand the flow of materials and individual important subfields in conjunction with the mutual interaction of water and the soil. The components of hydrology were discussed in Sections 3.1.1–3.1.5. The most important aspects of physics are the energy input (which is decisive for the processes of volatilisation and for the speed of the materials cycles), the rate of erosion, temperature effects, the sedimentation rate, and silting up. As the natural displacement of soil, i.e. transporting away particles of soil (Chapter 4) on the Earth's surface due to the effects of wind or water, erosion increases with the flow rate and decreases with the size of the transported particles and their bond strength in the soil. Sedimentation results in loose layers of soil in a standing body of water, which play an important role in their material cycle.

Due to the sedimentation of organogenic materials (as sludge or peat) in particular, standing or slowly flowing waters can ultimately close in from the shore out (the process of silting up). The chemistry of bodies of water is determined by the material cycles and by the equilibrium between soil and water. Important factors include pH changes, conversions such as that of the nitrogen cycle (Chapter 2), processes of humus formation (Chapter 4), the capability of being washed out (and often the associated removal) of materials from the solid phase and generally the processes of chemical weathering. In contrast to physical weathering, which describes a mechanical disintegration of rock (thermally, by frost or salt), chemical weathering includes four different processes. Solution weathering is chemical weathering in general, as the result of the reaction of rock with water or with CO_2, O_2, SO_3 or NO_x. Hydrolytic weathering (the most widely distributed type of chemical weathering) is the decomposition of minerals due to hydrolysis (especially due to the salts of weak acids, carbonates and silicates). Oxidative weathering involves considerable involvement by oxygen in addition to hydration and hydrolysis (with iron and manganese carbonates). Finally there is weathering due to the formation of complexes. In the last case, organic acids (acetic acid, citric acid, or fulvic acids – see the discussion of humic matter in Section 4.2) generated in the decomposition of biomass or as plant exudates contribute to the chemical weathering of rock and especially to the mobilisation of heavy metals. The chemistry of waters also determines their biology. High levels of nutrients raise the net productivity of organisms. Monocultures (of special algae, for example) can develop. Bioregulatory processes determine the ecology of a body of water, which also includes the niches (as a particular formation determined by ecological factors such as space, time, nutrients and temperature for a particular species). The ecological circuit includes physical and chemical factors as significant disturbance and quality variables. Finally, eutrophication is a result of the anthropogenically induced increase of the level of nutrients in natural waters, followed by an increase in productivity, which results in oxygen consumption in the water.

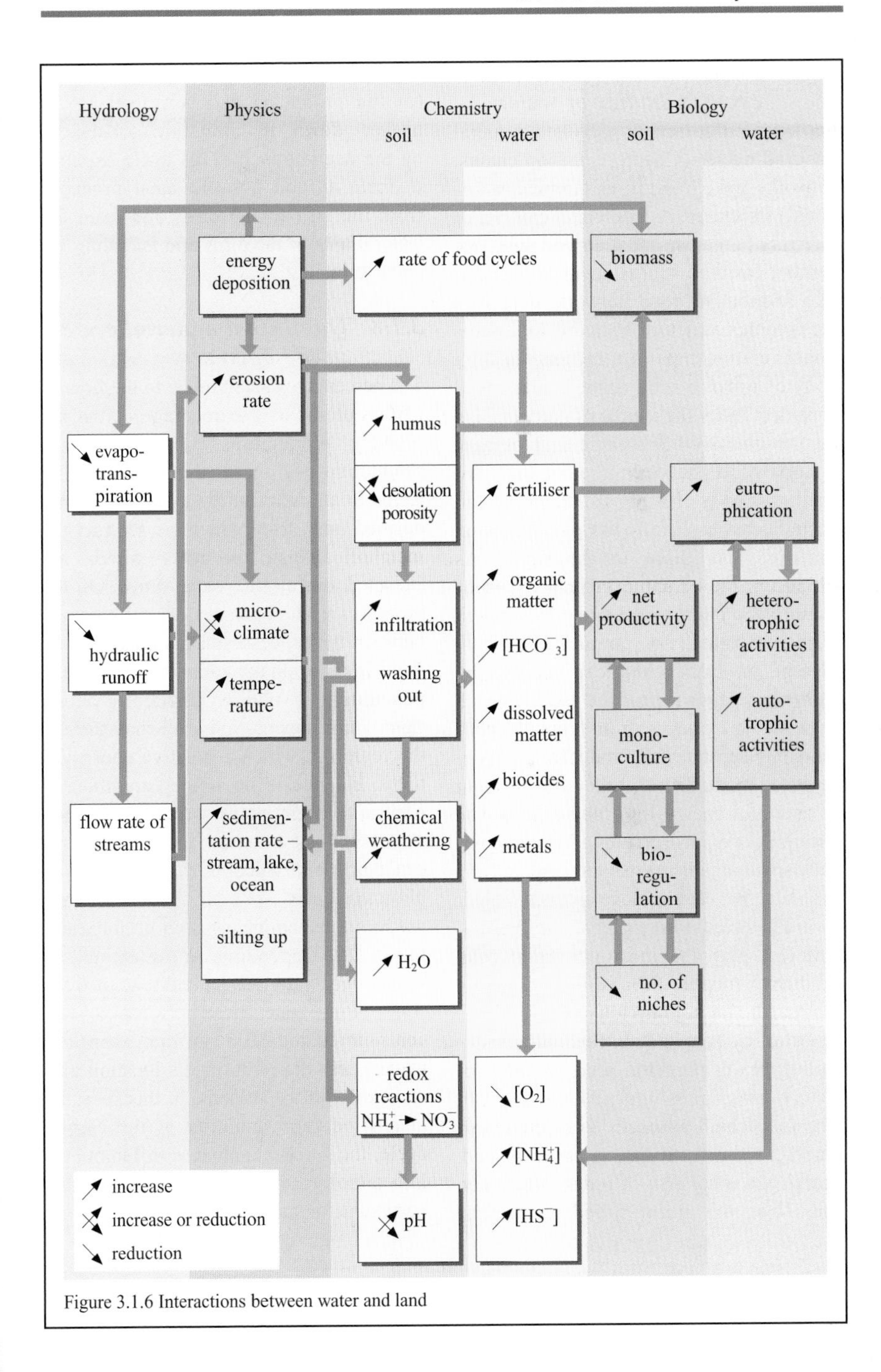

Figure 3.1.6 Interactions between water and land

3.1.7 *Food chain and material cycle in bodies of water*

Ecology differentiates between general material cycles (Chapter 1), food chains, networks and pyramids, and trophic (nutritional) structures. A food chain consists of a series of organisms that depend upon one another for their nourishment in a chain-like fashion. A 'food network' describes the complex structure of interwoven nutritional relationships in a biocenosis. In a body of water, a food chain is influenced considerably by the supply of nutrients, i.e. by the dissolved inorganic and organic materials. At the beginning of the food chain in a body of water are the plants and bacteria, but especially the phytoplankton (in the food chain to the right). As producers (autotrophic organisms), they acquire the energy necessary for growth from sunlight (via photosynthesis by plants) or from chemical energy (by bacteria which produce energy-rich organic materials from energy-poor inorganic substances). Producers are characterised by the fact that they convert light or chemical energy into food energy. The term plankton incorporates all microscopically small organisms that are suspended in the water but cannot move around on their own. Phytoplankton (e.g. diatoms, dinoflagellates) play the most important role, producing roughly 50% of the oxygen on the Earth. Phytoplankton are used as a foodstuff by zooplankton (including krill). Both types of plankton serve as food for fish, worms, crabs, mussels and starfish. The food chain proceeds from the plant-eaters to the meat-eaters, such as cod, sharks and other fish or birds. Altogether, they form the group called consumers. Finally, the dead organic material – detritus (of the producers and consumers) – is broken down to inorganic substances by the decomposers. This cycle expands to form a food network as it proceeds from the fish to the birds and from the water plants to the birds and including the small animals.

3.1.8 *Quantified material cycle in the open sea*

The percentages listed relate to the flow of carbon atoms as a percentage of the net primary production. By net primary production we mean the biomass that forms in an organism by assimilation in a defined time, from which we subtract the metabolic waste products which are excreted during the same time. Of the processes in the open sea, we differentiate between the processes in the deep sea and those in the euphotic zone, the zone that is penetrated by photosynthetically active light. The compensation level characterises the region in which a positive photosynthetic balance is no longer possible. In general the light intensity there is less than 1% of the light at the surface. In the course of a day, the production of biomass due to photosynthesis (P) by the primary producers is completely used up for respiration (R). Depending on the amount of suspended matter and plankton, in fresh water this zone lies between a few centimetres and 30 m; in the ocean it is usually less than 100 m. Its location also changes with the seasons. In this specially quantified representation of the carbon cycle, the carbon content is differentiated into dissolved organic carbon (DOC) and particulate organic carbon (POC).

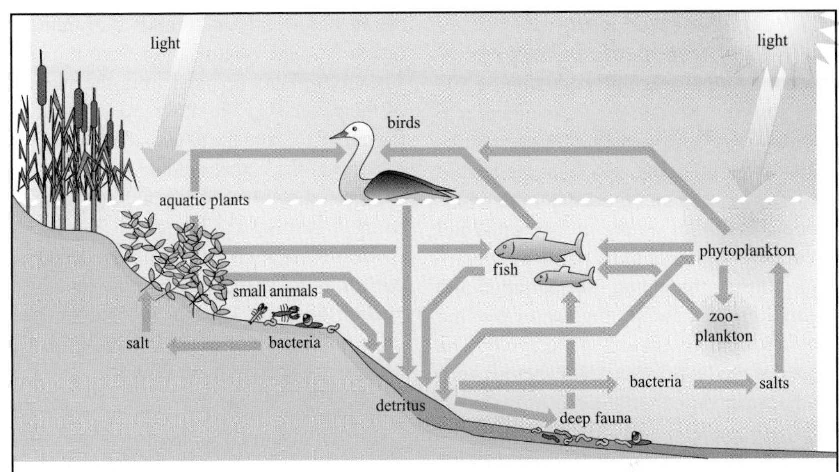

Figure 3.1.7 Food chain and material cycle in bodies of water

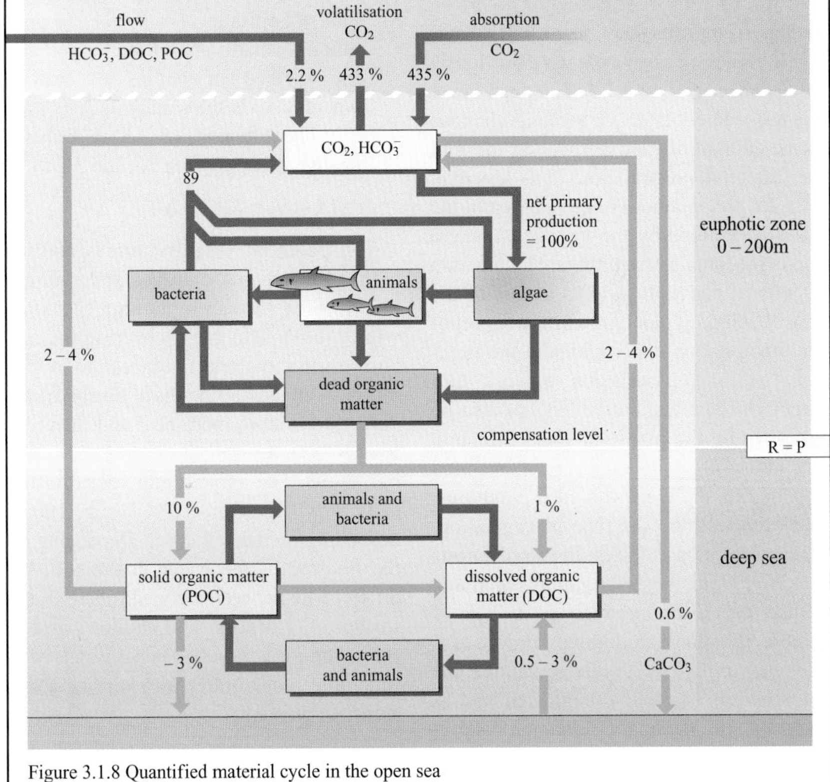

Figure 3.1.8 Quantified material cycle in the open sea

3.1.9 *Water cycle with anthropogenic influences*

Assuming an average precipitation of 803 mm per year (Section 3.1.5), in addition to the natural cycles (Chapter 1) of volatilisation from the ocean and from the Earth's surface, we show in Figure 3.1.9 the routes leading to surface waters, groundwater, and especially to households, small businesses and industry (including transportation and agriculture), including the use of cooling water for the generation of electricity. The width of the lines indicates the percentages of the water flow from the precipitation.

3.1.10 *Water cycle in the area of water–soil–vegetation*

The information in this section is drawn mainly from Kuntze *et al.* (1983). Water levels in the soil as a whole were previously discussed in section 3.1.4. The plant cover has the greatest effect on these levels. Due to the amount of growth covering the soil, the 'effective precipitation' N_e is less than the total precipitation N because part of the latter is retained by the vegetation shield and is precipitated from there. This fraction is called 'interception 1'. It is part of the water balance of plant growth as a subset of the precipitation that remains in the vegetation shield as a perfusion, which in turn largely evaporates, but which can also be taken up by the plants in small measure. The loss of precipitation due to interception depends on the meteorological conditions and especially on the type of vegetation. The interception is lowest in mixed forests at 15–30% (in stands of young beech and spruce trees it can be as low as 10–12% of the precipitation); in tropical rainforests it can reach as high as 70%. In agriculture, the interception is usually neglected and is included with the total evaporation volatilisation V. The latter is composed of transpiration V_T, and volatilisation from the soil V_{ES}, from plants V_{EP}, and from the surface of the water V_{EW}. We differentiate between the current V_a and the potential V_p, where V_a represents the actual amount of water lost by covered soil by volatilisation (measured using a lysimeter). The result is always volatilisation when the saturation deficit of the air is greater than the soil moisture tension of water in soils and plants. The following equations can be derived from the water cycle shown in the figure.

1. The water balance with $N = A + V + (R - B)$, where N consists of the total precipitation composed of N_e and I.
2. The total run off $A = A_o + A_b + A_u$; the total volatilisation $V = V_T + V_E$ (evaporation = volatilisation from the soil and the surface of bodies of water), whereby V_E is the sum of $V_{ES} + V_{EW} + V_{EP}$. The difference $(R - B)$ describes a change in supply (see Section 3.1.5):

$$(R - B) = \Delta W_B + \Delta W_G + \Delta W_O$$

The potential volatilisation is considerably greater in the spring and summer (32–40% of V) than in autumn (20%) or winter (8%). Mulching (covering the soil with organic material) reduces the evaporation V_E by about one-third. Furthermore, darker soils store more heat and therefore lose 30% more water to evaporation than light soils. The annual total volatilisation can vary between 20% and 100%, depending on the soil cover. Depending on the binding of the water in the soil, we differentiate between retained water (on the surface), adsorption and capillary water, percolating water, gravitational water, back water and groundwater with varying availability to plants.

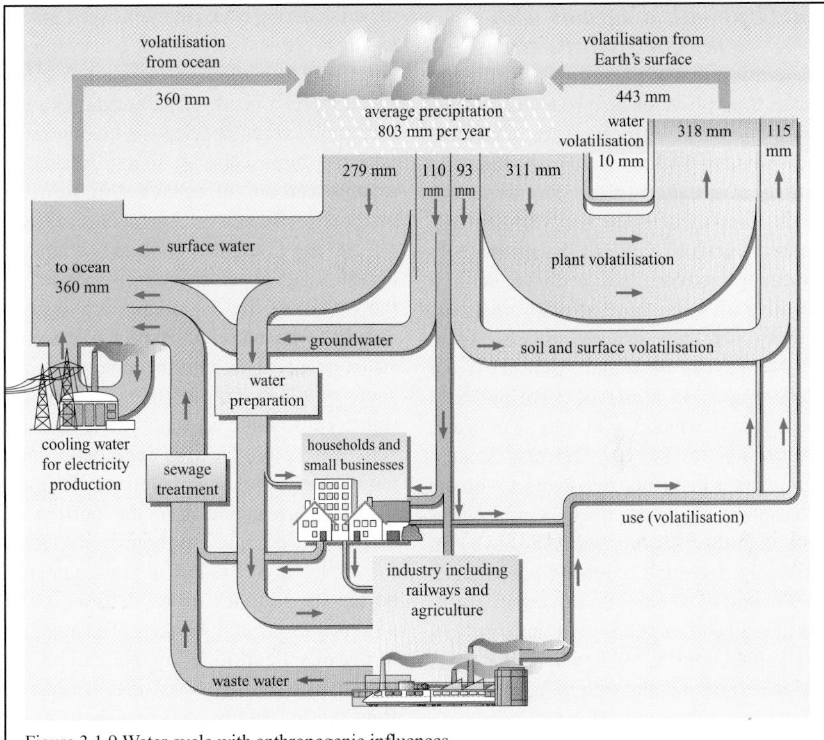

Figure 3.1.9 Water cycle with anthropogenic influences

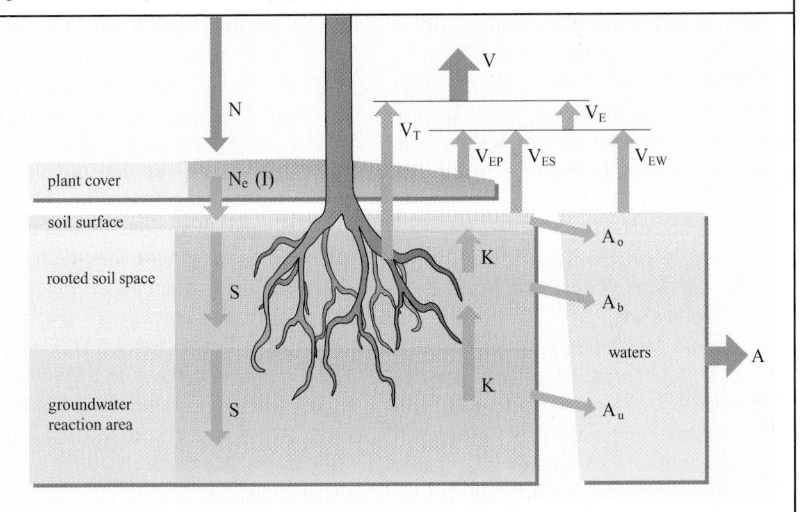

Figure 3.1.10 Water cycle in the area of water–soil–vegetation

3.1.11 *pH–pE diagrams with atmospheric influences*

In addition to the pH value in bodies of water, their pE value also has considerable economic significance. It represents a measurement for the oxidising or reducing activity of solutions (pE = –log *E* where *E* is the electrochemical redox potential). Solutions with pE values > 5 have strongly oxidising activity, while those with a negative pE value have strongly reducing activity. The light-coloured area in Figure 3.1.11 between pE 0 and 20 at pH 0 and about 6 and –15 at pH 14 corresponds to the region in which liquid water can occur. According to R. M. Garrels, system components that come into contact with the atmosphere (fog, rain, rivers, and pit water and aerated saline waters) have an oxidising potential; oxidising agents such as O_3 and H_2O_2 are dissolved in them. Nitric acid and sulphuric acid cause the low pH values. On the other hand, the areas isolated from the atmosphere usually have a reducing milieu and pH values near zero: due to microbial activity, the oxygen is largely used up in marsh waters with a large amount of organic material. Seeping groundwater, from which pH- and pE-determining substances are largely removed due to adsorption processes, lies between the extreme ranges described. The ecochemistry in bodies of water is largely determined by pH and pE values.

3.1.12 *Carbon species in bodies of water*

Cycles and concentrations of various inorganic and organic carbon species (CO_3^{2-}/HCO_3^-, dissolved CO_2, CH_4 and other organic compounds) are determined by photosynthesis (formation of organic material, during which nutrients such as P, N and trace elements are bound) and by mineralisation (formation of inorganic carbon species). As a result of photosynthesis the inorganic carbon species are incorporated into the biota (*bios* = living world) as organic carbon. In oceans, this primary production (Section 3.1.8) amounts to some 50–100 mg C m^{-2} d^{-1}, or several grams of organic material per square metre per day. In the course of the degradation of organic carbon compounds, substances of varying structure occur as intermediate products, some of which are also released into the water. Thus humic acid and fulvic acid are some of the products of degradation and natural polymerisation reactions (Chapter 4). They are included in the portion of dissolved organic carbon, or DOC. Incompletely degraded organic material occurs as particles or as detritus, or is adsorbed to mineral particles. The amounts of dissolved and total organic carbon are increased by the material that washes in from the soil and by the contents of waste waters. The weathering of carbonate rock adds to the increased levels of inorganic carbon. The organic material stored in sediments can be further mineralised there. If the available oxygen is used up (Section 3.1.11), carbon species and also methane can be produced as an end product as a result of microbial activity. Typical values for DOC and POC (particulate organic carbon) in the ocean are 0.5 and 0.05 mg L^{-1}, respectively, whereby the POC in the form of organisms corresponds to a POC of 0.005 mg L^{-1}. In river water, a DOC of 1–10 mg L^{-1} stands in contrast to a POC of 1–2 mg L^{-1}. The DOC values in groundwater and rainwater lie between 0.5 and 2.5 mg L^{-1}; in ponds they reach 10–50 mg L^{-1}.

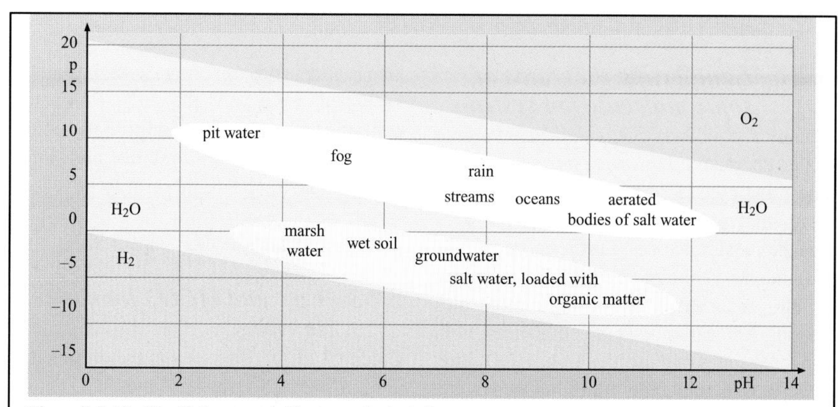

Figure 3.1.11 pH–pE diagrams with atmospheric influences

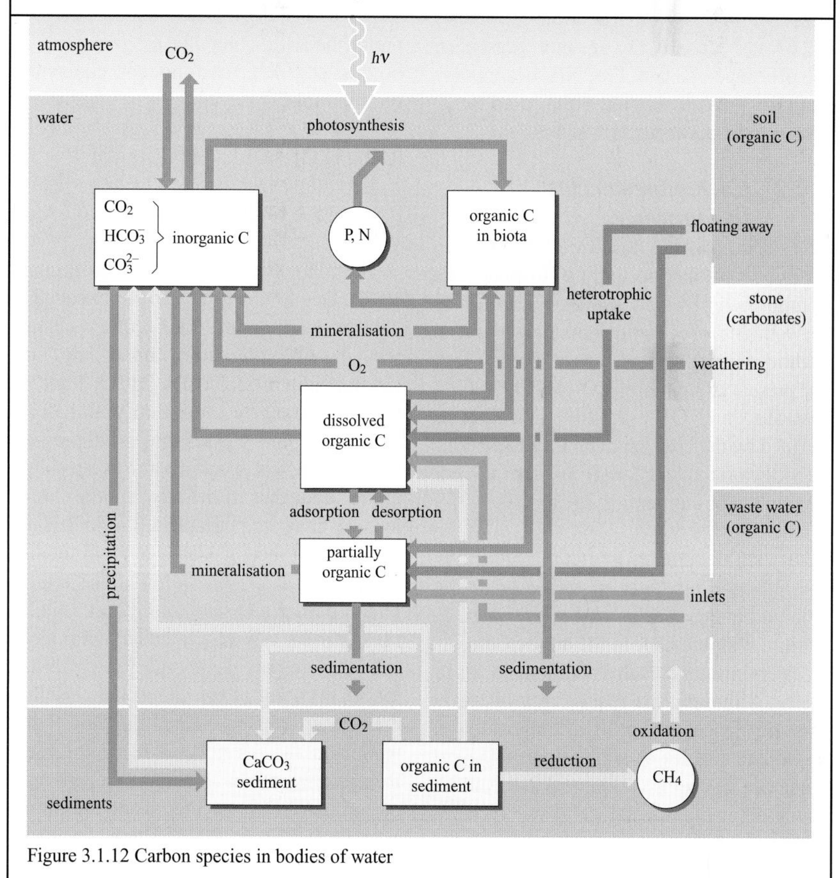

Figure 3.1.12 Carbon species in bodies of water

3.2 Chemistry in bodies of water

3.2.1 Equilibrium diagrams of some molecule/ion systems

Chemical equilibria play a significant role in water chemistry. However, in real samples, one must consider the high ion concentrations that are generally characteristic of ecochemical processes. Furthermore, the pH dependency is the most important influence parameter for the different states: molecules or ions. The carbonic acid equilibrium, with the three states of dissolved CO_2, HCO_3^- and CO_3^- ions, is frequently superimposed by the lime–carbonic acid equilibrium due to the presence of Ca^{2+} ions (Sections 3.2.2 and 3.2.3). CO_2 and HCO_3^- are both present in the pH range of 2 to 5. Free NH_3 occurs only at pH > 7. With sulphur, HSO_4^- and SO_4^{2-} ions coexist between pH 1 and 3.

3.2.2 Carbonate species in rainwater

The term pristine rainwater is used to describe rainwater in a pure atmosphere, the composition of which is determined only by the equilibrium of carbon dioxide without an additional acid or base. The following ions are present in solution: H_3O^+, HCO_3^-, CO_3^{2-}, $CO_2{\cdot}H_2O$ and OH^-. Pristine rainwater is defined by the electron neutrality conditions or by the proton balance $[H^+] \approx [HCO^-_3]$. For $c\gamma$ soluble the summation equation is

$$[CO_2 \cdot H_2O] + [HCO_3^-] + [CO_3^{2-}]$$

The carbonate species shown are in equilibrium with the CO_2 in the atmosphere. As a rule, rainwater contains additional acids or bases. Although this results in a different proton balance, the specific composition of carbonate species as a function of the pH value can be read from Figure 3.2.2. And for every other type of natural water in equilibrium with the CO_2 in the atmosphere, these depicted equilibria and diagrams are also valid, after reacting with bases (of rocks) as well as with acids. Rainwater contains principally additional acids, whereas fresh water contains additional bases. Due to the high salinity in ocean water, activity-correcting factors must be used to calculate the CO_3^{2-} species.

3.2.3 Ca^{2+} and HCO_3^- ions in rivers

Calcite ($CaCO_3$) dissolves in the presence of CO_2 in water to form Ca^{2+} and HCO_3^- ions. With a CO_2 content in the atmosphere of 3×10^{-2}%, and taking into account the charge or proton balance, computer programs can be used to calculate the following concentrations in mol L^{-1}:

$$[CO_2 \cdot H_2O] = 10^{-5}\ ;\ [HCO_3^-] = 10^{-3}$$

$$[Ca^{2+}] = 5 \times 10^{-4};\ [CO_3^{2-}] = 1.6 \times 10^{-5}$$

The pH value is 8.3 (Sigg and Stumm, 1994). Due to this lime–carbonic acid equilibrium, many liquids on Earth are characterised on the one hand by the electroneutrality $2[Ca^{2+}] = [HCO_3^-]$ and on the other hand by saturation with calcite. However, in many cases the partial pressure of CO_2, which is in equilibrium as a result, is higher than that in the atmosphere. Many low-salt rivers are unsaturated or undersaturated with respect to $CaCO_3$ (left of the line in Figure 3.2.3 – see below). The atmospheric partial pressure for CO_2 at 25 °C is $p(CO_2) = 10^{-3.5}$. Rivers saturated with calcite often have a partial pressure in equilibrium higher than that in the atmosphere, such as the Rhine, the Danube and the Don River, which on the other hand fulfil the electroneutrality rule with respect to carbonate or carbon dioxide species. They are on the lines for $[HCO_3^-] = 2[Ca^{2+}]$ in the figure.

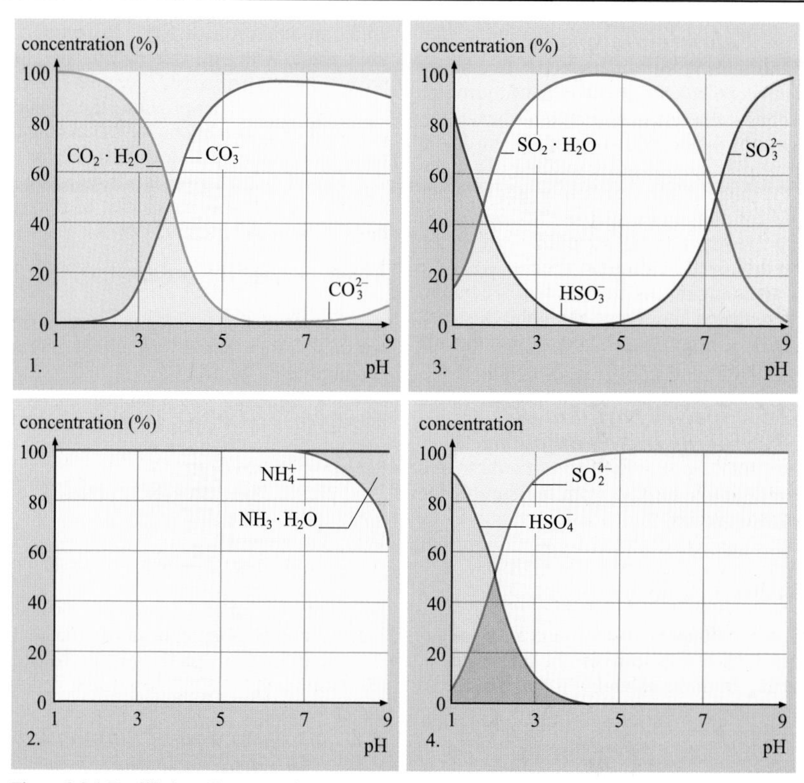

Figure 3.2.1 Equilibrium diagrams of some molecule/ion systems

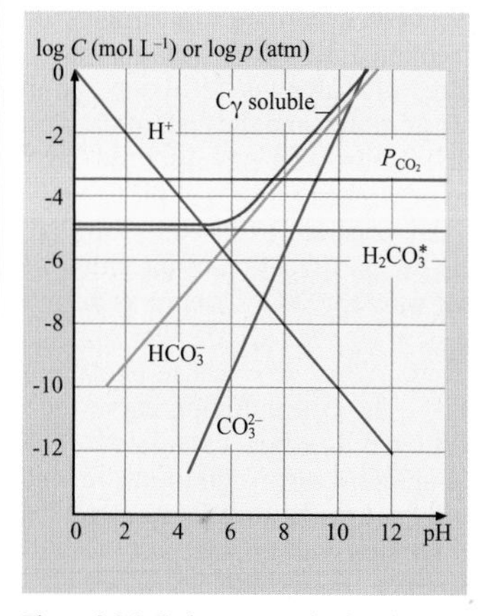

Figure 3.2.2 Carbonate species in rainwater

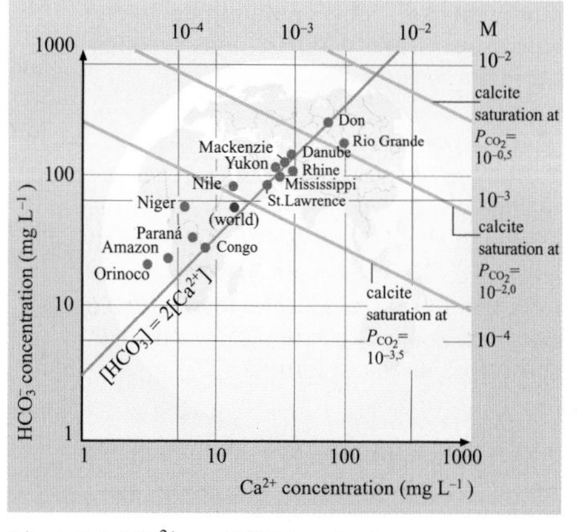

Figure 3.2.3 Ca^{2+} and HCO_3^- ions in rivers

3.2.4 *Solubility of aluminium species*

Aluminium is one of the most frequently occurring elements. It enters the atmosphere via the weathering of aluminium silicates. As an atmospheric element, aluminium has a strong pH dependence on solubility, whereby in the range of pH 4 to 8 in bodies of water different aluminium species (Section 3.2.5) with greatly varying solubility can occur. The dissolved aluminium species are toxic for some organisms such as fish. Therefore differentiated knowledge about the existence ranges of aluminium species and the pH dependence of solubility is very significant.

3.2.5 *Species partition of Al–hydroxo complexes*

The total concentration of dissolved aluminium is derived from the sum of the hydroxo species:

$$[Al]_{total,\ dissolved} = [Al]^{3+} + [Al(OH)^{2+}] + [Al(OH)^{+}_{2}] + [Al(OH)_4^-]$$

Each individual concentration can be determined as a function of the pH value – starting from the solubility product of the Al hydroxide:

$$K_L = [Al^{3+}]\,[OH^-]^3$$

$[OH^-] = [H^+]$ can be derived from the ion product of water K_W. Figure 3.2.5 demonstrates that Al^{3+} ions predominate in the acidic range, while $Al(OH)_4^-$ ions predominate in the basic range.

3.2.6 *Equilibria between nitrate and ammonium ions*

In addition to the pH dependencies shown in Sections 3.2.1 – 3.2.5, the redox potential also plays an important role in water chemistry (Section 3.2.3). Figure 3.2.6 shows the stability region of water with respect to oxidation to form O_2:

$$[2H_2O = 4e^- + 4H^+ + O_{2(g)}]$$

and to reduction to form H_2:

$$[2H^+ + 2O_{2(g)} = H_{2(g)}]$$

for pH 7.5. With a redox potential of >0.4 volts, an oxidation of NH_4^+ to form NO_3^- takes place. With a pE value < 5.5, NH^+_4 ions predominate, whereas NO_3^- ions do so with a pE value > 5.5. This provides the physico-chemical basic conditions for calculating the redox conditions, e.g. deep in a lake at pH 7.5. Figure 3.2.6 assumes a total concentration of both nitrogen species of 5×10^{-4} $molL^{-1}$ (Sigg and Stumm, 1994):

$$1/8NO_3^- + 5/4H^+ + e^- = 1/8NH_4^+ + 3/8H_2O$$

with a value for log K, the equilibrium constant, of 14.9 at 25 °C, yields a redox potential of

$$pE = 14.9 + 1/8\log[NO_3^-]/[NH_4^+] - 5/4pH$$

For pH = 7.5, we find

$$pE = 5.52 + 1/8\log[NO_3^-]/[NH_4^+]$$

Using the condition

$$[NH_4^+] + [NH_3] + [NO_3^-] = 5 \times 10^{-4}$$

the graphs in Figure 3.2.6 can be calculated. Similar correlations can be calculated for the occurrence of Fe(II) and Fe(III) ions (Section 3.2.8).

3.2.7 *Conversion of ammonium in running waters*

In bodies of water, NH_4^+ ions are oxidised via NO_2^- to form NO_3^- after a longer run of 25 km in this case, according to a sample calculation (Section 1.3.3). NH_4^+ gets into the soil and bodies of water via fertilisers, animal waste, waste water, and the path of biological nitrogen fixation and precipitation. In the presence of oxygen (pE > 6 at pH 7; Section 3.2.6), NH^+_4 ions are oxidised to form NO_3^- ions with the participation of bacteria (nitrification), whereby NO_2^- occurs as an intermediate step (via the bacteria *Nitrosomonas*). The bacteria *Nitrobacter* cause oxidation to form NO_3^-. The concentration curve for the three nitrogen species calculated in the example shows a clear maximum for NO_2^- after about 5 km. Both NH_3 and NO_2^- are nitrogen species with toxic activity for fish.

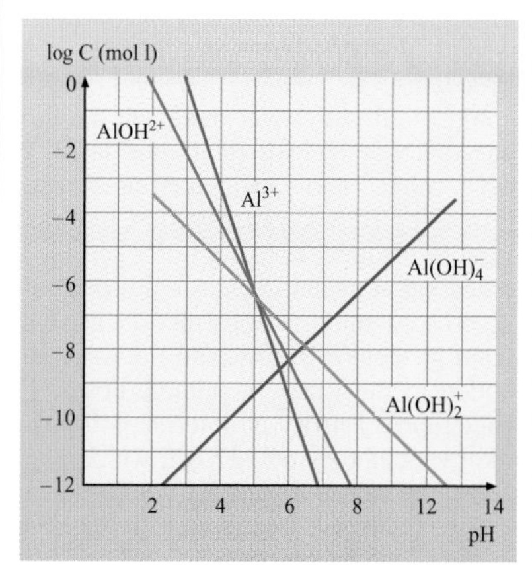

Figure 3.2.4 Solubility of aluminium species

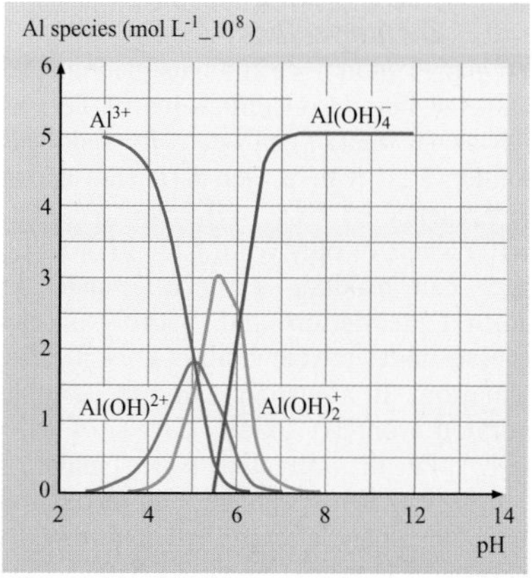

Al species (mol L^{-1}_10^8)
6
5
4
3
2
1
0
Al^{3+}
Al(OH)$_4^-$
Al(OH)$^{2+}$
Al(OH)$_2^+$
2
4
6
8
12
14
pH

Figure 3.2.5 Species partition of Al-hydroxo complexes

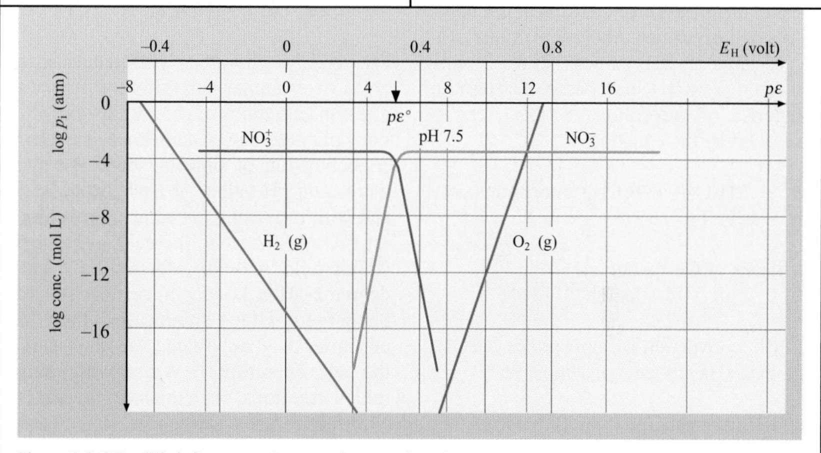

Figure 3.2.6 Equilibria between nitrate and ammonium ions

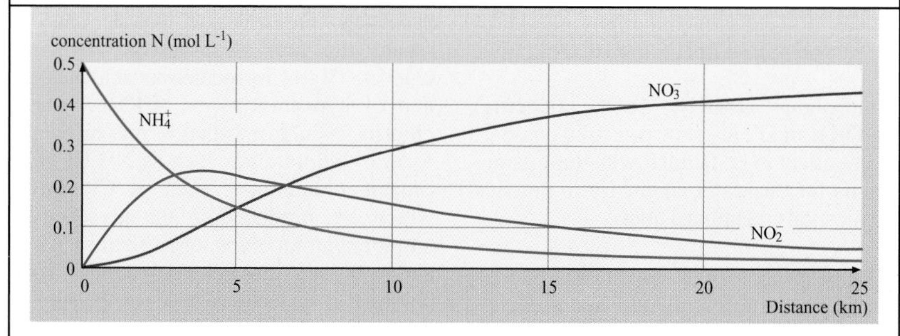

Figure 3.2.7 Conversion of ammonium in running waters

3.2.8 *pE–pH diagram for iron species*

Iron is present in trace amounts in almost all natural waters. Concentrations of Fe^{2+} as high as 10 $mg\,L^{-1}$ can occur in 'reduced' ground waters with oxygen deficiency and in the presence of CO_2, NH_4^+ and S^{2-} ions. Fe(III) levels of only 0.3 $mg\,L^{-1}$ in drinking water can produce a metallic taste. In addition, oxidation and hydrolysis can generate undesired clouding as a result of the formation of amorphic, slightly soluble hydrated iron(III) oxide. Levels of 0.15 $mg\,L^{-1}$ Fe in untreated water make it necessary to perform deferrisation (oxidation, coagulation of hydrated iron(III) oxide and filtration). Figure 3.2.8 shows the ranges in which the different iron species exist in the presence of CO_2 and as a function of pH and pE values. Above a pE value of 12 to 13 and at pH values below 3, only hydrated iron(III) ions occur. The area of existence of amorphous $Fe(OH)_3$ can be described by the equation

$$Fe^{2+} + 3H_2O = Fe(OH)_3(\text{amorphous, s}) + 3H^+ + e^-$$

by means of the function

$$pE = 16 - \log[Fe^{2+}] - 3pH$$

For the conversion of amorphous $Fe(OH)_3$ into iron(II) carbonate, we know that

$$Fe(OH)_3(\text{amorphous, s}) + 2H^+ + HCO_3^- + e^- = FeCO_3(s) + 3H_2O(s = \text{solid})$$

where

$$pE = 16 - 2pH + \log[HCO_3^-]$$

Amorphous $Fe(OH)_3$, $FeCO_3$ (siderite), $Fe(OH)_2$ and Fe are shown as solid phases. It is important to be familiar with these associations for the water chemistry in practical applications mentioned above.

3.2.9 *Concentration–pE diagrams for chlorine species*

Processes in natural waters and in conjunction with water treatment are influenced largely by hydronium ions, i.e. by the pH value and by electron-exchange processes (redox processes). Chlorine is used to disinfect drinking water and for the oxidation of matter in industrial waste water. To do so, a chlorine-water mixture is created from gaseous chlorine, and the mixture is added to the water in suitable doses. The bactericidal activity of chlorine is based on the hypochlorous acid HOCl, which is pH- and temperature-dependent. Altogether four different chlorine species occur concurrently according to the equation

$$Cl_2 + H_2O = HOCl + H^+ + Cl^-$$

Figure 3.2.9 shows the differences in the areas of existence for different pH values The concentrations of chlorine species at the same pE value can be compared. The relative concentration of chlorine decreases with increasing pH value. At a pK_s value of 7.5 and with increasing pH values, hypochlorite ions ClO^- form in increasing numbers, which help with disinfection to a small degree as HOCl, since in comparison with uncharged HOCl molecules they can penetrate only slowly into the inner part of the cell. Therefore, in waters with comparable fractions of microorganisms or oxidisable organic substances, an increase in pH value also means an increased need for chlorine, or a longer reaction time is necessary. 'Free chlorine' in water analysis means the sum of dissolved, elemental chlorine Cl_2, of hypochlorous acid (HOCl) and of hypochlorite ion (ClO^-). 'Bound chlorine' is chlorine that is present in the form of chloramines such as NH_2Cl or of organic chloramines such as CH_3NHCl, which are produced in the presence of ammonia/ammonium ions (Section 3.2.1) and/or organic nitrogen compounds. 'Total chlorine' is understood to mean the sum of the free and the bound chlorine.

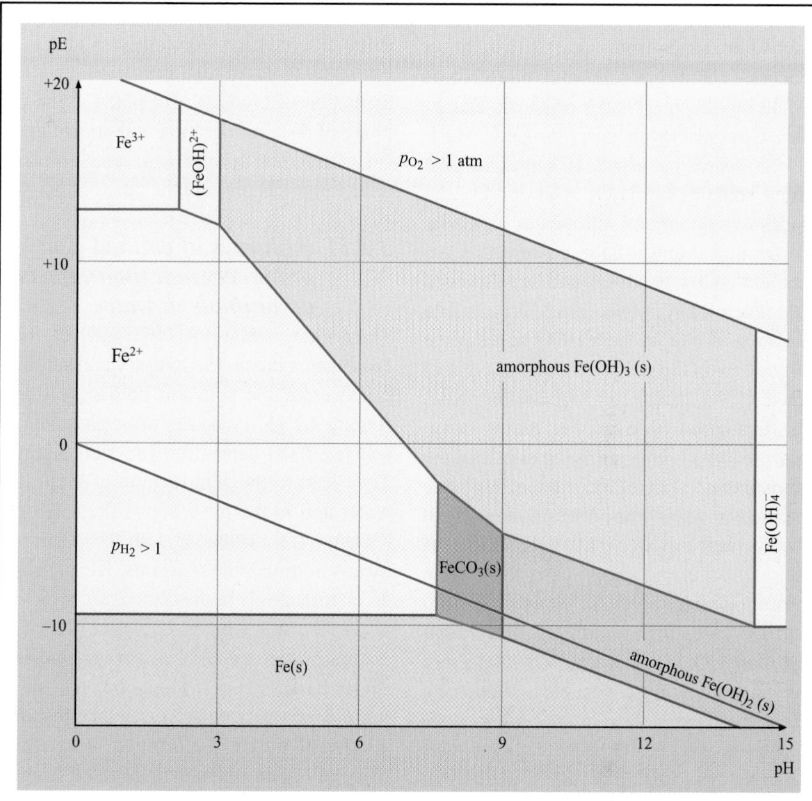

Figure 3.2.8 pE-pH diagram for iron species

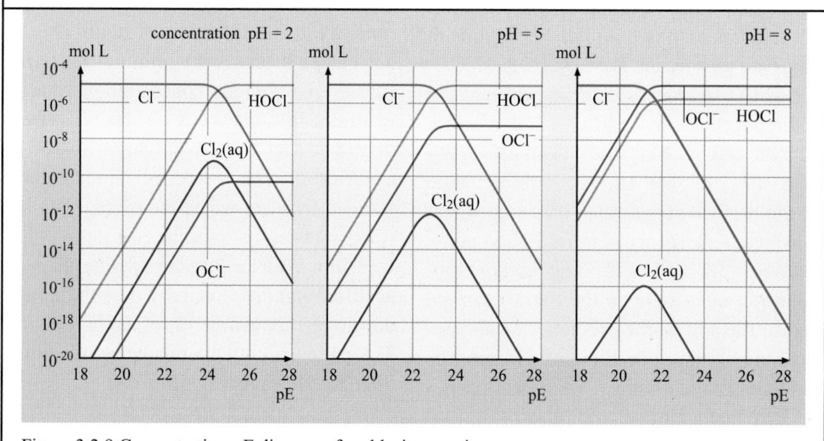

Figure 3.2.9 Concentration-pE diagrams for chlorine species

3.2.10 Genesis of rainwater

Rainwater comes from nature's distillation process (Section 1.1.5), and therefore it should be the purest water on Earth. Due to intensive contact with the atmosphere, which has an oxidising effect (Chapter 2) and which contains O_2, H_2O_2, $\cdot OH$ and O_3, the oxides of sulphur and nitrogen in particular are formed. Many of these processes are accelerated by catalysis and are induced photochemically (Section 2.2). The oxidation of NO_x to form HNO_3 takes place particularly in the gas phase, whereas that of SO_2 to H_2SO_4 takes place in the water phase. The formation of water via precipitation incorporates a highly purifying effect for the atmosphere. Therefore, materials are enriched in water from precipitation which dissolve well in water – like O_2 and CO_2 as naturally occurring gases – and those from anthropogenous sources, especially from exhaust gases from industry and transportation, CO, SO_2, nitrous gases, NH_3, soot and industrial dust particles containing heavy metals as well as organic compounds such as disulphides (RSSR). This includes ocean sprays (Section 2.1.8). Especially at the start of a rainstorm and after a long dry period, rainwater can contain concentrations of substances that would be found in domestic wastewater. Bases such as $MgCO_3$ and $CaCO_3$ get into the rain via geogenic and anthropogenous dust particles and, together with the anions of the acids mentioned along with NH_4^+, K^+, Na^+ and Al^{3+} ions, generate an ion balance in rainwater, the composition of which is shown at a pH of 4.3 (acid rain – Sections 2.4 and 4.2). The contents of rainwater contribute on the Earth's surface and in the soil to the processes of disintegration, in the formation of biomass as phytomass or as humus. They also participate in nitrification, denitrification and sulphate reduction (Sections 1.3 and 1.4). The processes of disintegration and the formation of biomass take place in the pH range of 4–5, nitrification and anoxic denitrification occur at pH 4, and sulphate reduction takes place in the range of 6.5–8.

3.2.11 Pathways of emitted acid producers and their effects on animals in water

The primary acid formers SO_2 and NO_x from combustion processes (industry and automobiles) enter the soil and bodies of water (Figure 3.2.11, 1) via dry (dust particles) or wet (rainfall) deposition (Section 3.2.10). They lead to the acidification of bodies of water and to the lowering of the soil's pH (Section 4.2) associated with the release of metals from sediments or soils. The pH plays an important role in the ecology of bodies of water. The first damage occurs at pH < 6.5. Natural waters are labelled as over-acidified if they have a pH of 5.5. And below a pH of 4.5–5.0, no living organisms will be found in a body of water. The different acid-sensitivity of organisms is demonstrated in their areas of existence. The lower animals such as crabs, snails and mussels are especially sensitive to the addition of acids (Figure 3.2.11, 2). Lakes with minimal buffering capacity, i.e. with low concentrations of acid-neutralising ions such as HCO_3^- (from dissolved $CaCO_3$) have a high degree of acidification. This is especially true for many lakes in Sweden, where the average pH has sunk as low as 4.5 since the end of the 1950s. One-fifth of all the lakes in Sweden are over-acidified and therefore have no fish. Similar developments can be observed in Canada. The previously great abundance of salmon there has degraded significantly.

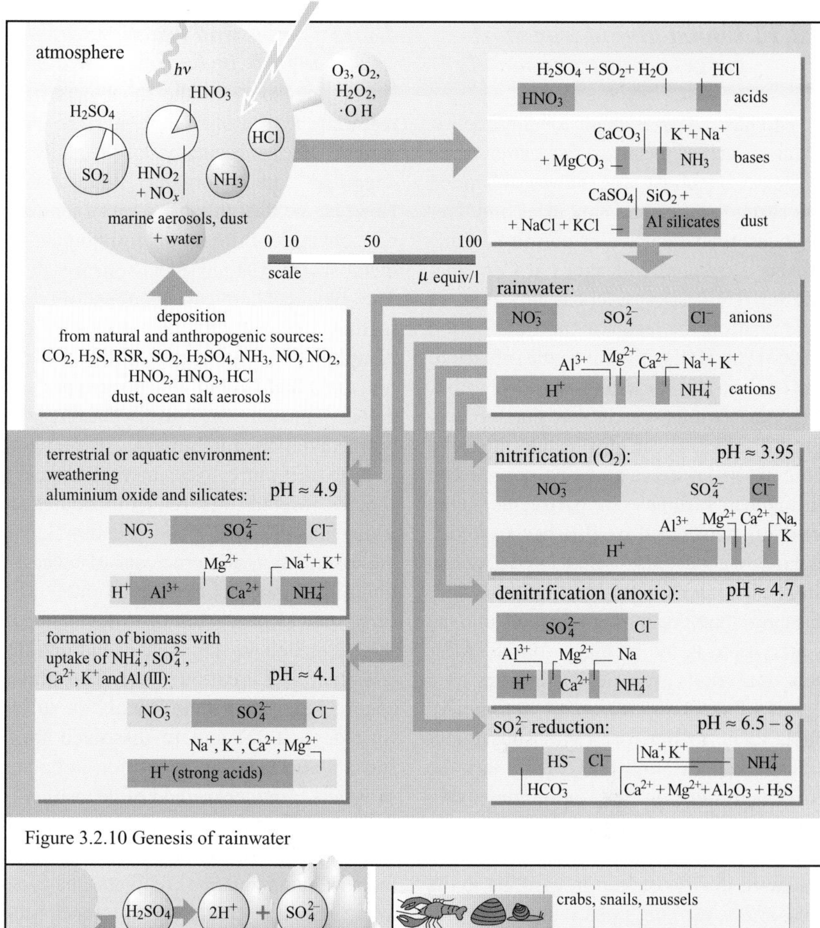

Figure 3.2.10 Genesis of rainwater

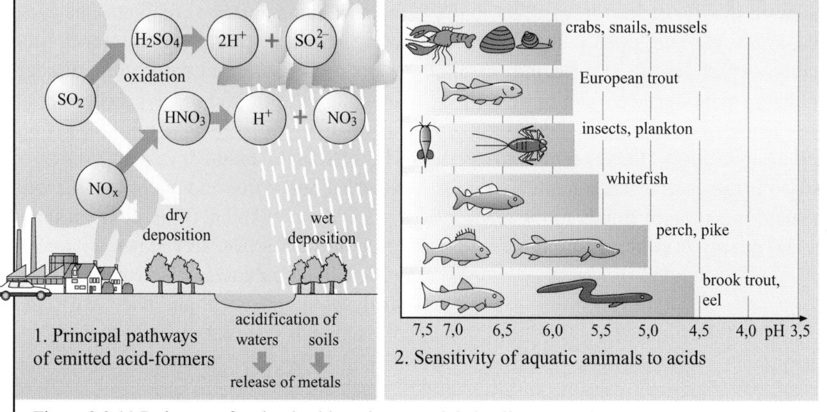

Figure 3.2.11 Pathways of emitted acid producers and their effects on animals in water

3.2.12 Emission and transport processes of trace metals in riverbeds

To produce an assessment for a particular species, such as metals in trace amounts in this case, one must take into consideration the current source and sink sizes and their changes over a particular period of time. Robert Ayres and colleagues at Carnegie-Mellon University selected this type of examination for lead for the time period 1880–1980 and applied it to the estuary of the Hudson and Raritan rivers in New York Harbour. Various sources (see upper row in Figure 3.2.12) can influence the lead content in an area, and one source can influence various compartments. One analysis performed on this basis yielded the following details (Graedel and Crutzen, 1994). During the time from 1880 to 1930, the emission of colouring pigments into the waterway was the primary source. After this, emissions from lead additives as a result of the combustion of automotive fuels came to the fore, with subsequent transfer into the atmosphere. After this, the total emissions of lead with increasing concentrations were distributed nearly globally. After the industrialised nations outlawed the use of lead in gasoline in the mid-1970s, air emissions reduced considerably, which is also reflected in the discharge in the surface water of a riverbed. The Nobel Prize winner Paul Crutzen (1995) considers these types of very difficult studies as a fundamental basis for the determination of trends in the environmental quality over large periods of time.

3.2.13 Cycles and reactions of metals in bodies of water

Metals are some of the most persistent materials in the environment since they cannot be biologically or chemically degraded like organic substances. However, on the other hand, metal ions can be converted into other compounds or metal species (as physicochemical states). The physicochemical changes of state include adsorption, precipitation and complexation processes. Biochemical processes that lead to new bond types and metal species are assimilation and biomethylation, for example, of mercury ions to the more toxic methyl mercury species. Metals transfer from bodies of water into the food chain (Section 3.1.7) via assimilation and processes of bioaccumulation and biomethylation. The ecochemical processes in water lead via interfaces into the atmosphere and into the sediment and create a cycle. The atmosphere helps distribute metals as solids (dry deposition) and in dissolved form (precipitation, as wet deposition in bodies of water). Another method of depositing is by means of inflow (Section 3.2.12), again, in dissolved form or as solids (suspended matter). Sedimentation completes the cycle at the sediment stage, which returns back to bodies of water via desorption, dissociation, dissolution and redox processes, as a result of mobilisation or remobilisation. The effects of metal levels in bodies of water on organisms and the behaviour with respect to the neighbouring compartments (evaporation into the atmosphere or precipitation in sediment) are determined decisively by the physical and chemical states of the metals (species).

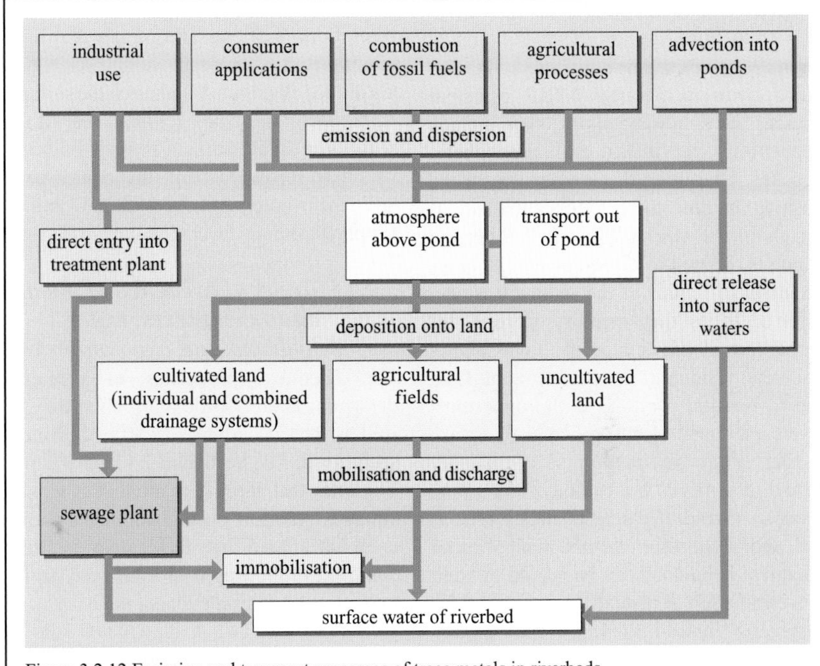

Figure 3.2.12 Emission and transport processes of trace metals in riverbeds

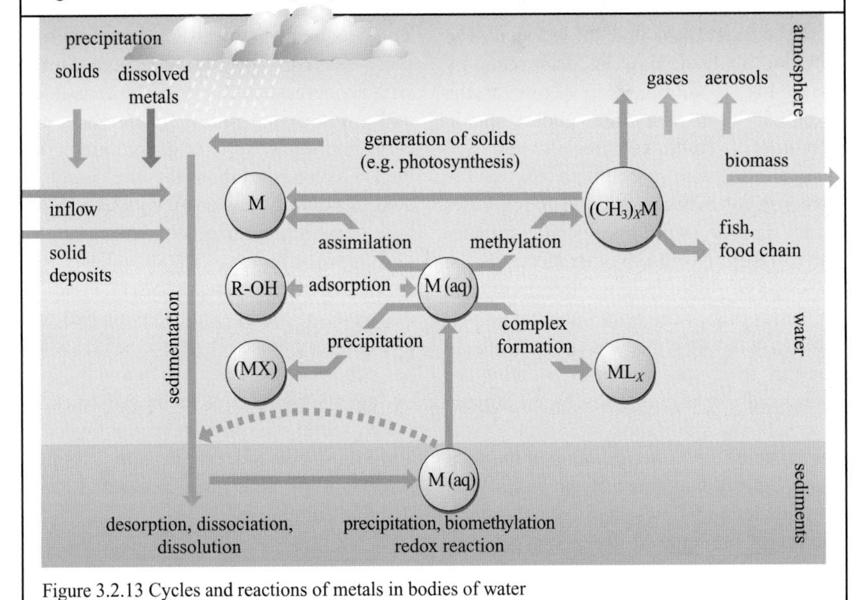

Figure 3.2.13 Cycles and reactions of metals in bodies of water

3.2.14 Reactions of metal ions in a lake

Figure 3.2.14 expands upon the more general view given in Section 3.2.13 principal sources for heavy metals in lakes include depositions via influx and atmospheric deposits. One thing that is decisive for the ensuing or the current concentration of metals in the water column is that portion which is retained in the sediment. The accumulation of metals in the sediment is also referred to as the 'memory storage' for earlier metal loads or deposits. If one follows the metal residue in the water column from top to bottom, portions in the uppermost layers are bound by adsorption or by uptake during algae production. The primary ligands for metals are carboxyl and amino groups. Instead of essential metals such as Zn and Fe, other metals with similar chemical behaviour can be bound as well; these include Cd instead of Zn, or AsO_4^{3-} (arsenate) instead of PO_4^{3-} (phosphate). Metals are also carried along when biological material sinks into deeper levels of the lake and finally into the sediment. The iron and manganese cycles determined by the redox conditions play a role at the sediment–water interface under anoxic conditions. Hydrated iron oxide and manganese oxide precipitate out at the border region between oxic and anoxic water layers. They present large specific surfaces and can therefore adsorb other metals. In the presence of sulphide ions, slightly soluble sulphides (e.g., FeS: black) precipitate out, which can in turn precipitate other metals. In contrast to rivers, processes in lakes are influenced very significantly by the various layers. In the springtime, the surface layer warms up by +4 °C (temperature of the deep water, density maximum of the water), and wind-generated flows lead to a thorough mixing of the body of water. Oxygen-rich water travels to the depths of the lake, whereupon nutrients and metals that have accumulated in the depths during stagnation are diluted, distributed, and reach the surface. As the water heats up over the course of the seasons, a stable surface layer is built up, which is separated from the colder deep layer, the hypolimnium, by a spatially limited temperature jump layer (metalimnium).

3.2.15 Areas of existence of metal–aquo complexes, water hydroxo and oxo complexes

The chemical properties of elements determine their coordination chemistry as well as their occurrence as defined element species (see also Sections 3.2.13 and 3.2.14). The rules that apply here are that cations in aqueous solution occur hydrated as aquo complexes and that they form hydroxy complexes following hydrolysis (and deprotonation). An example for M(+2) is:

$$Zn(H_2O)_6^{2+} \leftrightarrow [Zn(H_2O)_5OH]^+ + H^+$$

The tendency towards deprotonation increases for different aquo-complexes with an increase in charge of the central ion and a decreasing radius. If elements can carry multiple charges (e.g. chromium +6), then in aqueous solutions they are deprotonated multiple times, or as with chromium they form anionic oxo complexes (CrO^{2-}_4: chromate). Figure 3.2.15 shows an overview of the pH-dependent areas of existence of metal–aquo ions or hydroxo and oxo complexes. If no overacidification has occurred (Sections 3.2.10 and 3.2.11), in the pH range 7–9 in most bodies of water, metal ions are present as hydroxo and oxo complexes. Finally, slightly soluble hydroxides are generated from hydroxo complexes.

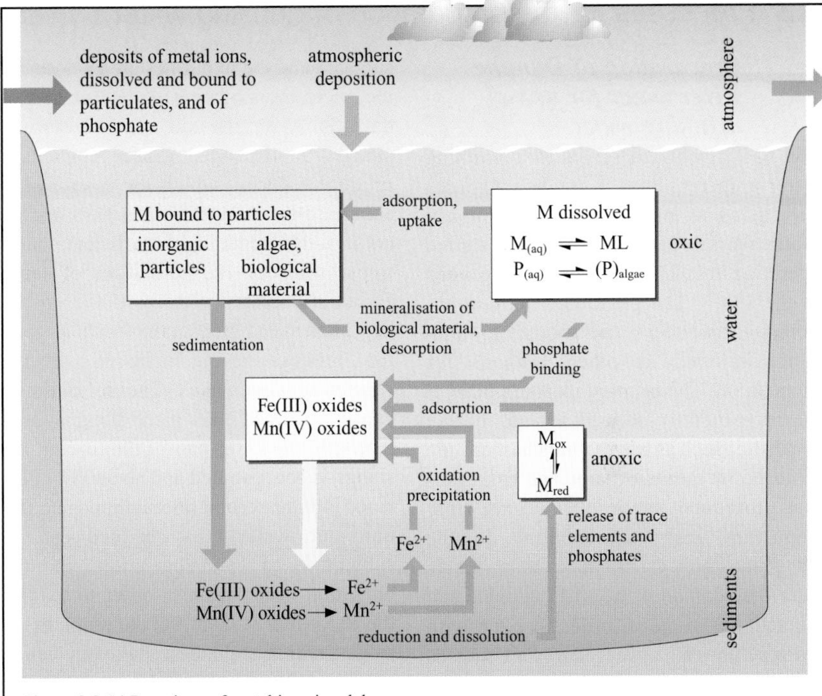

Figure 3.2.14 Reactions of metal ions in a lake

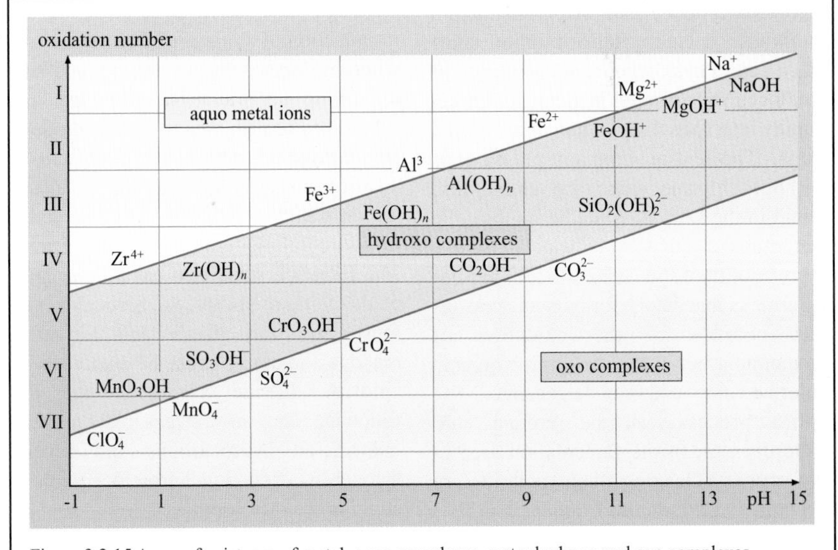

Figure 3.2.15 Areas of existence of metal–aquo complexes, water hydroxo and oxo complexes

3.3 Processes for the preparation of drinking water

3.3.1 Preparation of Danube River water for use as drinking water

The basic technical stages in a drinking water preparation plant are demonstrated here using as an example the Langenau Water Works (Bank, 1994), which started operating in 1973 and which was expanded in the 1980s. The plant uses the conventional ('classical') preparation of groundwater as well as other methods for preparation. The primary cleaning plant is where generally non-dissolved, coarse impurities such as leaves, branches, etc. are removed by means of large and fine rakes and subsequent sieving (to remove finer impurities such as plankton). In the Langenau plant, a first flocculation occurs after the addition of iron(III) sulphate and flocculation aids, in order to cope with extreme untreated water situations such as high water with special depositions of pollutants or else low water with an increased concentration of pollutants. On the whole, a compact flocculation takes place here. Physicochemical knowledge of the flocculation process is optimised procedurally by a spatial separation of individual steps. The entire preparation is done without chlorine since the ammonium found in the Danube could be biologically decomposed. For surface waters that are more polluted, in general the use of chlorine is mandatory for hygienic reasons and because of the oxidation of ammonium, but at the same time, organic chlorine compounds can be generated. The entire process can be broken into subprocesses. In the flocculation area in the primary cleaning plant, colloids are destabilised by the addition of iron(III) sulphate and the impurities are enclosed in the microflakes of iron(III) hydroxide that develop. With the aid of the returned sludge, a suspended matter contact is produced, during which the microflakes adhere to the sludge. Then a flocculation aid is added under highly turbulent conditions, whereby the formation of large flakes is prepared. Finally, the latter are deposited as sediment in the fourth stage of the process upon inclusion of the remaining fine flakes. The solid–liquid phase separation takes place along parallel plates; in the lower part of the reactor, the sludge is concentrated and about 90% of it is fed into the second flocculation area. The untreated water thus obtained can be subjected to preozonisation in the untreated water basin in order to oxidise the undesirable contents of the water. In the subsequent flocculation stage, in which lime and flocculation aid are added, the resultant sludge is separated once again. Next the ozonisation takes place via the introduction of a mixture of ozone and air, whereby pathogens are killed and any remaining organic materials, especially substances contributing to taste and odour, are oxidised. Then the water is subjected to a filtration stage, where residual pollutants are separated in a double-layer filter made of hydroanthracite and quartzous gravel. The filtration stage also has a filter layer made of activated charcoal. Since this layer must also adsorb disinfectants, for safety reasons the water must be treated with chlorine dioxide during storage. By following all of these stages in the process, chlorinated hydrocarbons and several frequently occurring herbicides such as atrazine are safely removed.

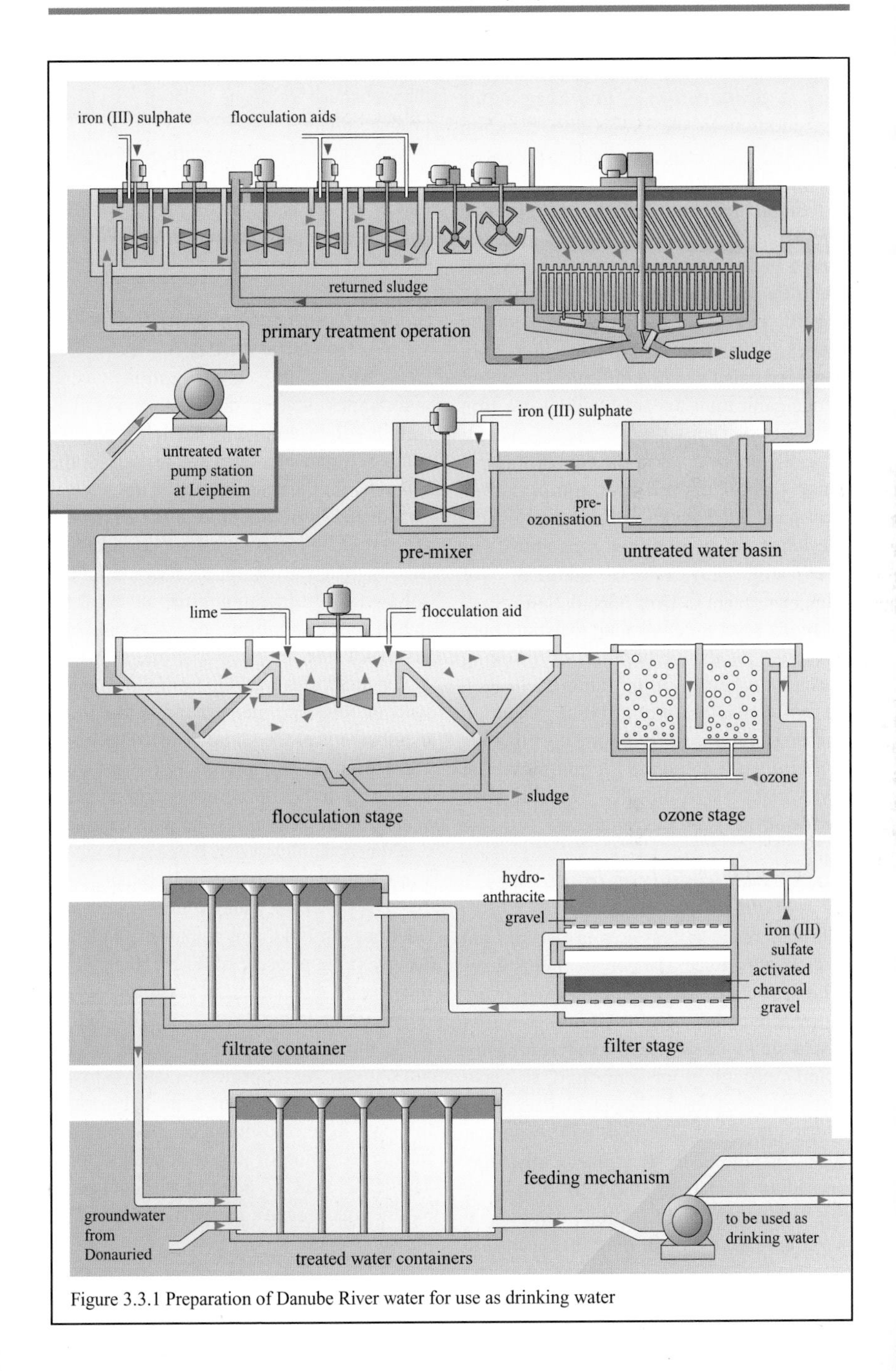

Figure 3.3.1 Preparation of Danube River water for use as drinking water

3.3.2 Classical processes for the preparation of drinking water

In addition to the procedures listed in Section 3.3.1, chemical–physical processes for the preparation of water include aeration (to enrich with atmospheric oxygen or to remove gases such as CO_2, H_2S and materials with a high vapour pressure); iron removal (with Fe levels in untreated water > 0.15 mg L^{-1}, as Fe^{2+} in low-oxygen waters) by means of aeration or with the aid of an oxidising agent or via a biological pathway; demanganisation (corresponding to iron removal starting at 0.07 mg L^{-1} Mn^{2+}); deacidification to adjust the equilibrium pH value (lime–carbonic acid equilibrium – Sections 3.2.2 and 3.2.3), performed mechanically by aeration (see above) or chemically by the addition of $MgCO_3/CaCO_3$, CaO or $Mg/Ca/Na_2CO_3$, all of which react in an alkaline fashion. Other processes in the preparation of drinking water are softening (using cation exchange); denitrification (removal of nitrate using ion exchange, reverse osmosis, or electrodialysis – Sections 3.3.3 and 3.3.4); and protective layer formation in order to protect against corrosion and scale formation.

3.3.3 Electrodialysis in a three-chamber cell

This membrane process uses an electric field as the driving force. The polymer membranes represent cation- (C) or anion-exchangers (A), whereby they become permselective (as diaphragms), i.e. they allow only cations or anions through at any one time. If heavy metal ions are present, they precipitate out as hydroxides due to the hydroxide ions that form on the cathode. Technically, this procedure is largely used to desalinate brackish water (slightly saline, non-potable water near the mouth of rivers in the ocean). After direct current has been applied, the three-chamber technique makes it possible to produce a largely deionised solution in the middle section. However, with increasing deionisation the electrical resistance in the middle section becomes very high, and then concentration polarisations develop on the surface of the membrane.

3.3.4 Desalination of water by reverse osmosis

In contrast to ultrafiltration using membranes to separate high-molecular-weight (often colloidal) particles, reverse osmosis (or counter-osmosis) is used for the separation of true solutions. This special membrane separation procedure works with pressures (29–100 bar) that are considerably higher than the osmotic pressure in the solution. The separation effect stems from the differing solubility of water and the dissolved ions in the membrane material. Water molecules transfer from the solution with the higher concentration through the membrane into the solution with the lower concentration. This procedure is used on a large scale to desalinate ocean and brackish water and in general for the preparation of drinking and utility water. If one starts with a 1 mol L^{-1} NaCl solution, which contains 58 g NaCl L^{-1}, then the osmotic pressure of this solution with 2 mol ions amounts to 44.8 bar. At a working pressure of 80 bar, there is an increase in concentration via the pressure difference of 35 bar. With a retention rate of 99%, the residual NaCl amounts to 600 mg L^{-1} in the permeate. Downstream mixed-bed ion exchangers remove the rest of the ions. Reverse osmosis can also be used for the reduction of nitrate concentrations. The efficiency is determined decisively by the membrane's properties and their measurements.

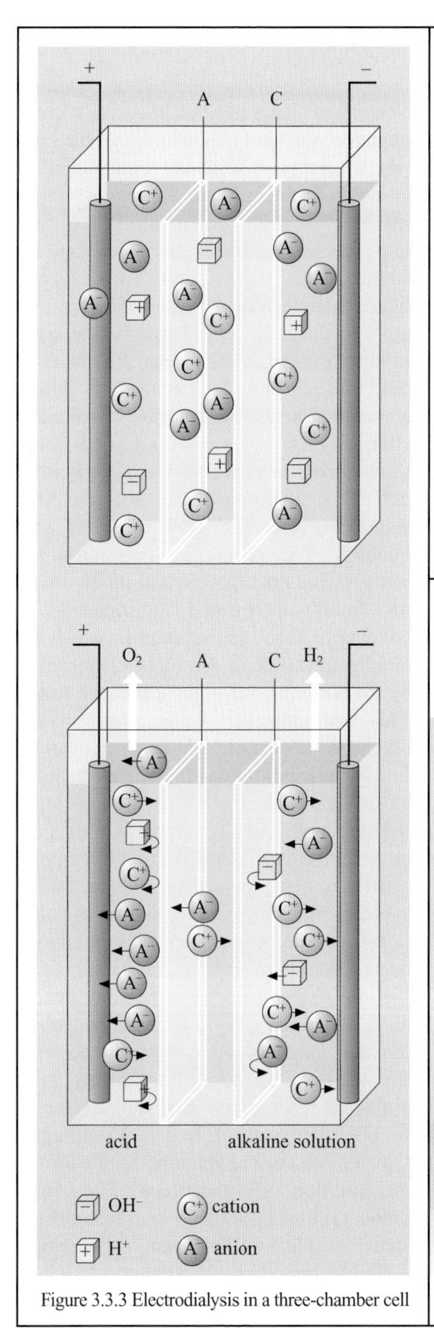

Figure 3.3.3 Electrodialysis in a three-chamber cell

Process	Measures
aeration	$+O_2$ removal from CO_2, H_2S, odoriferous substances
iron removal, demanganisation	oxidation with O_2, oxidant or biologically: precipitation of Fe(III)/Mn(IV) oxide hydrates
deacidification	+ CaO, MgO, Na, Mg, Ca carbonate: establishment of equilibrium pH
softening	ion exchange: removal of Ca^{2+}/Mg^{2+} ions
denitrification	ion exchange, electrodialysis, reverse osmosis: removal of nitrate
protective layer formation	addition of phosphates, silicic acid and its salts

Figure 3.3.2 Classical processes for the preparation of drinking water

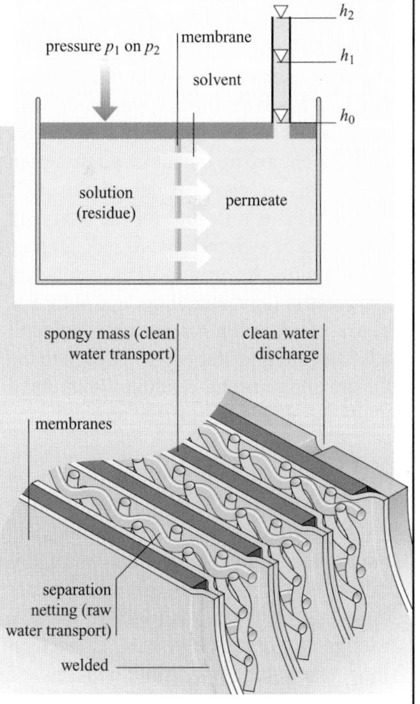

Figure 3.3.4 Desalination of water by reverse osmosis

3.4 Basic principles of waste water chemistry and sewage treatment

3.4.1 Local treatment plant as a direct discharger

The Water Content Law in Germany regulates the direct or indirect use of bodies of water. The requirements for the introduction of waste water into a body of water come from this law. Waste water is 'water which runs off and is changed due to use and all water which enters into the sewage system' (DIN 4045), which includes rainwater. We differentiate between indirect dischargers and direct dischargers: indirect dischargers are producers of waste water, the waters of which – along with other (e.g. domestic) waste waters – flow into a public (municipal) water treatment plant. Like a municipal water treatment plant, direct dischargers discharge their waste water directly into a body of water after appropriate purification. To do so, they require water rights permission and must pay waste water fees based on the diverted pollutants. A waste water plant is generally a facility for the diversion and treatment of waste water. Beyond this, a waste water purification plant, or WPP, is a facility for the treatment of waste water with the goal of reducing the contamination. As an important ecological component in the material cycle, a WPP has the task of returning water rich in mineral matter and of comparable quality back to the drainage ditch (natural running water) and a humus-like sludge back to the soil.

3.4.2 Layout of a mechanical–biological water treatment plant

Sewage water from homes, streets and entire subdivisions is fed into the waste water purification plant via a network of the city sewerage and drainage system, along with falling rainwater (as a combined sewage system). Mechanical purification (first stage) includes rakes for large-scale cleaning of pieces of wood, leaves and fabric remnants; the sand trap, which uses the sand which washes in with the water or other granular mineral substances to trap materials through settlement in long sand traps or in round basins; and an oil and fat capturing stage (for safety's sake, since these materials must be separated out previously by the producers). The flow rate in the sand trap is regulated so that the grains of sand, which are heavier than the finer sludgy contaminants, just barely settle. The primary clarification begins in a settling pond, where heavier materials separate out at the bottom due to a significantly slower flow rate. This step alone retains 30–35% of the organic material. The remaining two-thirds are organic contaminants present as dissolved or finely distributed (suspended or colloidally dissolved) substances, which are largely accessible to aerobic decomposition by the bacteria already present in the water. This decomposition takes place in the second stage, that of biological purification. The bacteria multiply in the activated sludge basin due to the rich nutrient source. With the introduction of large amounts of oxygen, biomass forms which is acquired in a post-clarification step as activated sludge and is subsequently subjected to sludge digestion (anaerobic decomposition: fermentation = decomposition of carbohydrates; and putrefaction: decomposition of protein, generation of sewage gas = methane and carbon dioxide). Biomass is a term used to describe the total of all living organisms in a biological reactor (activated sludge basin or trickling filter setup). Iron and aluminium salts are used as precipitating agents for the precipitation of phosphate. The final product is treated sludge, which can be dried and utilised as fertiliser among other things.

For further detail, see Section 3.4.18.

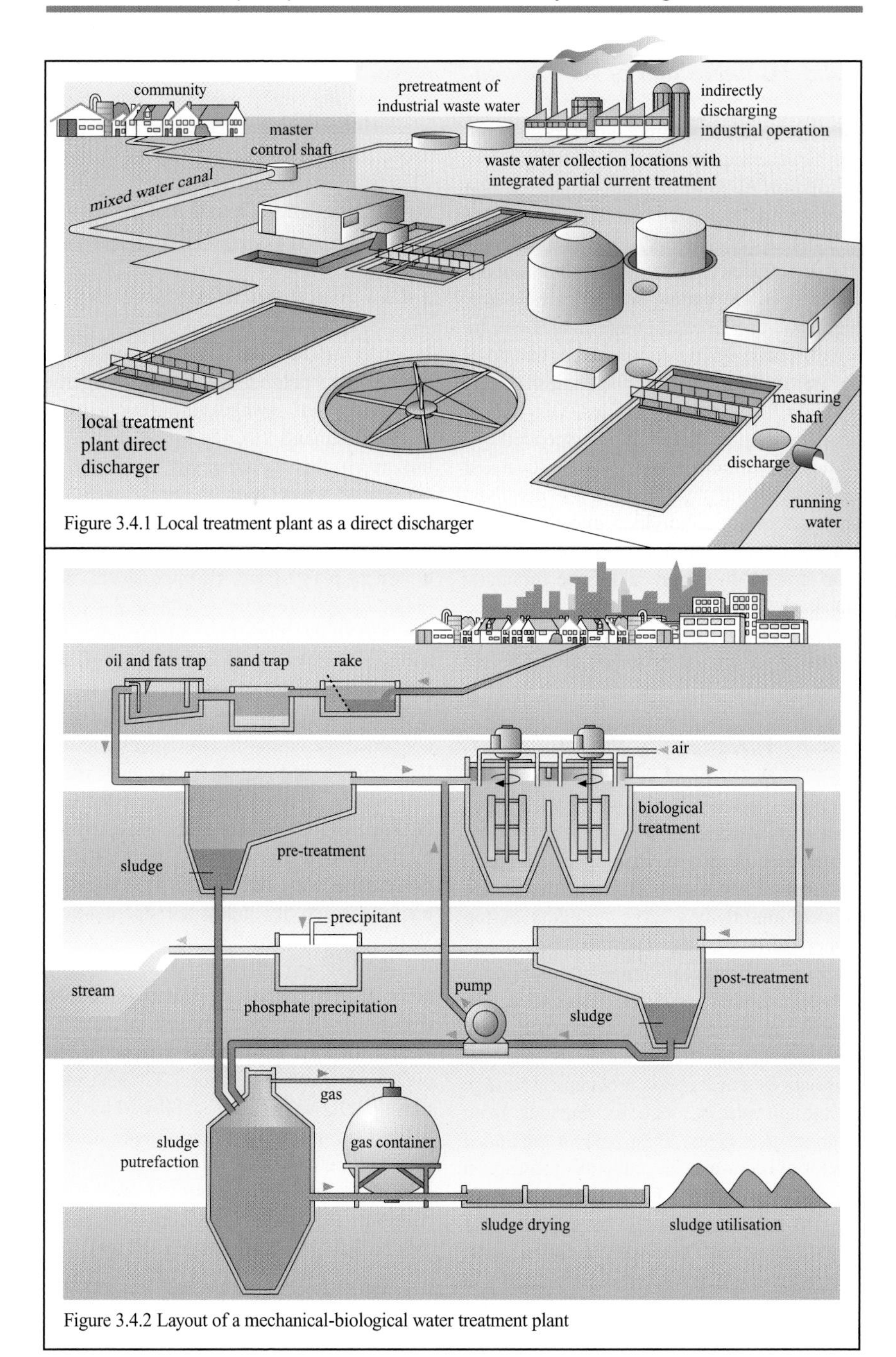

Figure 3.4.1 Local treatment plant as a direct discharger

Figure 3.4.2 Layout of a mechanical-biological water treatment plant

3.4.3 Composition of sewage and its possible treatment

Sewage treatment procedures utilise the physicochemical processes of concentration, formation of solid or gaseous phases, and their separation from the water. Sewage itself presents a two-phase system: one differentiates between undissolved and dissolved substances. Depending on their particle size, undissolved (filterable) substances can be divided once again into substances that do or do not settle out. The non-settling materials include 'swimming' particles, very finely dispersed particles, and suspended or colloidal substances, which can be separated analytically via ultrafiltration. The dissolved substances are subdivided into three categories: organic and degradable; organic and non- (or hardly) degradable; and inorganic substances. The two latter groups can only be removed from the sewage after a physicochemical treatment, for which very different methods are available (Kunz, 1992).

3.4.4 Variations in the amount of municipal sewage

Domestic sewage contains largely organic materials derived from urine, faeces, and water used in rinsing, cleaning and laundry. Inorganic substances (salts) enter the sewage system mainly via large amounts of tap water. Sewage contains a broad spectrum of various substances: organic compounds (carbohydrates, proteins, fats) that are readily broken down by bacteria, urea from excrement, salts and sand, as well as substances which create problems in sewage treatment plants, such as tensides from laundry detergents. Even without the added factor of rainwater, the amounts of sewage in a city (shown with 50 000 inhabitants in Figure 3.4.4) can change by a factor of 3 within 24 hours. These daily variations must be taken into account when planning a sewage treatment plant, whereby one does not use the daily average, but rather a weighted average that is roughly 30% above the 24-hour average for calculations. The amount of settling material has a very different curve from that of the amount of sewage.

3.4.5 Biochemical oxygen demand, BOD

In order to be able to characterise the biological degradability of organic substances in sewage, the biochemical oxygen demand (BOD) is defined as a measure of the amount of oxygen (in mg O_2 per litre of water) used up during oxidation by aerobic bacteria. The BOD is an important value in the planning of sewage treatment plants (size, cleaning stages, and biological requirements – Sections 3.4.1 and 3.4.2). A BOD value of 60 g O_2 is established as the resident equivalent factor (REF) for 200 L sewage per resident (Section 3.4.4) in order to break down the organic substances contained therein. With sufficient O_2 levels, readily degraded materials are quickly and completely mineralised, i.e. converted into H_2O and CO_2 (substrate respiration enzymic) by microorganisms within 5 days (BOD_5). Pollutants with inhibitory (usually for enzymes) activity delay the degradation. With substances such as tensides (Section 5.5.7) that are difficult to degrade, in addition to endogenous respiration (cellular respiration) by the primary consumers (which live off autotrophic producers like algae), nitrification (Section 1.3.3) also takes place. BOD values are established by lawmakers as minimum requirements for the introduction of sewage into bodies of water, and they form the basis of the 'Law regarding fees for the introduction of sewage into bodies of water (Sewage Fee Law)'.

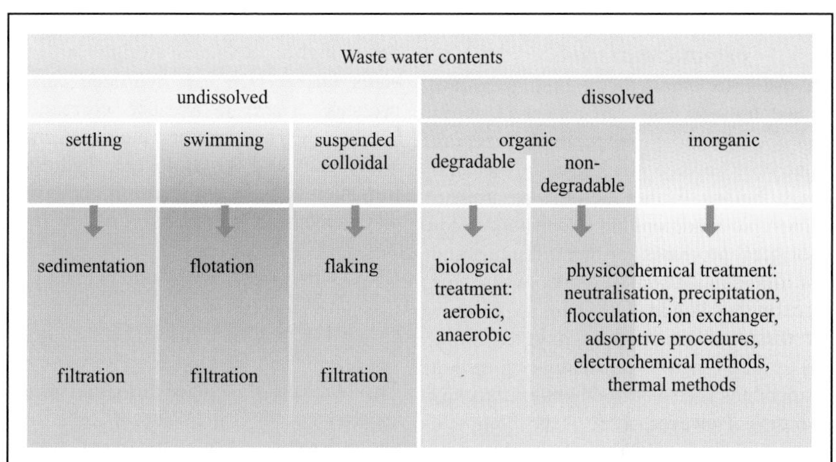

Figure 3.4.3 Composition of sewage and its possible treatment

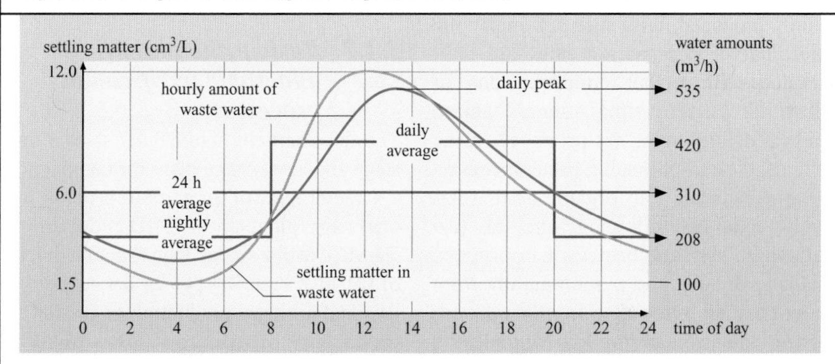

Figure 3.4.4 Variations in the amount of municipal sewage

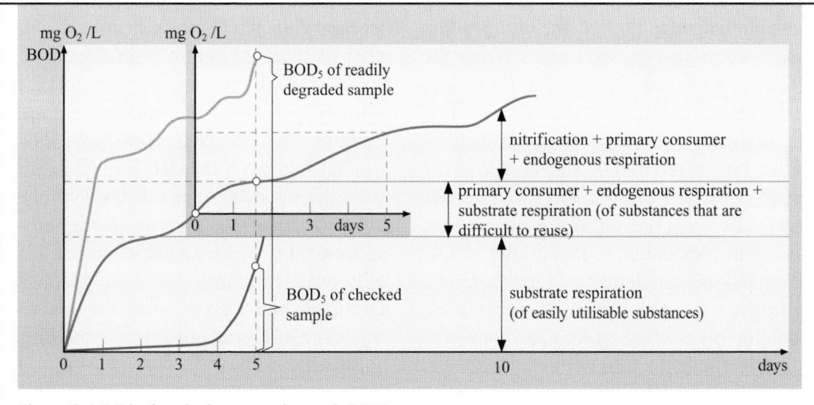

Figure 3.4.5 Biochemical oxygen demand, BOD

3.4.6 Anaerobic degradation of organic materials

In order to break down high substrate concentrations, waste treatment plants in the food industry in particular use an anaerobic sewage treatment as the first stage before further aerobic treatment. Anaerobic conditions are also necessary for biological processes for denitrification and for the elimination of phosphorus (see Sections 3.4.14 and 3.4.16). In the presence of dissolved (or bound) oxygen, some bacteria can metabolise energy-rich carbon compounds such as carbohydrates, fats and proteins. However, there is no complete mineralisation at a relatively low conversion rate, but rather methane is mainly produced. With such low metabolic rates, useful degradation rates are only attained with higher concentrations of substrate, whereby simple sugars, amino acids and fatty acids are produced in the first stage. Acid-producing bacteria convert these substances into organic acids relatively quickly, following one of two pathways. Methane bacteria can produce methane directly and predominantly from short-chained acids and from methanol. Longer-chained acids are first degraded by acetogenic bacteria to form methanogenic substances, which are then metabolised by methane bacteria. Finally, the end product biogas contains CH_4, NH_3, H_2S (from the proteins), CO_2 and H_2 from the portion of complete mineralisation. The degradation of complex (usually high-molecular-weight) organic compounds shown in the upper right of Figure 3.4.6 includes the energy flow (%) of an anaerobic sewage treatment, calculated using the COD (chemical oxygen demand). First there is a hydrolysis of the substrate molecules; insoluble organic substances are converted into water-soluble substances by extracellular enzymes. Acetogenic bacteria break down higher molecular weight organic acids further at a low hydrogen partial pressure. There is a close correlation between the H_2-consuming methane generation (fast) and the acetogenic hydrogen formation (slow). Two reactions come into play in methane formation:

$$CH_3COOH \rightarrow CH_4 + CO_2 \text{ (slow)}$$

$$CO_2 + 4\,H_2 \rightarrow CH_4 + 2\,H_2O \text{ (fast)}$$

The inhibition of methane formation occurs especially at lower pH values. Biogas from an anaerobic sewage treatment has a heating value between 20 and 25 MJm^{-3}.

3.4.7 Basic principles of anaerobic purification processes

The flowcharts in Figure 3.4.7 show a one-stage and a two-stage plant for comparison. Characteristics of an aerobic process in a one-stage plant are a COD reduction of 50–90% with a sludge load (Section 3.4.12) of 0.1 to 2 kg COD per kg dry matter per day, and a biogas production of 0.2-0.4 m^3 for each kg of reduced COD with a CH_4 portion of 65–85 and 10–25% vol. CO_2. In order to maintain the optimal temperatures necessary for the bacteria, heat exchangers must be installed before or in the reactors. Furthermore, it is necessary to maintain the optimal pH with the aid of a buffer container. A two-stage plant provides the optimal solution if the CH_4 formation is the rate-limiting step. At pH 4, the hydrolysis and acidification phases described in Section 3.4.6 are separated. However, if the hydrolysis is running slow or if the sewage has a large buffering capacity, the two-stage procedure has no advantage over the single-stage method.

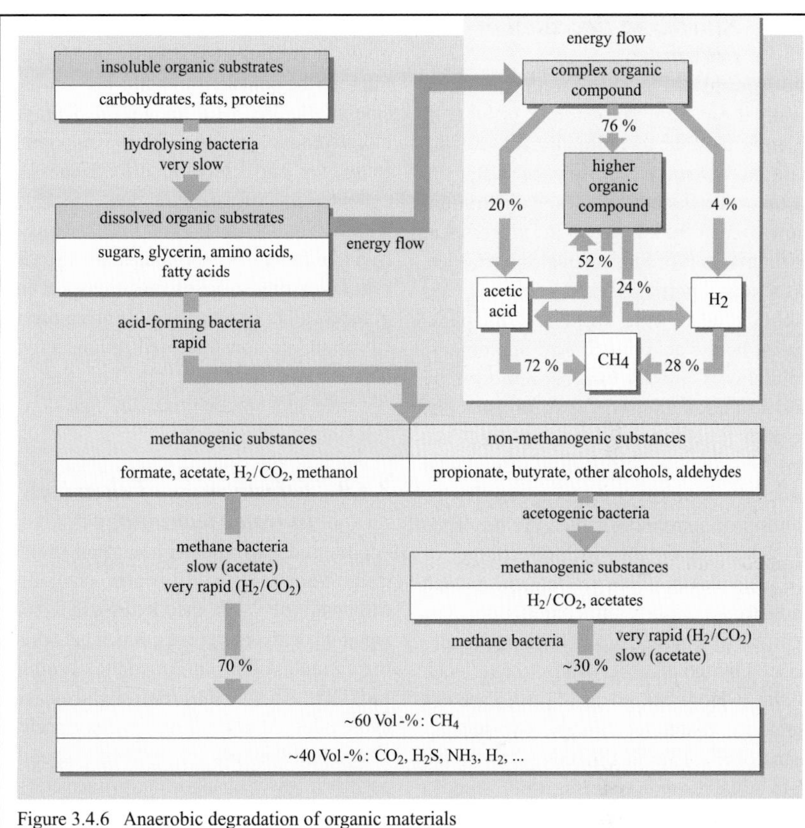

Figure 3.4.6 Anaerobic degradation of organic materials

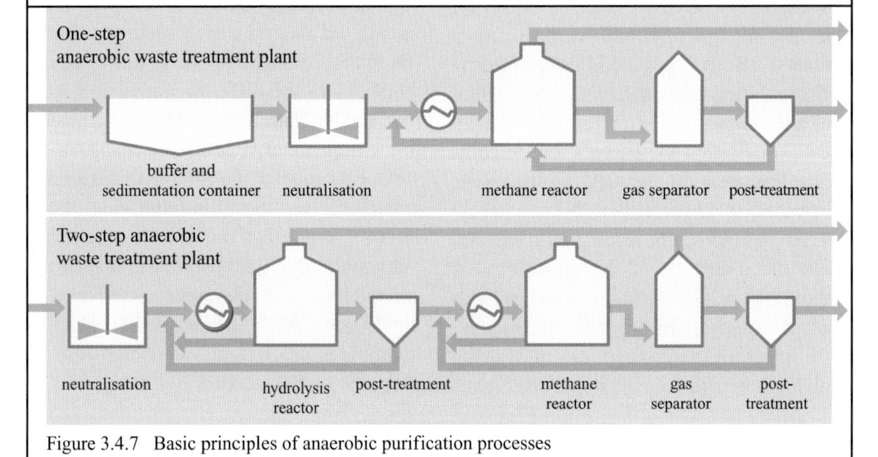

Figure 3.4.7 Basic principles of anaerobic purification processes

3.4.8 *Kinetics of flocculation and filtration*

The information in this section is drawn mainly from Sigg and Stumm (1994). In order to be able to optimally remove the materials dissolved in sewage water by means of flocculation and filtration, knowledge of the kinetics of the processes at the interfaces is of fundamental significance. Suspended particles and colloids are important adsorbents for metal ions, phosphates, humic acid, fulvic acid and organic pollutants. Colloids can be removed by means of separation only if the diameter of the particles is increased via agglomeration. Flocculation is an aggregation of colloidal particles to form larger agglomerates. By the addition of substances which are adsorbed on the surface, the surface charge of colloidal particles can be reduced, which leads to an adsorption coagulation. The stability of charged colloids is generally reduced by the addition of electrolytes such as NaCl. However, such additives are not usable in sewage technology. Two mechanisms play a role in this coagulation: the rapid destabilisation process, which leads to an aggregation of neighbouring particles, and the slower transport stage, by which the particles develop mutual contact, as a collision frequency, which is determined by diffusion, Brownian movement, shear forces and rate gradients. The collision effectivity factor α describes the success rate of collisions dependent on chemical adhesiveness, where $\alpha = 10^{-4}$ indicates one effective impact per 10^4 collisions. These simplified kinetics yield the reduction of the number N of colloids as a second-order rate law with $-dN/dt = k_p \alpha N^2$, where k_p is the rate constant.

In water and sewage treatment systems with high particle concentrations, the coagulation is optimised by selecting a high rate gradient (turbulence) and by additives to improve the collision strength. The left-hand side of Figure 3.4.8 shows the filtration effectiveness as a product of contact frequency and collision effectiveness. A comparison of natural and technical filtration systems showed that in spite of having different filtration rates, the filtration process during groundwater transport in a groundwater carrier, or groundwater filtration, demonstrates filtration activity (see above) comparable to that of commercial filtration processes with slow or fast filters.

3.4.9 *Effectiveness of chemicals in water technology*

Al(III) and Fe(III) salts are often used as coagulation agents. In water containing carbonate (pH 7–9), they hydrolyse to form metal–hydroxo complexes, which network to form metastable intermediate products $Me_x(OH)^{n+}_y$ before they form slightly soluble hydroxides. At pH < 8 they carry a positive charge and they are adsorbed by suspended particles, which changes the surface charge. The constant addition of an Al(III) salt leads first to a charge neutralisation (destabilisation) and then to a reversal of the charge (restabilisation). The left-hand side of Figure 3.4.9 shows the differing activity of Al^{3+}, Ca^{2+} and Na^+ ions, which induce coagulation due to the electrolyte activity (a change in the surface potential of the colloids or suspended particles – see the right-hand side of the Figure), and that of the hydroxylated aluminium described previously and of the precipitation with aluminium hydroxide. At higher doses of Al(III) and Fe(III) salts, the hydroxides precipitate out in the form of precipitation flocculation.

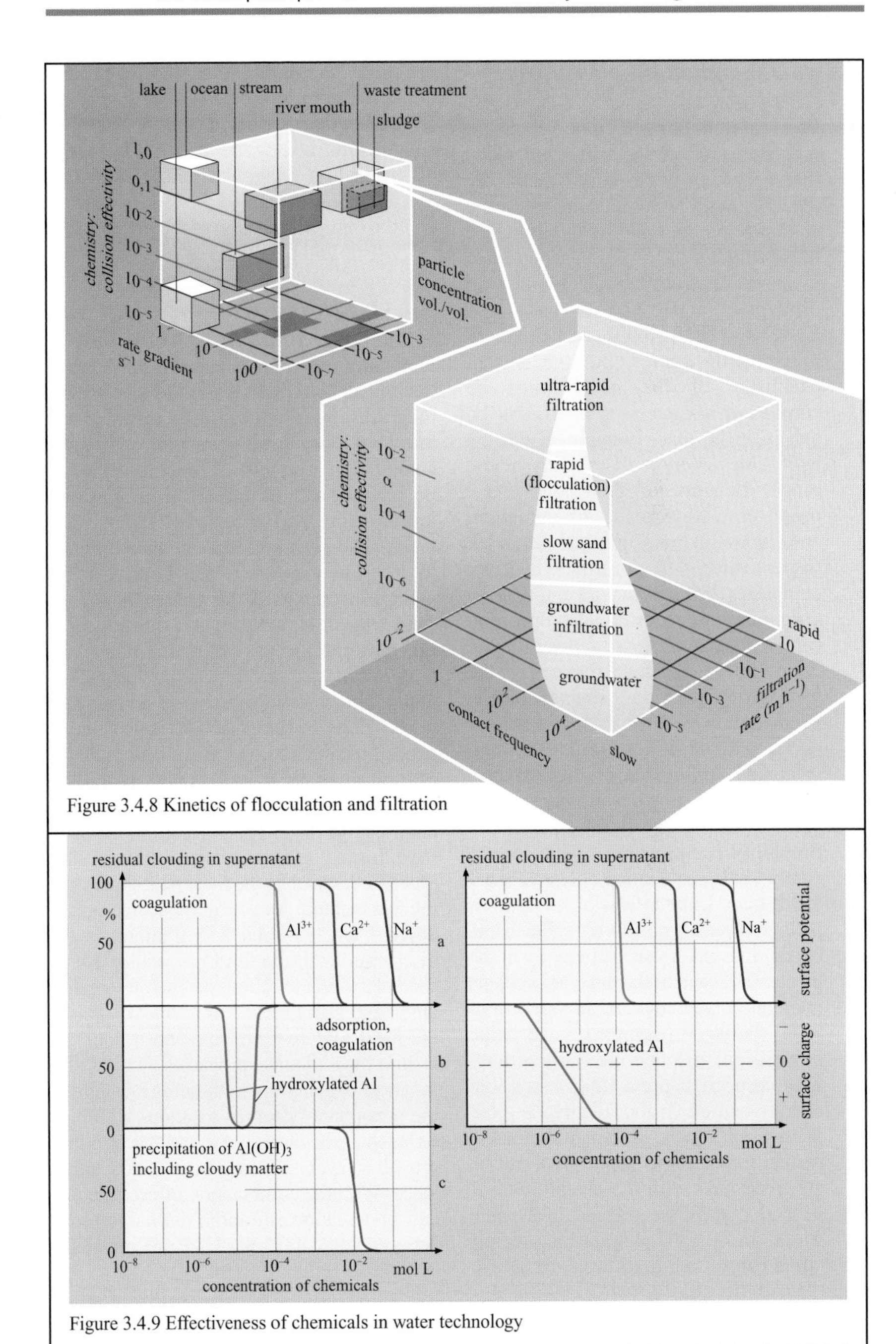

Figure 3.4.8 Kinetics of flocculation and filtration

Figure 3.4.9 Effectiveness of chemicals in water technology

3.4.10 Precipitation–pH ranges of metals

1. Precipitation ranges. The chemical procedures for sewage treatment include the separation of dissolved metals by precipitation, which is determined by the solubility product and by the influence of the solubility. The pH range permitted in waste water lies between 6 and 9. In principle, NaOH, Na_2CO_3 (soda) and $CaCO_3/Ca(OH)_2$ (lime) can be used as precipitating agents, but the special chemistry of the metal must be considered in each case: Fe(III), Al(III) and Cr(III) ions precipitate out as hydroxides even in the slightly acidic range. Because of their amphoteric character, the hydroxides of aluminium and chromium break up at pH > 8 and 9, respectively, to form hydroxy complexes. With chromium and amphoteric zinc, this renewed decomposition can be suppressed by precipitation with lime since the calcium compounds of the hydroxy complexes are slightly soluble. Precipitation with Na_2CO_3 is advantageous with zinc, lead and cadmium, since the carbonates of these metals precipitate out as the hydroxides at low pH.
2. Solubility curves. With Fe(III) ions, a nearly total precipitation occurs at a pH lower than 4, regardless of the type of precipitating agent. On the other hand, Cr(III) ions precipitate differently in the presence of each of the three precipitants mentioned, whereby CaO in the form of lime milk (calcium hydroxide suspension) displays the best activity with respect to preventing a possible decomposition in the alkaline region. With Cd(II) ions, precipitation in the neutral region up to about pH 8 can be achieved only with the use of soda. If several metals are present in sewage water, an overall quantitative precipitation can be achieved even at low pH, since the metal ions that are more difficult to precipitate (i.e. at higher pH values) or that precipitate incompletely are adsorbed by the more easily precipitable hydroxides like iron(III) hydroxide and are precipitated as well. In practice the optimal selection of the precipitating agents and pH ranges is very important.

3.4.11 Sewage treatment plant with precipitation of phosphate

Not only do phosphates reach sewage treatment plants directly from sewage waters, they are also produced as a result of the microbial breakdown of organic substances containing phosphorus. Phosphate ions can be removed with the aid of chemical precipitation using Fe(III) and Al(III) ions. The precipitation can take place at three different locations in a sewage treatment plant. If the precipitant is added at the stage of the aerated sand trap or directly in the inflow to the preclarification basin, then the phosphate portion that was brought in is precipitated out along with organic substances, relieving an impact from the biological phase. In simultaneous precipitation, the precipitant is added at the stage of the activated sludge basin. The flake separation takes place in the secondary clarification stage. The settling properties of the activated sludge are improved using this procedure, which is the most widespread and reliable method. An average concentration of about 1 mg L^{-1} total P is attained in the discharge. With the third procedure, that of secondary precipitation, a separate precipitation and settling basin is necessary. This is the most effective method, and it does not disturb the biological processes, but it is also the most expensive variation. The dosage of the precipitant occurs at locations with high turbulence (Sections 3.4.8 and 3.4.9). Flake growth is favoured in regions with low turbulence. With the combination of precipitation and flocculation filtration, discharge levels between 0.2 and 0.3 mg L^{-1} P are attained.

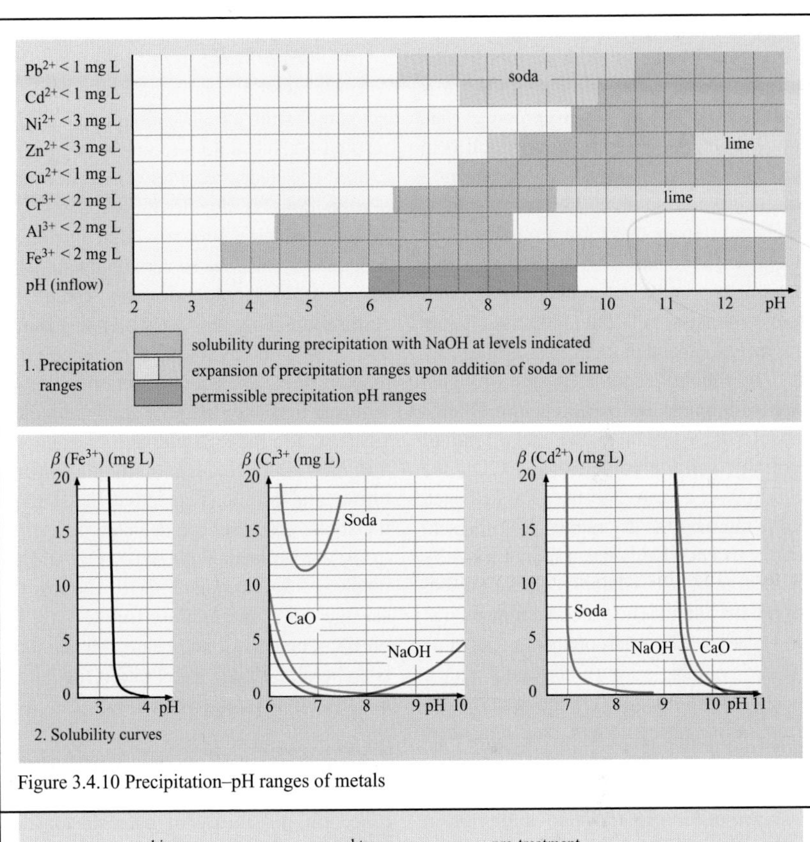

Figure 3.4.10 Precipitation–pH ranges of metals

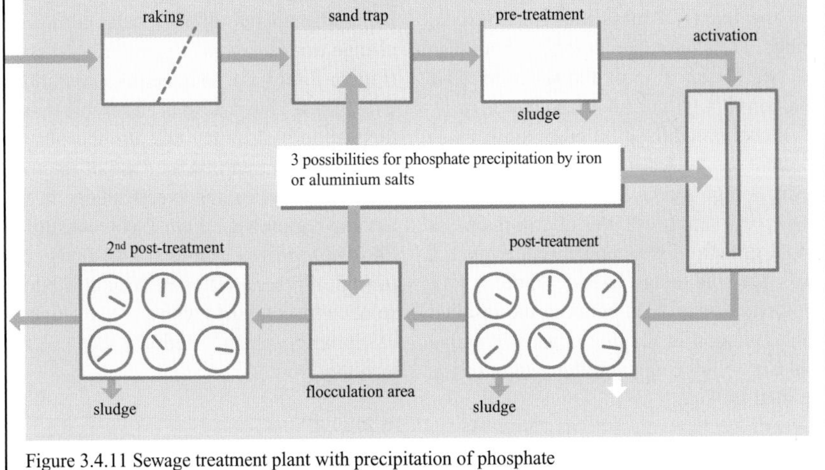

Figure 3.4.11 Sewage treatment plant with precipitation of phosphate

3.4.12 Sludge formation and BOD

In treatment plants for the biological purification of waste water, biological sludge is produced in an activation process. The sludge bacteria (*Pseudomonas*, coliform bacteria, *Nitrosomonas*, *Nitrobacter*, etc.) excrete water-soluble, high-molecular-weight substances that form a mucilaginous envelope around them and that aggregate the bacteria to produce flocculation. Activated sludge consists of the biomass, slimy substances, silt and inorganic materials (only 0.5% of which is dry matter, and of which 30% is inorganic and 70% organic). Sludge impact (B_{DM}) refers to the daily load of respirable sewage contents (as BOD_5, see Section 3.4.5) relative to the total activated sludge biomass (as dry matter, DM) present in an activated sludge basin (Section 3.4.2). The degree of purification is all the greater, the smaller the mass is of contaminants that can be utilised by microorganisms per time and per available mass of activated sludge. The age of the sludge is an important factor, particularly for bacteria with long generation times, such as nitric bacteria. The age is calculated as the quotient of the bacterial mass present in the activated sludge basin and the bacterial mass minus the excess sludge. The sludge age must be greater than the generation time of the nitric bacteria (Sections 3.4.15 and 3.4.16) so that as many of these specialised microorganisms as possible can be fed back into the activation basin. A high sludge impact generally leads to a good supply for the microorganisms (rapid growth of the population); in such a case, the sludge age is low. Figure 3.4.12 shows the correlation between the residual contamination in the discharge, the sludge build-up, sludge age and the causes of the residual BOD_5.

3.4.13 Oxygen consumption for metabolic processes

Various metabolic processes take place in activated sludge: an endogenous sludge respiration, substrate respiration and nitrification. The sludge respiration in an activated sludge flake is due to the symbiosis of aerobic bacteria, protozoa, and in part of fungi and yeasts. The bacterial biomass takes in the usable contents of the sewage for respiration. Therefore, in order to balance this growth in biomass, a corresponding amount of excess sludge (Section 3.4.12) is removed from the sludge cycle. During nitrification, the nitric bacteria *Nitrosomonas* and *Nitrobacter* oxidise ammonia to form nitrite and nitrate, respectively. All three metabolic processes are depicted in relation to the sludge load. With BOD_5 kg^{-1} DM d^{-1} below 0.3 kg, complete purification of the sewage water is possible. Below 0.15 kg, nitrification occurs, and below 0.05 kg there is an aerobic sludge stabilisation (i.e. elimination of the putrefactive ability).

3.4.14 Elimination of phosphate

The biological elimination of phosphorus is based on the ability of bacteria to store increasing amounts of phosphates during the change from anaerobic to aerobic conditions (Figure 3.4.14, 1): in anaerobic conditions they release phosphates; in an aerobic milieu (activation basin) they take up phosphates. As a result of optimisation (use of returned sludge and cycle), the excess sludge in the aerobic region has a high P content. In the Phostrip process (Figure 3.4.14, 2), the biological process (after the anaerobic return solution) is combined with a precipitation as calcium phosphate (Section 3.4.11) in the auxiliary flow.

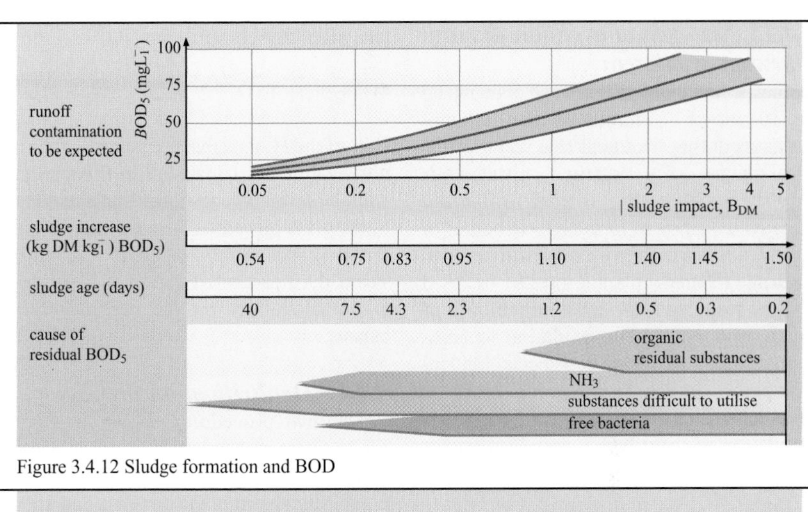

Figure 3.4.12 Sludge formation and BOD

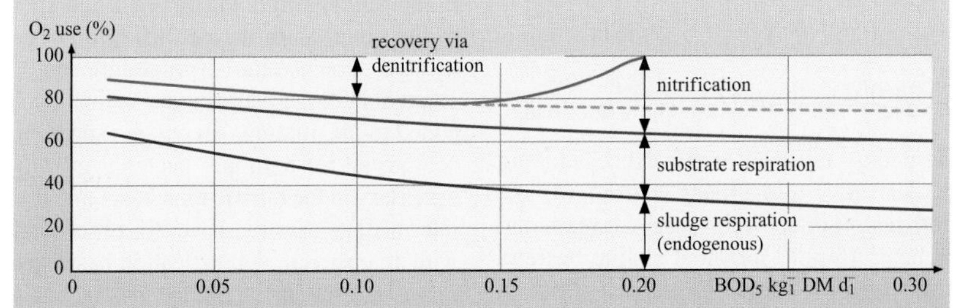

Figure 3.4.13 Oxygen consumption for metabolic processes

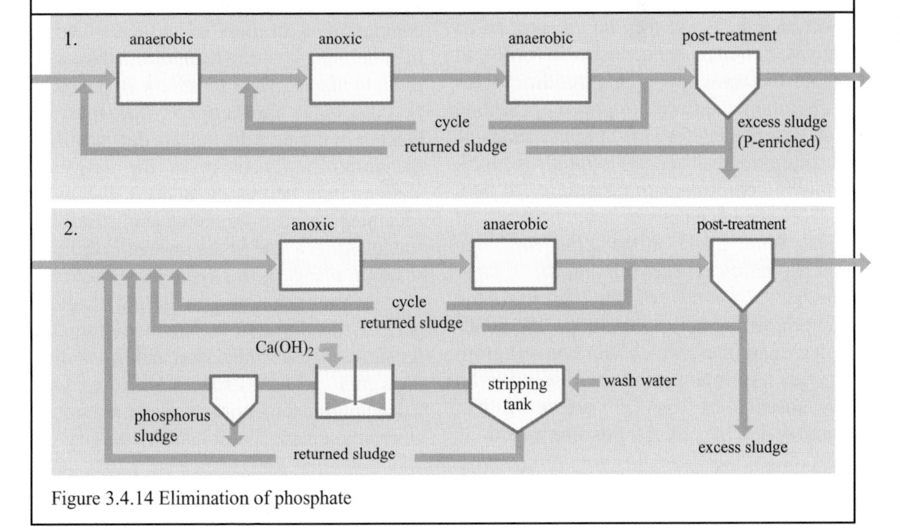

Figure 3.4.14 Elimination of phosphate

3.4.15 Nitrogen conversion during sewage treatment

Domestic and municipal sewage contains up to 30 mg L^{-1} ammonia; this amount increases during treatment as a result of the breakdown of organic nitrogenous substances. Since NH_4^+ is toxic to fish and since its oxidation to form NO_3^- in bodies of water leads to a high oxygen consumption, in a sewage treatment plant it must be oxidised and the resulting NO_3^- must be removed as much as possible. The oxidation to NO_3^- (nitrification) occurs via two special lithoautotrophic (utilising inorganic carbon) microorganisms of the genera *Nitrosomonas* (oxidation to form NO_2^-) and *Nitrobacter* (oxidation from NO_2^- to NO_3^-):

$$NH_4^+ + 1.5O_2 \rightarrow NO_2^- + H_2O + 2H^+$$

$$NO_2^- + 0.5O_2 \text{ (from } H_2O) \rightarrow NO_3^-$$

The energy released in this process (about 350 kJ mol^{-1} altogether) is used for the construction of cell substance. Compared to the heterotrophic activated sludge bacteria (utilisation of organic carbon, Section 3.4.13), these bacteria grow more slowly by a factor of 10. Only after a certain sludge age (Section 3.4.12) can they be accepted in an activated sludge biocenosis. In order to create the most favourable conditions for nitrification, the sludge impact B_{DM} (Section 3.4.12) (and therefore the nutrients available for heterotrophic bacteria) is greatly reduced. Optimal conditions are considered to be a sludge age of 7 days at a temperature of 15 °C, 2 mg L^{-1} O_2 in the activated sludge and pH values between 7.5 and 8.6. A large number (70–90%) of the heterotrophic bacteria in the activated sludge are facultative anaerobes, which can convert from oxygen respiration to nitrate respiration in the absence of oxygen (anoxic milieu; Section 3.4.14). Decomposition reactions take place that consume the NO_3^--oxygen:

$$4H^+ + 5C + 4NO_3^- \rightarrow 5CO_2 + 2N_2 + 2H_2O$$

Half of the H^+ ions that are generated during nitrification are consumed in this process, with the favourable response that a reduction of the pH is prevented. Nitrification requires 2 mol O_2 per mol N, whereas 1.25 mol is saved during denitrification, which has a favourable effect on the overall oxygen balance.

3.4.16 Denitrification process

The two procedures shown in Figure 3.4.16 differ in the placement of the denitrification basin, which is located before the nitrification basin in the one case and after it in the other. With the denitrification basin located downstream (procedure 1), an external hydrogen donor such as methanol is added to the mixture. An oxygen supply that is adjusted for optimum bacterial growth is necessary in the nitrification basin and in the intermediate aeration basin. In procedure 2, with an upstream denitrification basin, there is no intermediate aeration basin, but the sewage must be circulated around instead. The processes in denitrification are largely comparable with those of aerobic bacterial metabolism. The most important factor at close to neutral pH is to have a BOD_5/NO_x-N ratio > 3. Furthermore, the reaction kinetics of nitrate reduction are dependent on the respiration activity of the activated sludge. It is necessary to have the most accurate possible information about the nitrifiable nitrogen load in the sewage in order to scale up plants of this type. One must take organic nitrogen compounds into account, part of the nitrogen is removed from the sewage with the sludge, and another portion does not get oxidised. Therefore, load balancing for the optimisation of a nitrogen elimination plant is absolutely necessary.

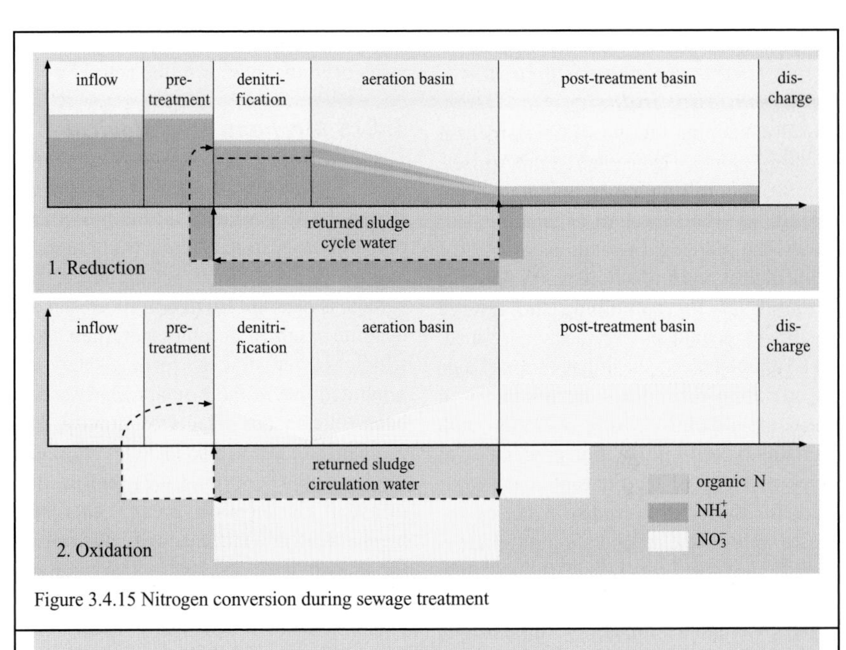

Figure 3.4.15 Nitrogen conversion during sewage treatment

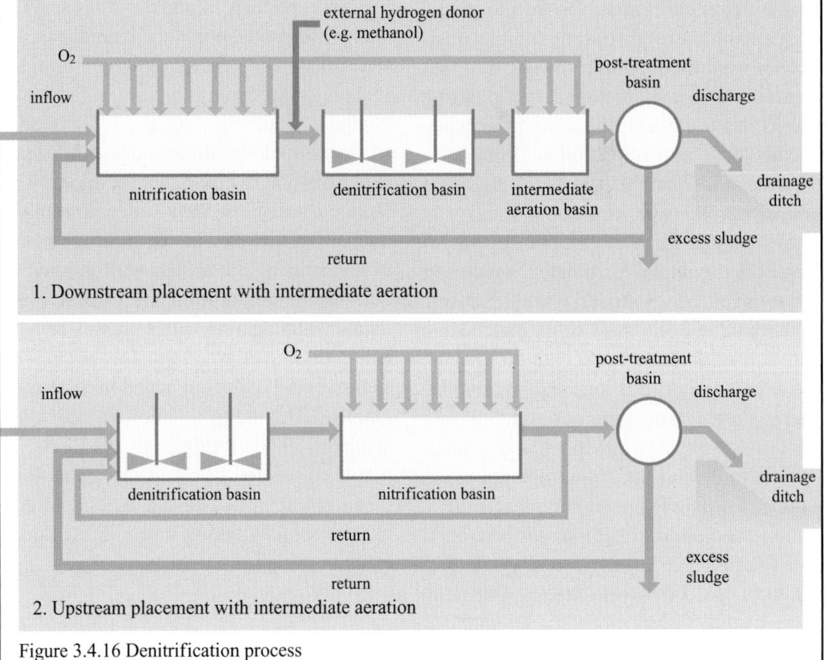

Figure 3.4.16 Denitrification process

3.4.17 *Waste water treatment in the metal-working (automobile) industry*

Metal-containing sewage accrues largely in pickling plants, electroplating enterprises, steelworks, rolling mills and hardening operations. The flowchart in Figure 3.4.17 shows the individual operations associated with special types of sewage as well as the possibilities for combining the sewage treatment techniques previously described. The figure makes it clear that various kinds of operating procedures are necessary in the sample metal-working operation with different types of production processes. An important part of the concept – as shown here for the central control room of the waste water disposal operation of an automobile manufacturer (Kunz, 1992) – is the return of treated material flows into production and the linking of stages (units) in the treatment. Often, few of the procedures shown are sufficient for the businesses described. Acidic sewage and that containing heavy metals from pickling plants has to be neutralised and the heavy metals must be precipitated as hydroxides. Electrolysis can be used for possible recycling. Sewage containing acid and heavy metals from electroplating operations often contains chromates, which can be reduced using iron(II) salts. Sewage containing cyanide must be oxidised. Here too it is reasonable to recycle the metals. Residual heavy metal ions can be removed using ion exchange or carburising processes. For sewage from steelworks, rolling mills and hardening operations, in addition to the physicochemical purification processes, in some cases microbial purification processes also play a role, wherein the dissolved metals from the sewage can be removed via storage on cell walls or in the interior of the cell.

3.4.18 *Mechanical–biological–chemical sewage treatment plant with sludge treatment*

Figure 3.4.18 combines all the procedures of a conventional industrial sewage treatment plant. In the first stage, the sewage is neutralised; dissolved substances become slightly soluble materials from which larger flakes form. They are separated out in the primary clarification basin due to the effects of gravity. The mechanical portion also includes rakes and sand traps (Section 3.4.2) before the primary clarification basin. Then the mechanically cleaned sewage flows into the activated sludge basin. The microorganisms present in the sewage are provided with oxygen both here and in a part of the biomass (return sludge – as seeding sludge) separated out in the secondary clarification. The excess sludge biomass (excess or secondary sludge) is fed into a sludge treatment area. The sludge treatment includes thickening, stabilisation (Section 3.4.13) and dehydration. Raw sludge consists of 90% water. The thickening step involves increasing the amount of solids using the influence of gravity; an additional dehydration step takes place using centrifugation, filter presses or heat. Lime (hydrate) is added for homogenisation. Fe(III) salts are added in the sludge conditioning stage to improve the water removal. The treated sludge can be disposed of by using it in agriculture (requirements as per the treated sludge regulations), by taking it to a refuse dump, or by burning it.

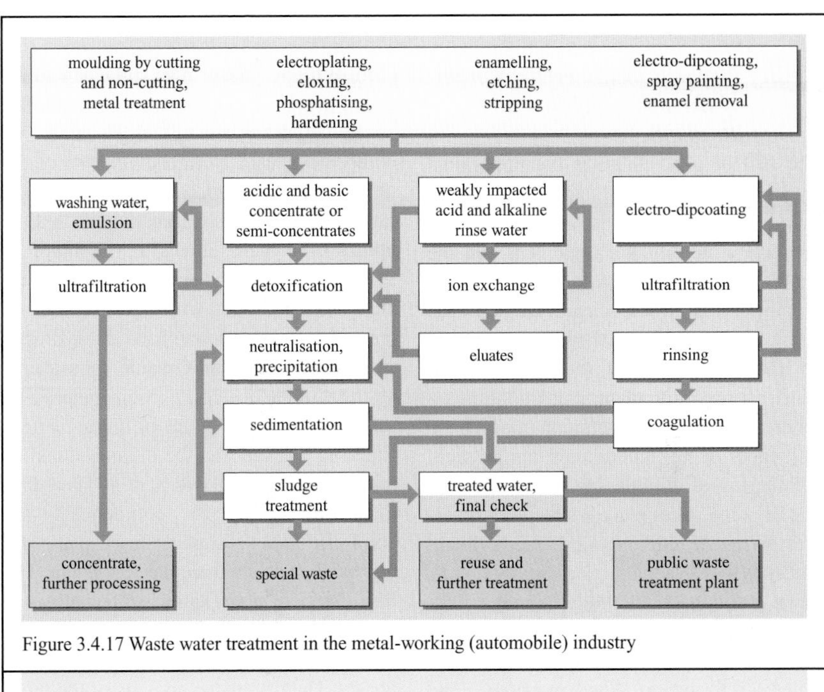

Figure 3.4.17 Waste water treatment in the metal-working (automobile) industry

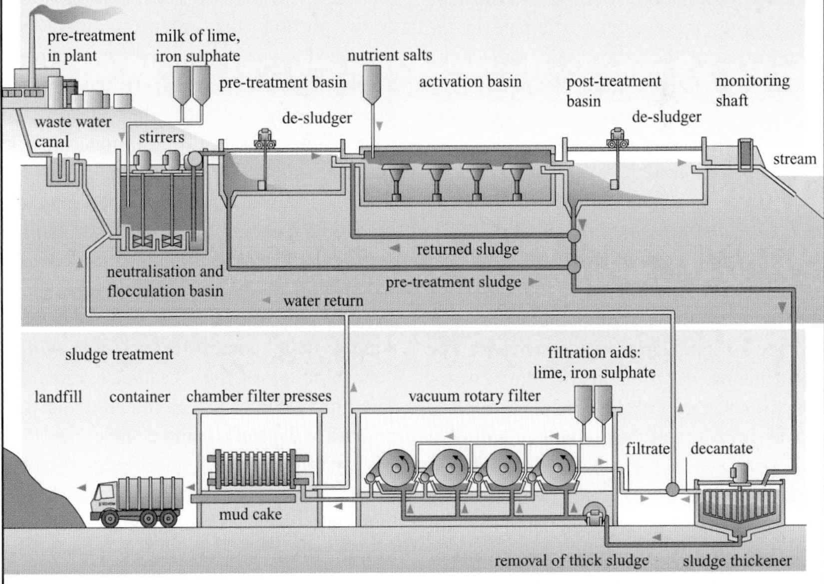

Figure 3.4.18 Mechanical–biological–chemical sewage treatment plant with sludge treatment

4 Soil

4.1 Basic principles of soil science

4.1.1. The soil in the environment of the litho-, hydro- and atmosphere

The soil or pedosphere is the uppermost layer of the Earth's crust that is inhabited by living organisms and is weathered. It is altered by the climate, organisms and the people who cultivate it. According to the definition used in soil science, the soil ends at the transition into unaltered stone. Based on the sphere model (as the spheroid construction of the planet Earth), the pedosphere is a sphere of penetration in which the influences of all four spheres can be seen. Weathering of the lithospheric component, that of stone, takes place in the pedosphere. Weathering includes physical (mechanical), chemical and biological processes that break down and convert the body of rock of the lithosphere. Changes in the body of rock in which all three major spheres participate are called soil formation. Where one of the three other spheres does not participate in the penetration of the lithosphere, there is no pedosphere. Examples are the alpine rock region = weather sphere, and the water bottom with no influence by the atmosphere = diagenesis sphere.

4.1.2 Components of the litho-, bio-, hydro- and atmosphere

The soil consists of lithogenic, biogenic, atmogenic and hydrogenic components. The percentages contributed by the various spheres can vary greatly from place to place in the three-dimensional space of soil. The soil structure is formed from the spatial arrangement of the solid components of the soil and the soil pore component. The soil structure directly determines the amount of water, air and heat. The gaseous phase in all parts of the volume of pore space of a soil is called the soil atmosphere. Since it is influenced largely by biological processes, its composition usually varies with that of the atmosphere. It is determined by soil respiration, the release of carbon dioxide and the uptake of oxygen, i.e. via the respiration of soil organisms and plant roots. Under anaerobic conditions, with high soil moisture content, gases such as methane and hydrogen sulphide collect here. Organic components as a part of the biogenic portion comprise all the dead matter of plant or animal origin found in the soil and the organic transformation products of this matter. The lithogenic and biogenic components are combined as a solid phase, so that the soil can be shown as a three-phase mixture. The basic step in quantitative soil analysis is considered to be the determination of the weight and volume portions of the three phases. The dry matter consists of the lithogenic portion, the organisms that have converted to a dry state (died), and the postmortal organic material. If the organic material makes up more than 30% of the whole, one uses the description 'organic soil'. In Central Europe, this figure is 3–10%. The water content is the difference between the mass of the original soil sample and the dry material. Differentiation between the mineral and organic portions takes place via annealing loss (at 500 °C with an oxygen supply) or by combustion analysis with a determination of the CO_2 that is released. The air volume is calculated from the volume of the soil column cylinder (removal of a soil sample with a soil column cylinder) minus the volume of solid and liquid materials. With normal humus-containing soil, the size of the pore space is around 43% by volume.

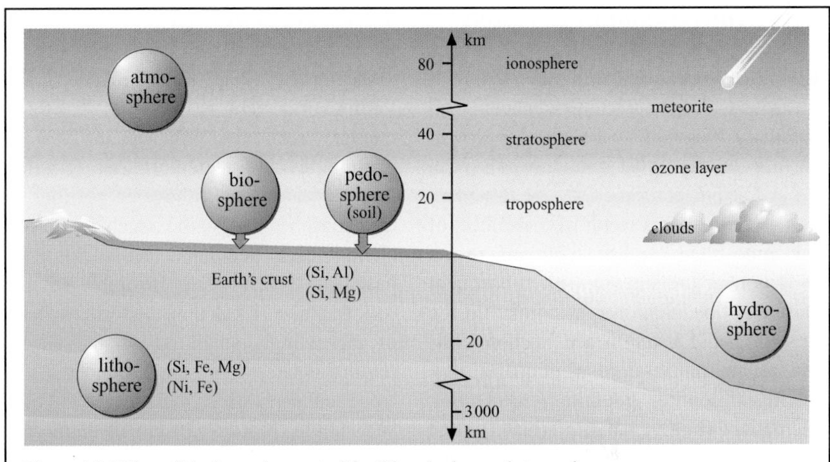

Figure 4.1.1 The soil in the enviroment of the litho-, hydro- and atmosphere

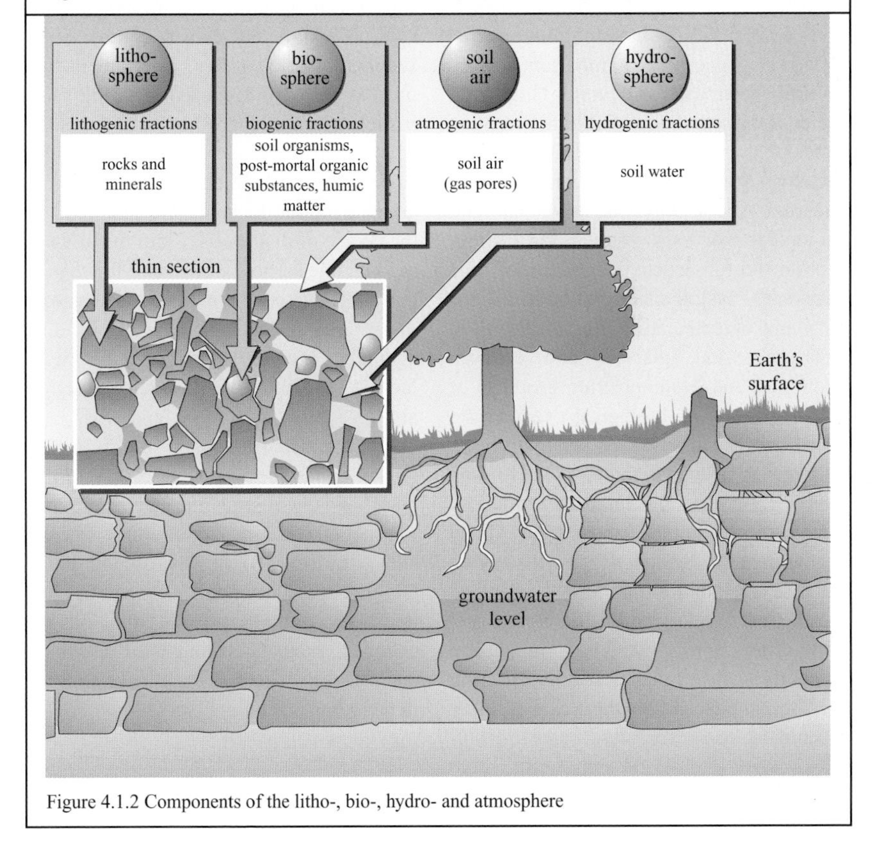

Figure 4.1.2 Components of the litho-, bio-, hydro- and atmosphere

4.1.3 Interrelationships among types of rock

The processes of weathering and sedimentation are of special significance in soil chemistry. Magmatic rock makes up 95% of Earth's external crust, metamorphic rock 4% and sedimentary rock 1%. Sedimentary rock is formed as a result of the effects of the atmosphere and/or hydrosphere on regions near the surface with subsequent sedimentation. Such regions make up the largest part of Earth's surface (Section 4.1.4) and contain mineral deposits of quartz, clay, calcite, dolomite, goethite (α-FeOOH), haematite (Fe_2O_3), halite (NaCl) and gypsum. Sedimentary rocks include shale, sandstone and limestone. Gneiss, marble and quartzite are metamorphic rocks. Weathering is critically important for the natural generation of soil. This term includes decomposition and transformation processes resulting from interactions with the atmosphere, hydrosphere and biosphere (Section 4.1.1). Physical, chemical and biological processes result in the disintegration and subsequent distribution of solid materials. Physical disintegration is induced by jumps in temperature (frost), changes in pressure, wind, glaciers and erosion. Excretion and decomposition products of organisms cause biological weathering. Chemical processes such as hydrolysis, carbonising (reactions among CO_2, H_2O and especially $CaCO_3$), reduction, oxidation, complex formation (e.g. of Al, Fe and Mn by humic matter – Section 4.2.15), dissolving out and crystallisation all lead to chemical weathering. Weathering is a process of soil generation during which soil-specific minerals such as clay minerals (Section 4.2.7) are produced from the rock. For futher discussion see Sections 1.1.2 and 1.1.4.

4.1.4 Types of rock

In contrast to the Earth's crust (Section 4.1.3 – the top layer, up to 16 km deep), the Earth's surface consists of up to 75% sedimentary rock and only up to 25% of magmatites and metamorphites (Section 4.1.1), meaning the lithogenic portions of the pedosphere (A.) come largely from sedimentary rock. Sediments can be classified based on their genesis. Insoluble silicate rock fragments break down to form particles of increasingly small size – the clastic (*klastos* is Greek for 'split') sediments – as loose sediments which include crushed rock and rubble (blocks >20 cm in diameter), gravel types (pebbles, stones, gravel and detritus; 20–0.2 cm), sands (2000–60 μm), silt (60-2 μm) and clay (<2 μm). Solid sediment rocks (after a solidification process called diagenesis) with the same particle size are labelled breccia, conglomerates, sandstones, silt rocks (loess) and clay stone. Thus, the hard clastic sediments have arisen from the loose clastic sediments in the course of diagenesis. From the grain of sand on, the particle size is used for characterisation, and not the particle shape as with the coarsely clastic sediments. Binding agents in the sandstones and silt rocks include lime (in chalky sandstone), clay (in argillaceous sandstone), and SiO_2 (in quartzites, sandstones in the narrower sense). Sandstones contain mostly feldspar, quartz and mica. Sands and clays are usually transported and deposited in water, whereas silt is transported mostly by the wind. Bioliths are substances such as coal and peat; shell limestone is one of the chemical–biogenic sediments. Limestones and salt rocks occur as chemical sediments in the ocean and other places.

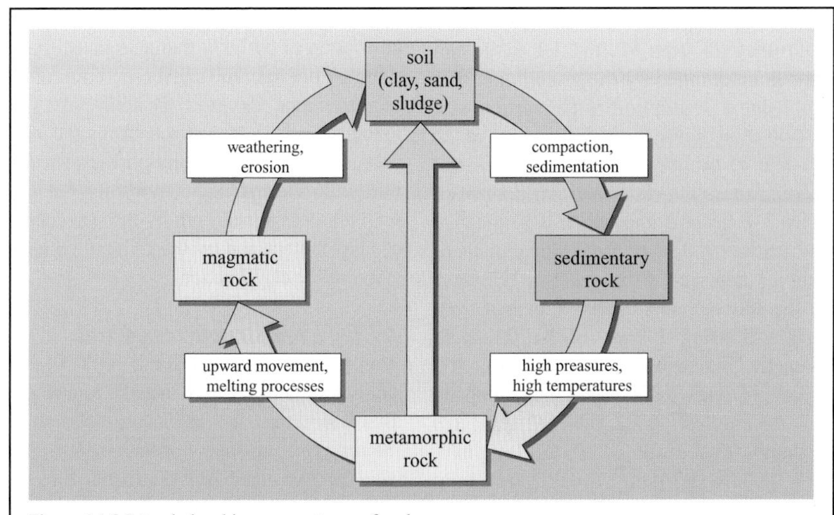

Figure 4.1.3 Interelationships among types of rock

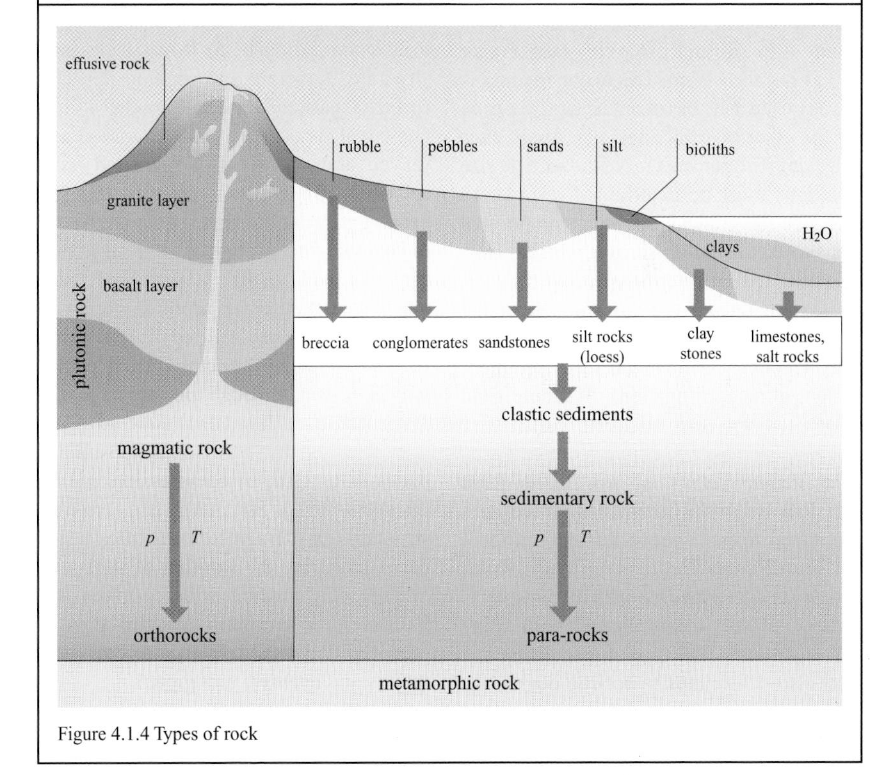

Figure 4.1.4 Types of rock

4.1.5 Soil granulation classes

Following on from Section 4.1.4, the soil type is determined by the granulation of a soil using a description of the particle size composition. It influences factors such as the particle distribution in the soil pores and therefore also the water and gas content of soils. The granulation, which is also called soil texture, is derived from the amounts of clay, silt and sand. Coarse soil (in the sense of the soil skeleton) is distinguished from fine soil using 2 mm as the threshold diameter. The main fractions of the fine soil are sand (coarse to fine sand internationally 0.2–2 mm; DIN 4022 0.063–2 mm; US system 0.05–2 mm), silt (0.002–0.02 mm, DIN 0.002–0.063 mm; US 0.002–0.05 mm) and clay (<0.002 mm). If one of these fractions is predominant, then the soil type is named after it. A soil comprised of 40% sand, 40% silt and 20% clay (see Figure 4.1.5) is called loam. The major fraction is given an attribute based on the larger portion of the other two fractions: e.g. sandy clay, silty clay. The percentages of the particle size fractions must be determined in order to describe the granulation. Sand components can be determined via sieving, whereas finer fractions can be determined using the sedimentation rate of the particles under the influence of gravity and by other methods (nephelometry, centrifugal force sedimentation, ultracentrifugation). A simple field procedure for soil scientists consists of rubbing the moist soil between one's fingers. The characteristics of the three main fractions are sand S (grainy, easy to see, does not adhere to the ridges of the fingers, cannot be shaped), silt U (velvety-floury, rough smear surfaces, does not adhere) and clay C (sticky, plastic, easily shaped with shiny smear surfaces). The type of granulation of a soil also determines to a large degree the ecochemical behaviour of pollutants, their adsorption or mobility and also transformations of organic substances or binding to inorganic ions. However, the ability of a soil to be adsorbed by cations from the soil solution (equilibrium solution between the lithogenic and hydrogenic portion – Section 4.1.1) is determined both by the percentage of clay content and by the mineral composition of the clay fraction.

4.1.6 Clod structures of soil

With respect to soil chemistry properties of soil types, their structure, usually referred to as soil structure, is significant. It describes the nature of the spatial arrangement of the solid soil particles. The soil structure directly influences the levels of water, air and heat and indirectly affects the biological activity, the soil development, yield and susceptibility to erosion (Section 4.1.5). In the single-grain structure, minerals and organic particles (primary particles) are not bonded to one another. This occurs in low-clay sand and gravel and in freshly deposited silty sediments. In the coherent structure the primary particles are kept together by means of cohesive forces. They occur in silty soils, clay soils and loam soils in the subsoil; in the upper soil, they break up into aggregates after drying out. Aggregate structures include the prism structure (from prisms with 3–6 usually rough side surfaces and 10–300 mm in diameter), plate structure (characterised by horizontally positioned plates as a result of compression, with a thickness of 1–50 mm), and crumble structure (as a fragment structure, which develops during the working of soils with average clay content with optimum soil moisture). A crumble structure then is produced under the influence of biological activity and intensive root growth.

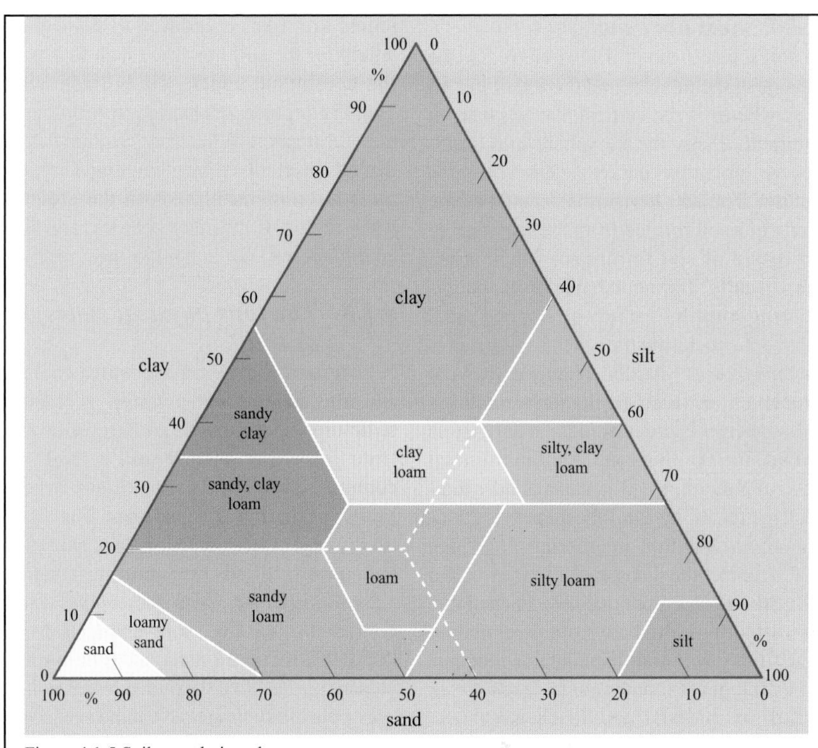

Figure 4.1.5 Soil granulation classes

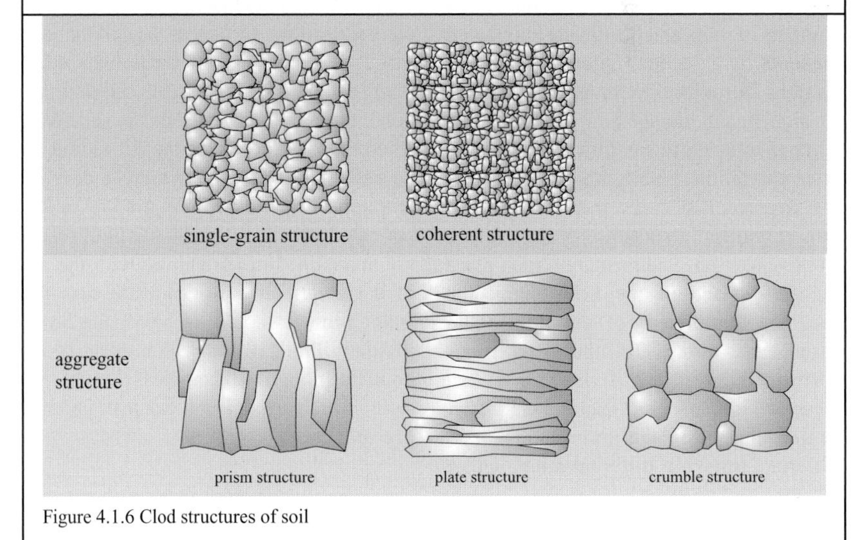

Figure 4.1.6 Clod structures of soil

4.1.7 Soil horizons

In soil science, the profile is viewed as the change in the mixture and the activity of all four spheres in the vertical plane as a depth gradient. Thus, the biosphere and atmosphere portions decrease from top to bottom. The soil horizon is defined as the strata of a soil formed from the substrate by processes of soil formation, which strata are visually distinguishable from neighbouring ones by a fault or stepped area. Above the rock substrate is the stone that is unchanged or hardly changed by the processes of soil formation (retaining lithospheric characteristics) and that is called the C horizon. As the mineral subsoil horizon, the B horizon is only indirectly affected by the lithosphere or atmosphere via solution weathering. It is here that organic substances and oxides of Fe, Mn, Al and clay minerals are enriched due to weathering. Moisture from precipitation that sinks down and that contains biogenic and atmogenic components leads to an influx of material and to chemical and structural changes. The mineral upper soil horizon, the A horizon, contains large amounts of roots and burrowing animals. It presents dark humus materials and light sections due to root excretion and is subject to significant changes in its lithogenic components due to the loss of substances after eluviation. Finally, the O horizon is the organic soil horizon as a layer of mineral soils. It consists of dead refuse from vegetation. Sedimentation took place on the surface of the lithosphere; the biogenic component of the lithosphere formed as humus on the lithogenous surface. Figure 4.1.7, 1 shows a model, whereas Figure 4.1.7, 2 shows a characterisation that uses the term humus and different tints for differentiation. Other letters are used in addition to the capital letters for the soil horizons, such as Aa/A_A for a mineral upper soil horizon with 15–30% organic substance, or Ap/A_P as a mineral upper soil horizon (Section 4.1.8) that has been mixed by ploughing. G represents the mineral horizon in the groundwater region, R indicates bedrock, and H is an organic soil horizon out of peat residue.

4.1.8 The four physical states of soil

A correlation between the water content and the degree of cohesion (coherent structure – Section 4.1.6) determine the ‘four physical states’ of a soil. A compressed soil exists especially where heavy agricultural equipment has been used. If the water content is low, hardened soils occur naturally (as aggregate structure – Section 4.1.6). With increasing water content, a cloddy soil changes to a kneadable soil and finally into a fluid state. Depending on the cohesion, a lower and an upper plasticity threshold can be specified. The purpose of working the soil in this connection is to create aggregates of sufficient size on the surface of the soil that can make water and oxygen available so that seeds can germinate and the seedlings can break through the surface of the soil. With reference to root growth, loosening is necessary by which hollow areas develop for aeration and for improving water permeability. Fine-crumb structural forms are maintained in agriculture with the aid of harrows. When working the soil, the water content should lie below the lower plasticity limit. If the water content is too high, the plough will compress the soil into clods, which are difficult to break up even in a dried state.

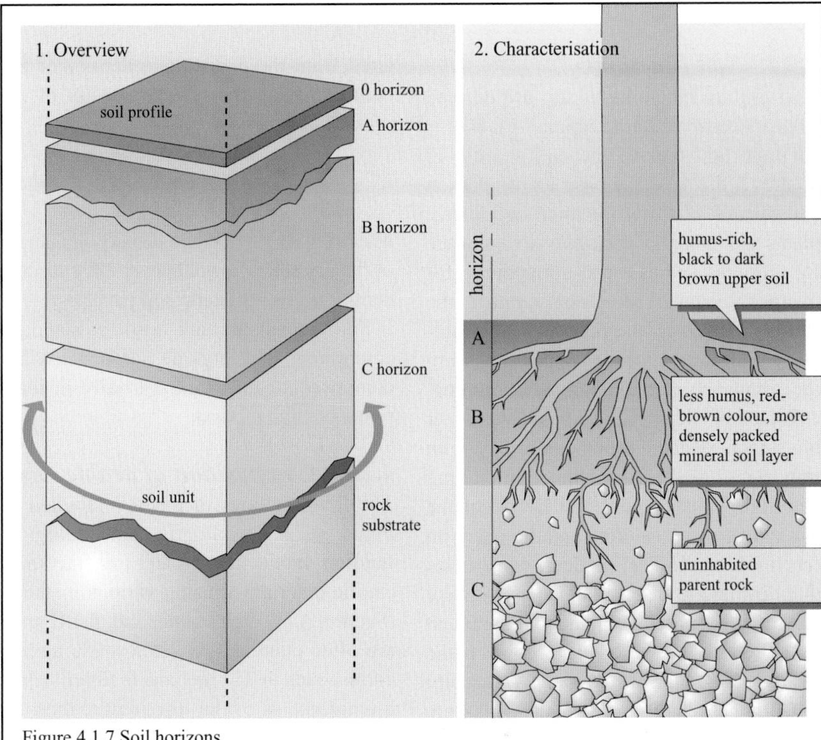

Figure 4.1.7 Soil horizons

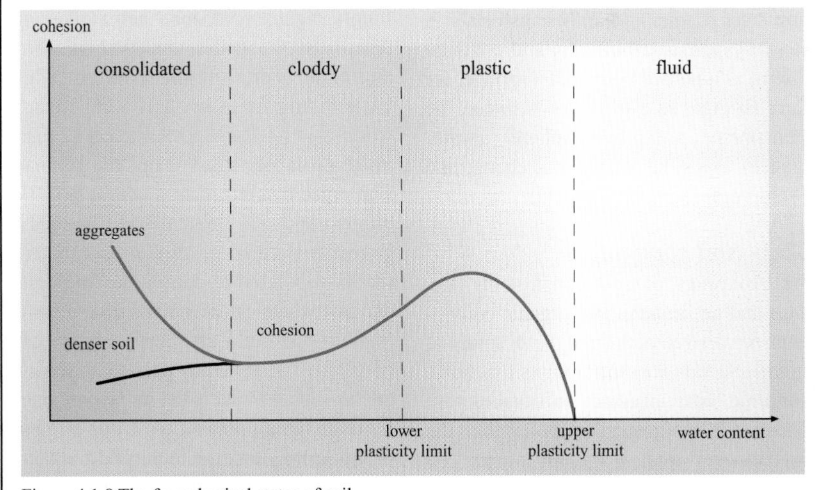

Figure 4.1.8 The four physical states of soil

4.2 Physico- and biogeochemical processes

4.2.1 Soil functions

The various functions of soil are derived from its composition (Section 4.1). Soil is an important part of the entire cycle of nature; it provides the base for vegetation and overall a habitat for microorganisms, plants and animals. It acts as a storage place for nutrients and as a culture medium (for microorganisms). The nutrient supply in the soil is changed by the processes of weathering and mineralisation, deposition from the atmosphere, eluviation and seeping, erosion and extraction due to plants. Other physicochemical processes play an important role in this habitat, including aeration, water content, the course of the soil temperature, filtering and the retention (chemical or physical binding) of particles that flow into it, from either the atmosphere or the hydrosphere. A functional approach to soils was used in a series of lectures in the Institute for Soil Science at the University of Göttingen. Additional soil functions were derived from this activity: soil as an accumulator and converter of biogenic substances (humus), as an N transformer in the ecosystem, as a buffering and colloidal system, as a porous body and a sponge, for water storage, as a mediator between the atmosphere and hydrosphere (water balance), as a system capable of change and development (soil development).

4.2.2 Soil components

The groupings of soil constituents, i.e. water and air, mineral and organic components, are derived both from the structure of the soil (Section 4.1) and from its functions. Inorganic and organic substances are dissolved in the groundwater; just like the air, they are found in the soil pores. The pore space of a soil, the porosity e, is calculated from the dry storage density of the entire soil and the storage density of the mineral soil particles:

$$e = 1 - \delta_b/\delta_s$$

where $\delta = m/V$ = storage density, m = mass of the dry soil, V = volume, b = dry storage density, and s = mineral soil particles.

Soil organisms are also a standard component. The organic components are composed of dead and partially decomposed organic material.

4.2.3 Composition of arable land

The four fractions of a soil – labelled in Figure 4.2.3 as solid fraction, biological fraction, liquid fraction and gas fraction – can be determined using various methods (Section 4.1). With arable land, the figures are often cited as t ha^{-1}. Thus, 2% humus corresponds to 74 t ha^{-1} and 0.13% of (dry) biomaterial is 5 t ha^{-1}(or if wet, then 27 t ha^{-1} with a water content of 80%). The biomaterial contains the mass of bacteria, fungi, algae, protozoa, nematodes and mites, insects, beetles, spiders, earthworms and plant matter (straw, roots), of which bacteria and fungi (with 10 t ha^{-1}), earthworms (4 t ha^{-1}) and plant matter (82 t ha^{-1}) make up more than 90% of the biomass. The organic solid fraction consists of 93% humus and 7% biomaterial. Humus is generally defined as all the dead material that is subjected to continuous biological and pedochemical decomposition, transformation and build-up. However, methodologically it is not really possible to separate the biological fraction, the living organisms, from the humus portion. When determining the annealing loss, the biological fraction is included with the organic solid fraction.

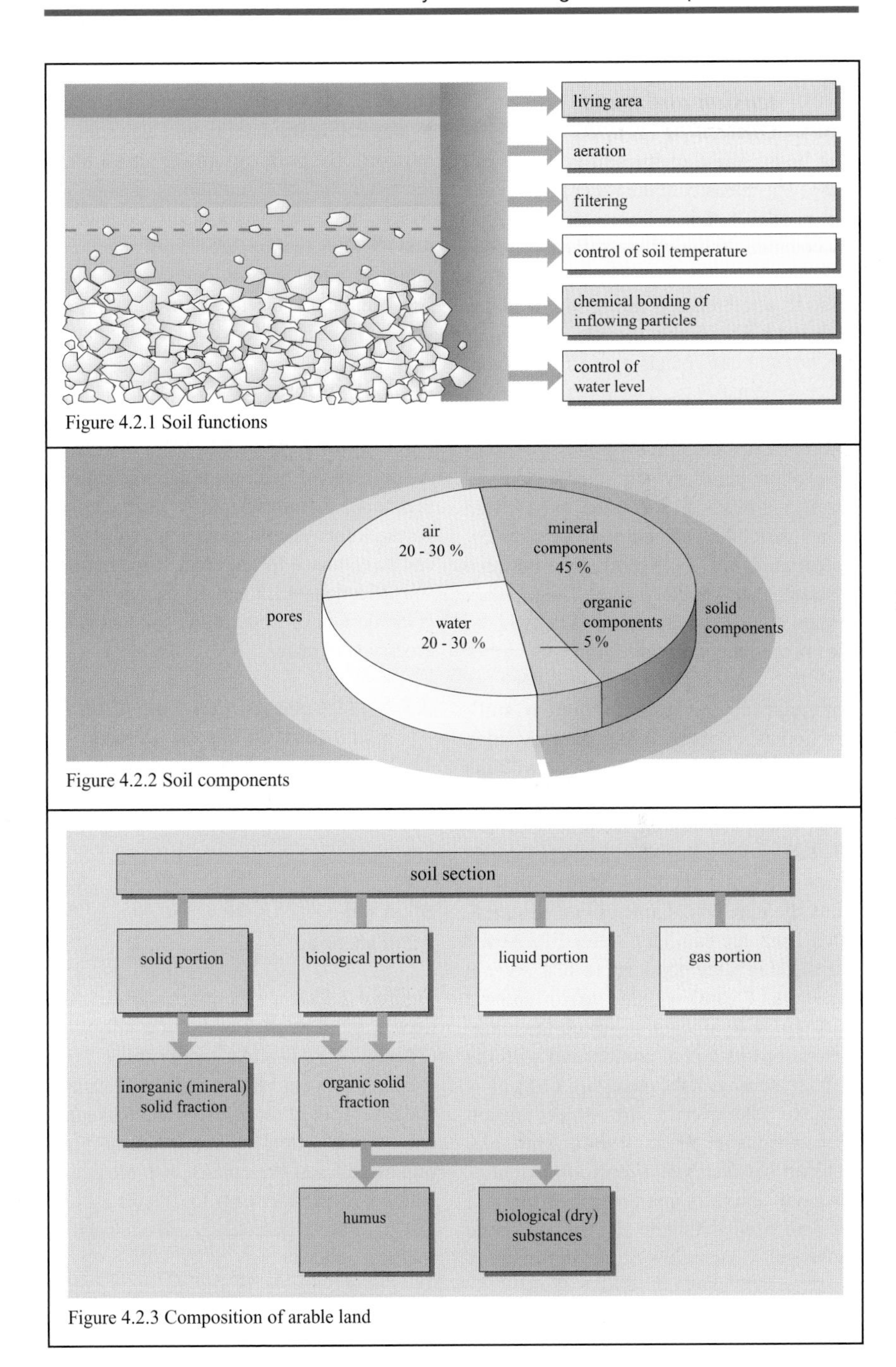

Figure 4.2.1 Soil functions

Figure 4.2.2 Soil components

Figure 4.2.3 Composition of arable land

4.2.4 *Relationship between water tension and water content:pF graphs*

The liquid phase of the soil is called soil water. The intensity of the water binding in a soil is calculated using a potential concept. By potential we mean the work necessary to transport one unit of water from a given point to a reference point. A soil saturated with water has a highly negative potential. If it is brought into contact with a free body of water (potential = 0), then water will rise in the soil (capillary action) until an equilibrium with all active forces (forces of adsorption, capillary action, cohesion and gravity) is reached. Soil water moves from points with high potential to those with low potential. We differentiate between adsorption and capillary water. The capillary tension is cited as a numerical value (without the preceding negative sign) as water tension. The capillary law, which is important in soil science due to soil's function as a sponge, results from equating air pressure and capillary tension (with the acceleration due to gravity and surface tension as determining factors) with $h = 2970/d$ (h = the lift of water in the soil capillaries in cm, d = capillary diameter in μm). Thus the capillary diameter can be calculated from the capillary water lift. At the basis of this is the physical law that water is drawn in a capillary due to differential pressure up to a height at which the atmospheric pressure is in equilibrium with the capillary tension at the meniscus. Log_{10} pF is used to characterise the water tension (Schachtschabel *et al.* 1984). The relationship between water tension (or matrix potential) and the water content depends on the pore size distribution and the pore space in the soil. Conclusions can be drawn about storage properties, rate of drainage and availability for plants (pF < 4.2 as a soil constant) by comparing sandy, silty and clay soils (A horizon). By field capacity (FC) we mean the water content of a naturally situated soil which has established itself counter to gravity after two to three days of precipitation at one location (saturation). Water between the permanent wilting point (PWP) and the field capacity is available for plants.

4.2.5 *Volumes of water, air and substances as a factor of soil type*

A summary of Sections 4.2.2–4.2.4 yields Figure 4.2.5 from which the characteristic volumes for water, air and solid substances can be extracted for the sand, loam and clay soils (Sections 4.1.2 and 4.1.5). Dead water is the portion of water in the soil that is not accessible to plants.

4.2.6 *Oxygen and carbon dioxide content in the air in soil*

The average composition (% vol.) of the soil atmosphere in comparison with atmospheric air amounts to:

	O_2	N_2	CO_2
Air in soil	20.6	0.30	79.1
Atmospheric air	21.0	0.03	78.9

Figure 4.2.6 shows the graphs at different depths for a sandy loam or a silty clay (Section 4.1.5) under apple trees. Gas exchange occurs via the open soil surface and a network of macropores and thus guarantees respiration in subterranean plant organs and soil organisms. A biologically active soil produces up to 16 000 kg ha^{-1} CO_2 (two-thirds microbially and one-third by root respiration).

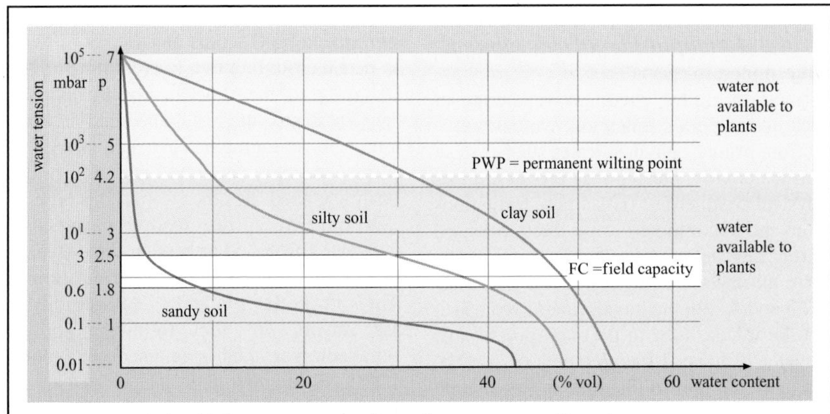

Figure 4.2.4 Relationship between water tension and water content: pF graphs

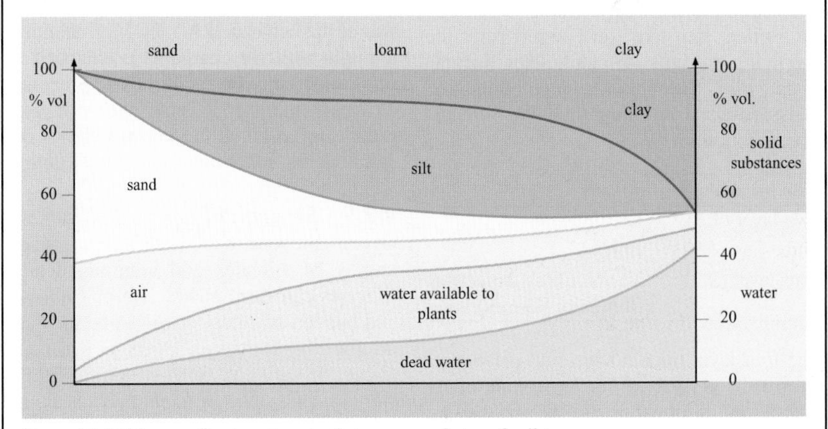

Figure 4.2.5 Volumes of water, air and substances as a factor of soil type

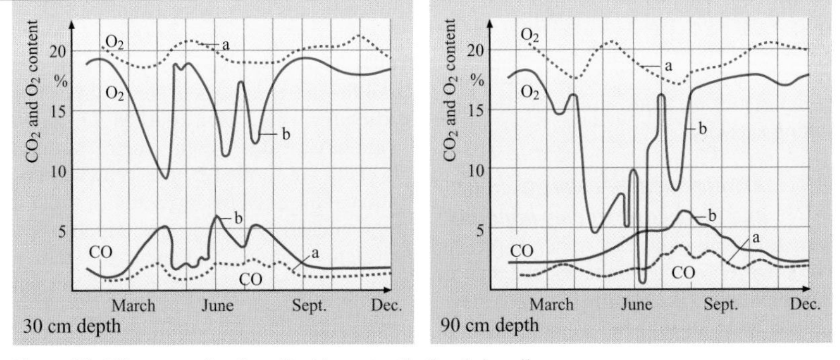

Figure 4.2.6 Oxygen and carbon dioxide content in the air in soil

4.2.7 *Formation, transformation and decomposition of clay minerals*

Clay minerals (particle size < 2 μm) are formed during the chemical weathering of rock and minerals following two pathways: (i) from primary phyllosilicates such as mica via principally mechanical processing, or physical weathering; (ii) as new formations from the decomposition products released during the chemical weathering of feldspars. Clay minerals are phyllosilicates (Sections 4.2.8 and 4.2.9), the transformation of which can be traced back to the incorporation or transfer of ions of the elements Al, K, Na, Ca, and Mg and to changes in the silicate groups. In the first pathway, after the grinding of mica minerals down to colloidal size, cation and anion exchange reactions take place that are very significant for supplying nutrients to plants. New formations from the breakdown products of feldspars take a very long time. We differentiate as follows:

Clay minerals	
Two-layer (1:1)	Kaolinite
Three-layer (2:1)	Illites
Four-layer (2:2)	Vermiculites, smectites (montmorillonites)

The four-layer minerals have an octahedral intermediate layer of Mg hydroxide or Al hydroxide. Allophanes are paracrystalline and amorphous aluminium silicates, the unit structures of which are combined with hardly any recognisable order. In addition to the layered construction and the small particle size with correspondingly large surface, one characteristic of clay minerals is their swelling capacity.

4.2.8 *Arrangement of elements in two- and three-layered clay minerals*

The two-layered (1:1) clay minerals present an ordered series of a tetrahedral and then an octahedral layer. The tetrahedron consists of a silicon atom surrounded by O atoms in a tetrahedral shape. Al, Mg or Fe atoms surrounded by O atoms or OH groups form an octahedron. In a two-layered mineral, the OH groups of the Al octahedron are opposite the O atoms of the Si tetrahedron. H bonds form, the stratographic sequence interval is correspondingly small and is therefore not easily changed. In other words, this type of clay mineral is not readily capable of swelling. On the other hand, a three-layered (1:2) mineral such as the montmorillonite $Al_2(Si_4O_{10})(OH)_2 \cdot nH_2O$ does swell readily. It has layers in the following order: Tetrahedron/octahedron/tetrahedron (Section 4.2.7). In the tetrahedral layer Si can be substituted by Al, and in the octahedral layer Al can be substituted by Mg or Fe, whereby the substitution of tetrahedral Si^{4+} by Al^{3+} or that of the octahedral Al^{3+} by Mg^{2+} produces a single negative charge. It is balanced by one cation between the Si layers. Smectites have particularly good cation exchange properties: they swell up in moist conditions and shrink in dry conditions quite significantly.

4.2.9 *Structural types of silicates*

Overall, the rock in the Earth's crust is made largely of silicates and aluminosilicates. Silicates can form chains (fibrous silicates and banded silicates) or rings and they can arrange themselves in layers as described above, or they can form structural frameworks as in zeoliths and feldspars (as almost exclusively alkaline or alkaline-earth aluminosilicates). Well-known examples of orthosilicates are olivine $(Mg,Fe)_2SiO_4$ and zircon $ZrSiO_4$; biotite $K(Mg,Fe)_3(AlSi_3O_{10})(OH)_2$ is an example of phyllosilicates; and zeoliths are examples of three-dimensional structures, e.g. $NaCa_2(Al_5Si_{13}O_{36}) \cdot nH_2O$. Aluminosilicates with their structural diversity are formed by the stepwise substitution of the Si atoms by Al atoms, which are approximately the same size. Monovalent to trivalent cations compensate for the increased negative charge of the framework due to the exchange.

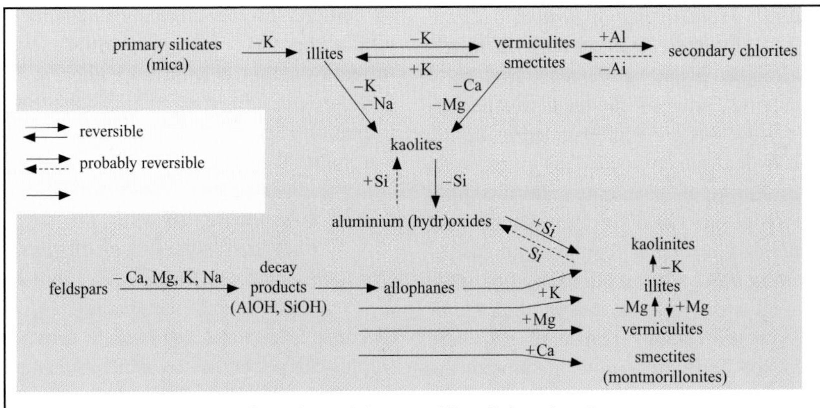

Figure 4.2.7 Formation, transformation and decomposition of clay minerals

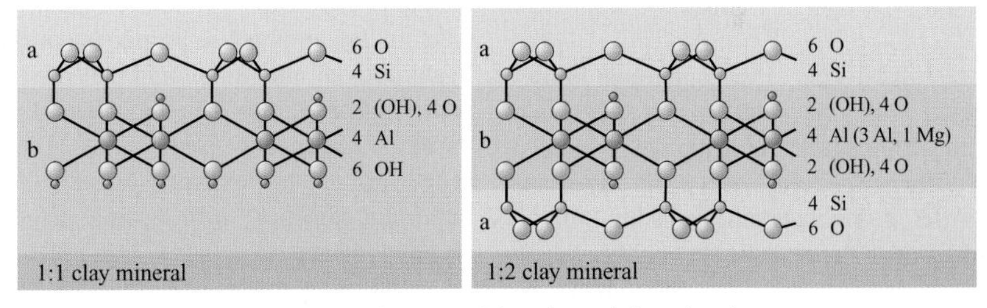

Figure 4.2.8 Arrangement of elements in two- and three-layered clay minerals

Structure	Formula	Description	Si-O ratio
	SiO_4^{4-}	orthosilicates	0.25
	$Si_2O_7^{6-}$	disilicates	0.29
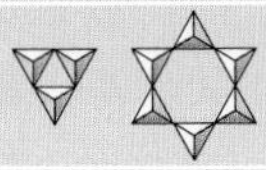	$Si_3O_9^{6-}$ $Si_6O_{18}^{12-}$	cyclosilicates	0.33
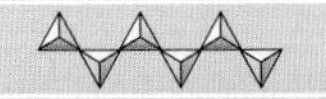	$(SiO_3^{2-})_n$	pyroxenes	0.33
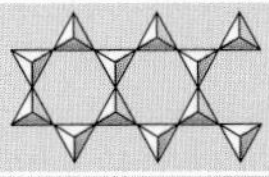	$(Si_4O_{11}^{6-})_n$	amphiboles	0.36
	$(Si_4O_{10}^{4-})_n$	phyllosilicates	0.40
	$(Si_{4-x}AB_xO_8^x)_n$	3D structures (tectosilicates)	< 0.50
	$(SiO_2)_n$	kieselguhr	0.50

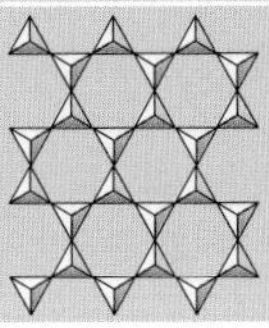

Figure 4.2.9 Structural types of silicates

4.2.10 *Weathering of potash feldspar to form clay minerals*

Hydrolysis is one of the most frequently occurring forms of chemical weathering. The processes are demonstrated using the potash feldspar orthoclase as an example (with a share of 20% of the mineral composition in the Earth's crust). The 'fresh' mineral is viewed as the potassium salt of the very weak silicic acid. In the first stage of weathering, first the ions on the edge of the crystal lattice (especially K) are loosened up by hydration; the ions of the base KOH form in solution via hydrolysis:

$$KAlSi_3O_8 + H_2O \rightarrow HAlSi_3O_8 + K^+ + OH^-$$

Potash lye forms in the outermost border region of the crystal undergoing weathering, with the generation of a type of 'hydrogen feldspar' (a watery slurry of orthoclase powder at this stage reacts alkaline). In the subsequent steps of weathering, bonds are split between O and Al or Si atoms in the crystal lattice. Finally, the structure of the orthoclase is destroyed with the formation of aluminium hydroxide and silicic acid:

$$HAlSi_3O_8 + 7H_2O \rightarrow Al(OH)_3 + 3H_4SiO_4$$

Kaolinite can be produced in the new growth pathway as a result of the recrystallisation of the solution partners (Section 4.2.7). The ions released from aluminosilicates in general are removed from the soil by seepage water, or are taken up by plant roots. The rate of weathering is increased markedly by low pH values (as a result of dissolved CO_2, organic acids as plant exudates, acid rain, or anaerobic decomposition) and by increased temperatures in the weathering zone due to solar radiation. In contrast to the mechanical–chemical weathering of mica (detritite clay formation), this chemical weathering of feldspar is described as an authigenic clay formation (Section 4.2.7; see Ziechmann and Müller-Wegener, 1990).

4.2.11 *Clay minerals as polyfunctional exchangers*

The isomorphic substitution of cations in the tetrahedral and octahedral layers (Sections 4.2.8 and 4.2.9) leads to negative surplus charges in the clay minerals that are described as a permanent charge. A 'variable charge', dependent on pH and the material concentration in the soil solution, rests on the amphoteric properties of the hydroxides of aluminium and that of iron somewhat, insofar as they are localised on the side surfaces of the clay minerals. These properties of clay minerals represent an important function with respect to the bonding of metal ions as well as their release. It is by means of the water of the soil solution that the occupation state of soil (ion) exchangers, i.e. the composition of the cations and ions, sorbs to balance the charge and is changed. Shaking distilled water (DIN elution S4) over soil causes an activity equilibrium between sorbed ions and the H^+ and OH^- ions from the water to be created. This process can be repeated until there is a complete exchange. On average, 66% (of the total amount) of Ca^{2+}, 26% of Mg^{2+}, 5% of K^+ and 3% of Na^+ (but 0% of Al^{3+}) can be exchanged in agricultural soils. If, as in Sweden, the soils have a pH < 5, the exchangeable amount of Ca^{2+} is reduced to 48%, although 33% of Al^{3+} can be exchanged. In overly acidic forest soils or soils loaded with heavy metals, heavy metal ions with toxic activity are released as well.

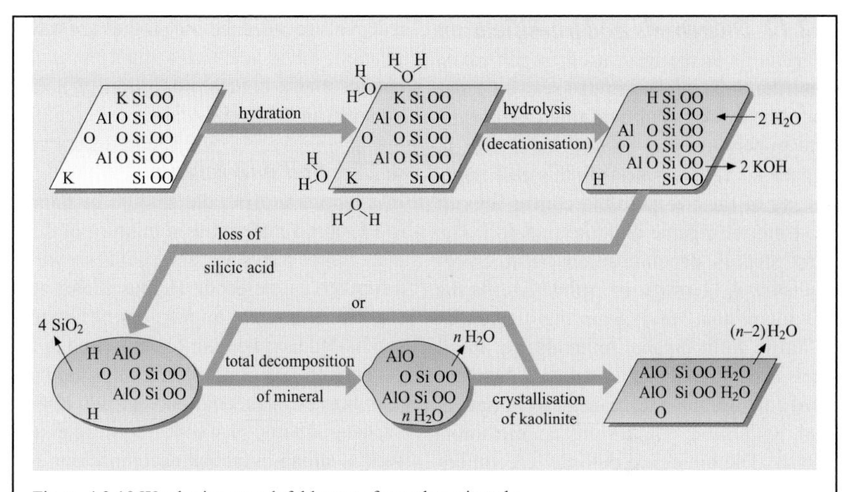

Figure 4.2.10 Weathering potash feldspar to form clay minerals

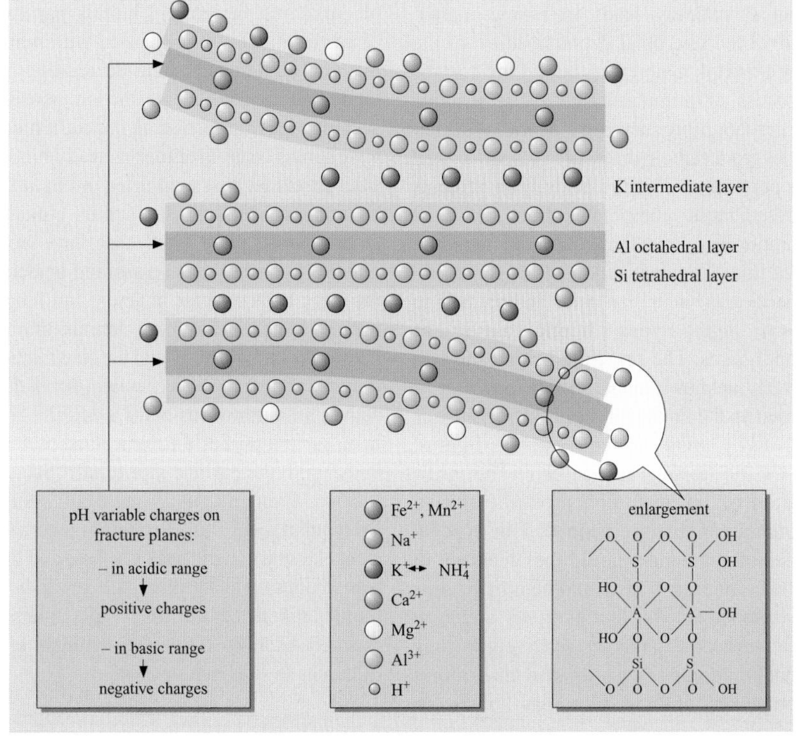

Figure 4.2.11 Clay minerals as polyfunctional exchangers

4.2.12 Diagenesis and humification

Diagenesis is the term used in general to describe those processes that convert loose rock into a more solid state, often chemically and mineralogically as well, as shown in Figure 4.2.12. Diagenesis in this case refers to one of the two possible conversions of post-mortal organic material in a soil. The first step is decomposition by microorganisms. A (geological) pathway leads the fragments that result from this decomposition to form organic minerals, i.e. fossil fuels (humic coal or mineral coal, lignite – hard coal – anthracite), called carbonisation, and with fluid conversions to metamorphosis. The biological pathway (decomposition) releases CO_2, H_2O, NH_3 and minerals from the organic phase. Finally, the pedological pathway leads to humic matter. Advanced microbial decomposition of the post-mortal material is necessary for the process of humification so that reactive degradation products such as monosaccharides from carbohydrates are present, as well as peptides and amino acids from proteins and phenolic compounds from cell wall components. Next, there occurs in a mixture like this a polymerisation of the monosaccharides, cyclic amino acids and phenols to form high-polymer humic matter as copolymers. The principle behind the still largely unknown humic matter formation is based on the linking of cyclic basic materials already present in plants, such as lignins, dyes and tanning agents, or on the cyclisation of linear fission products of ring compounds that are formed. This type of chemical reaction could be decisive in acidic, infertile mineral and high moor soils (Section 4.2.14). Humic matter and clay minerals can undergo relatively strong bonding in soils; they are called clay–humus complexes. Humins (not very soluble or reactive) are called the end products of humification; fulvic acids have small molecular masses and a higher proportion of acidic groups than the humic acids.

4.2.13 Soil and humus

In a humus profile, the amount of humus (post-mortal material as a mixture of dead matter from plants and animals) is shown as a function of the depth. Humus occurs as a superficial layer of humus – as O horizon, subdivided based on ash content – and in the A_H horizon (Section 4.1.7). The O_L horizon (L = litter) contains morphologically largely unchanged parts, such as leaves, needles and bark; a strong microbial decomposition has occurred in the O_F horizon (F = fermentation); the humification horizon O_H consists of dark mite faeces and humus particles (diameter < 0.2 mm) encrusted with humic matter. The humus content decreases from top to bottom; the mineral fraction increases accordingly (with increasing mineralisation). Raw humus consists of mechanically broken plant parts that are mummified or encrusted by organic acids. Musty humus contains largely those plant parts that have been crushed by arthropods (crabs and spiders), the dung balls of these animals, and fungal hyphae. In contrast to musty humus, in mild humus the mineral body and the clay fraction create a relatively solid chemical bond. The activity of earthworms plays a decisive role here, since it pushes the humus and mineral body up to the particle size of silt (Section 4.1.5) through its intestinal tract. Prerequisites for mull formation are sufficient clay and materials such as lime and free iron oxides, which foster the binding of humic substances to the clay minerals (Section 4.2.12). Mull is the dominant type of humus in soils rich in clay.

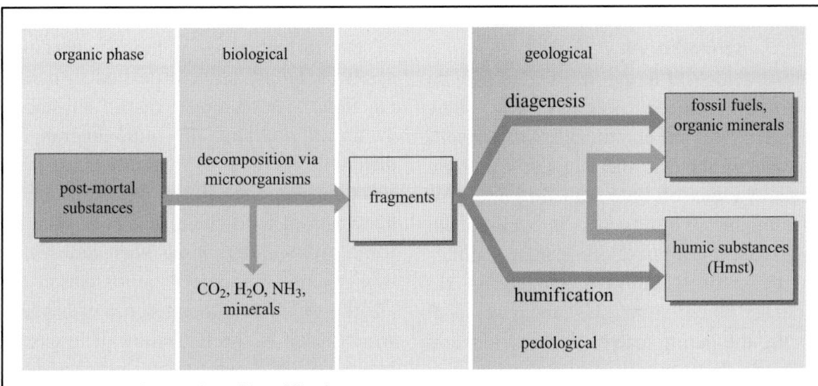

Figure 4.2.12 Diagenesis and humification

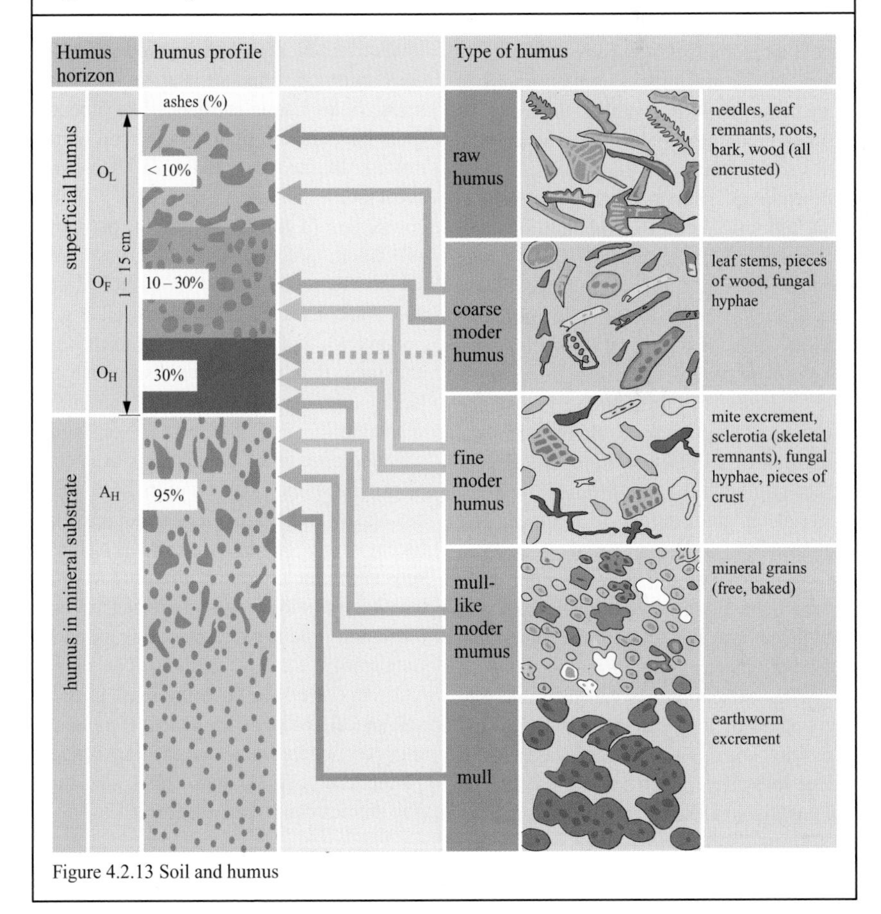

Figure 4.2.13 Soil and humus

4.2.14 *Biogenesis of humic substances*

Humic acids and humic substances play an important role in polluted soils: they complex with heavy metals, they adsorb organic substances, and they can also have decisive effect on their transformation. W. Ziechmann (Göttingen), a well-known researcher of humic materials, offers a differentiated representation of biogenesis. He includes humic substance precursors HsP among the humic materials, the acidic and relatively stable humic acids Ha and humins Hu as end products of the humification process (Section 4.2.13). According to Ziechmann, masses are converted in this important natural process which compare with those of photosynthesis in terms of quantity. According to the current state of the art, various stages in biogenesis can be differentiated. In the metabolic stage there is a partial microbial degradation of high molecular weight substances. This is where the humifiable material is generated. If one considers the aromatic branch, an introductory stage of humification begins with the aromatic natural substances from plants. The genesis of the humic substance precursors HsP begins, along with the formation of radicals. It is not possible to detect an uptake of non-aromatic starting materials in the conformation stage. The non-aromatic starting materials stem from carbohydrates, fats and proteins. In the stage in which the entire very complex process is completed with the actual formation of the humic substance system, contaminants such as metals or xenobiotics can infiltrate into the system or they can be drawn into the reactions. For this reason the biogenesis of humic substances is currently still of interest not only for soil science but also for environmental sciences as a whole. So far the stage of the formation of a humic substance system can only be described using model reactions. These model humic substance syntheses include the autoxidation of different phenols to radicals and their reactions, as well as the Maillard action known from food chemistry. For example, using hydroquinone it has been shown that this phenol undergoes autoxidation in alkaline solution in the presence of oxygen, which leads to the formation of intensely brown-coloured non-uniform products with properties similar to humic substances. Even the browning reaction, known as the Maillard reaction, that is observed during the conversion of reducing sugar with amino acids, yields humic substance-like products (plus flavours and pigments in food preparation). In the first phase of the humic substance system, numerous stable complexes of humic materials (or humic substance precursors) with non-humic substances such as phenols, carbohydrates and amino acids are detected, as well as with polycyclic hydrocarbons, steroids and enzymes. It is possible to distinguish the terms 'humic substances' and 'non-humic substances' as follows. Non-humic substances are all of the materials from dead plants and animals that occur in the phase of biological and abiological decomposition (decay) and in transformation. Humic substances include all of the products that have been abiologically synthesized as transformation products and synthesis products. Lithospheric fractions of the soil (minerals, oxide hydrates) are also involved in these abiological synthesis processes. If inorganic soil components are included in organic synthetic processes, then stable, structural clay–humus complexes are formed.

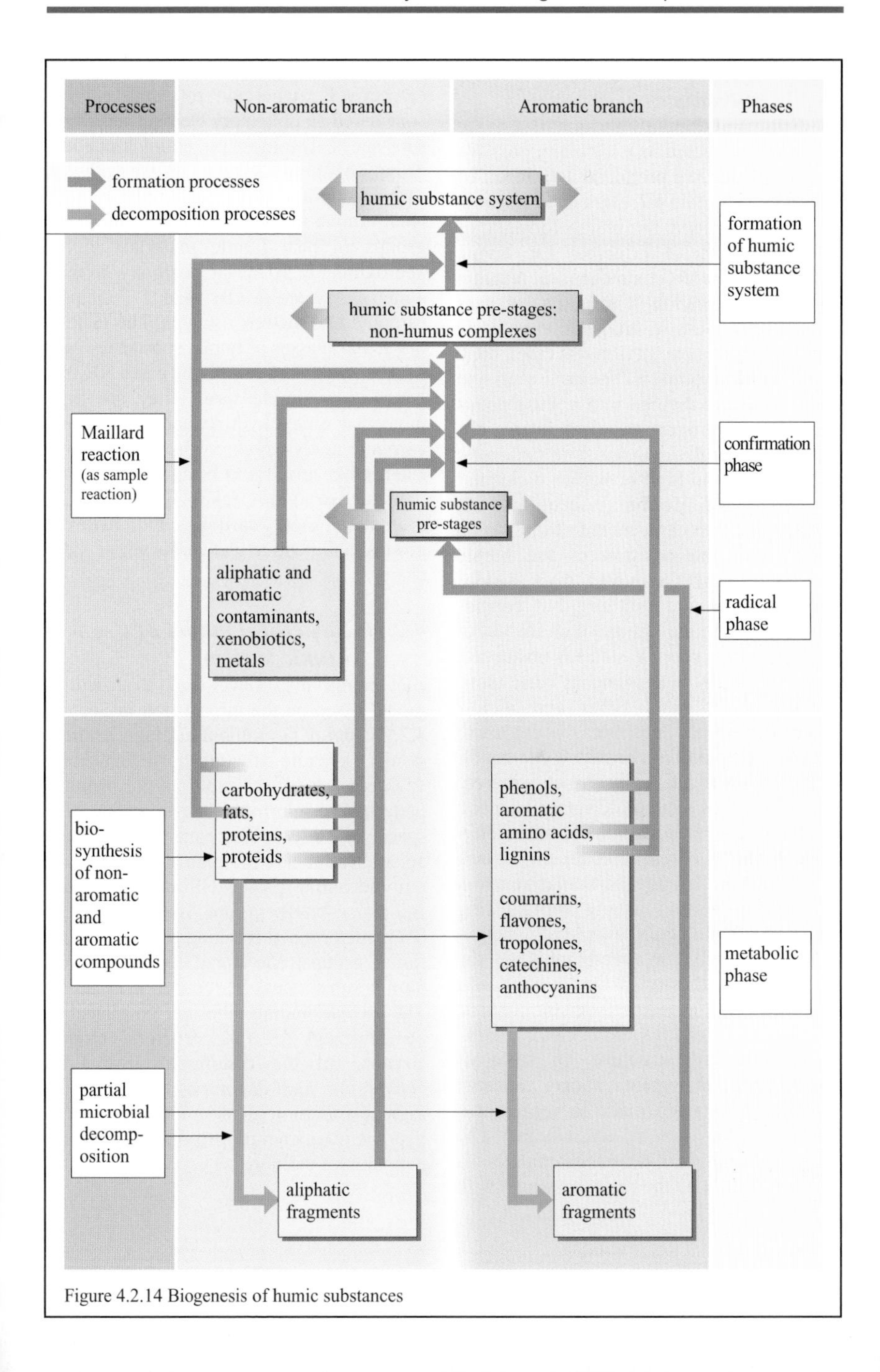

Figure 4.2.14 Biogenesis of humic substances

4.2.15 Synthesis of humic substances

1. Structural chemical position. Due to the complex composition with many possible variations that are produced as a result of biogenesis (Section 4.2.13), it is not possible to cite a defined structural formula or uniform synthesis principle for humic substances. The structural chemical diagram for the classification of natural substances generally consists of the parameters basic unit U (for the monomers in this case), bond types B and substrate S. The basic units and bond types are divided into homogeneous (ho) and heterogeneous (he) forms; the substrates are divided into low molecular weight (lm) and high or higher molecular weight (hm) forms. According to Ziechmann, one can conclude from Figure 4.2.15 that the position of the humic substances is far removed from that of carbohydrates and proteins, but considerably closer to the lignins. With the aid of this diagram, a specific status is postulated, characterised by heterogeneous basic units, heterogeneous bond types and a low molecular weight substrate. In contrast to classical preparative chemistry, the conditions here are regulated by chance, a prerequisite for chaos analysis. Chemical aspects for the chaos phenomenon are an indeterminable variety of reaction partners, the lack of dominance by any particular reaction mechanism, and no driving or organising forces in the soil (such as enzymes, energy-rich compounds or membranes) as the controlling authority of humic substance synthesis.

2. Hypothetical structure. In spite of Ziechmann's statements above, repeated attempts have been made to represent a structure for humic substances. The elemental analysis of humic materials yields C, O, H and N as the main elements, with averages of 54%, 33%, 4.5% and 2% respectively whereby nitrogen is not considered an obligatory element for humic substances. The most important functional groups of humic substances are carboxy, carbonyl, amino, imino and hydroxy groups. Quantitative analysis of the functional groups in peat humic materials yields approximately 10 meq g^{-1} carboxy groups, 4 meq g^{-1} phenolic OH^- and 1–2 meq g^{-1} carbonyl and methoxy groups. The relative molecular masses of humic substances vary between 1000 and in extreme cases 500 000 g mol^{-1}. In simple terms they are high molecular weight hydroxy and polyhydroxy carbonic acids which are linked via C–C, ester, ether and imino bridges. The hypothetical structure takes into account aromatic cores, carboxy and hydroxy groups, and peptide and carbohydrate side-chains.

4.2.16 Structural model of a humic system

Ziechmann also developed a structural model based on the information in Section 4.2.15 (point 1), without adding chemical details. Spheroid structures formed by main valences are recognisable as the primary pattern. The intermolecular interactions in space yield a secondary pattern. Other characteristics include the planar aromatic and quinoid border groups (IS inner structure, BG planar border groups, Hs hydrate shell, SB predetermined breaking point, FG functional groups, Me metal ions, NH non-humic substances, BS border structures). The special binding ratios are postulated to be dominant for this structural model. Because of the continuous formation, conversion and decomposition in abiological and biological processes, even a hypothetical structure can only approach a temporary condition.

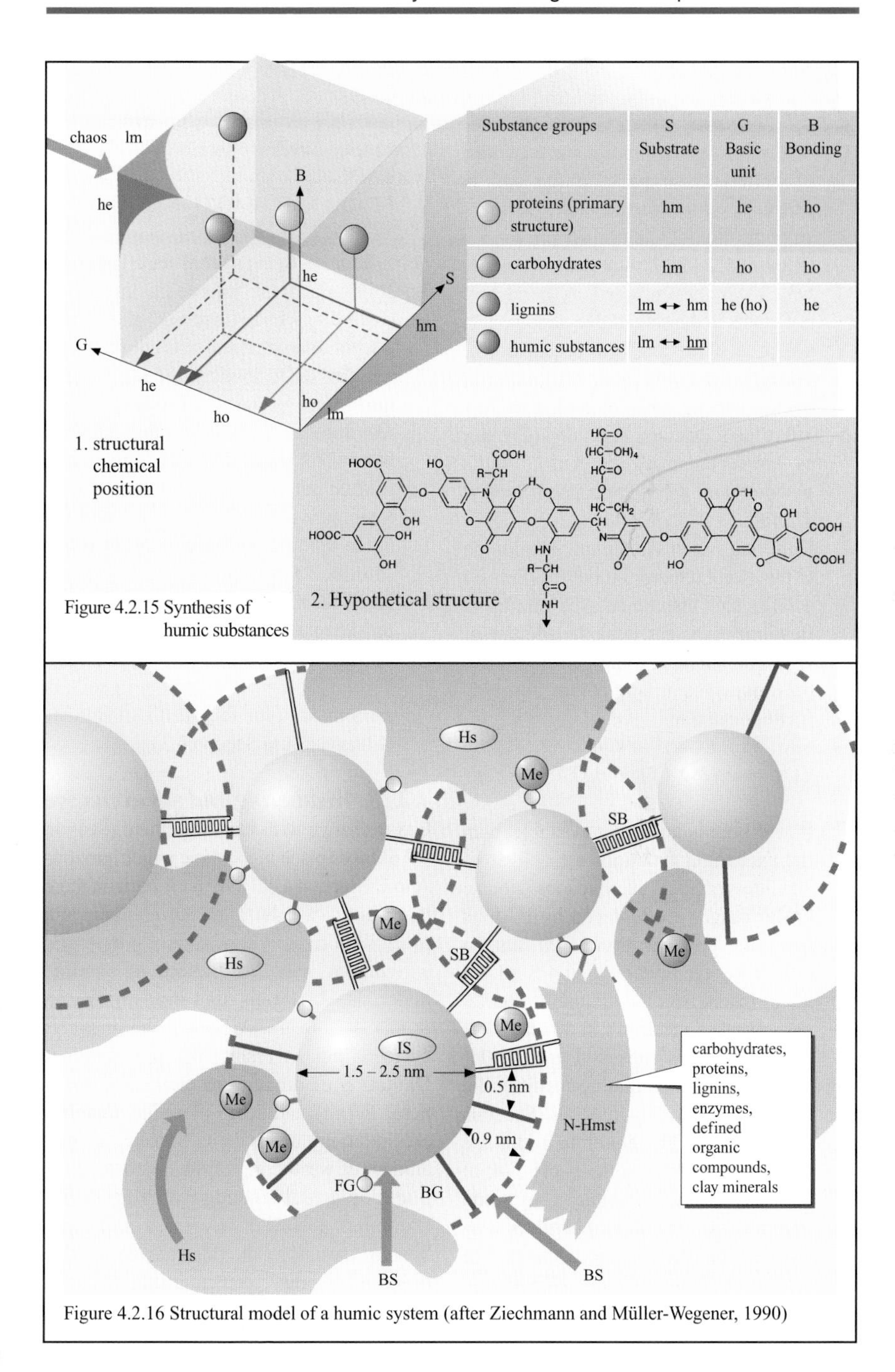

Figure 4.2.15 Synthesis of humic substances

Figure 4.2.16 Structural model of a humic system (after Ziechmann and Müller-Wegener, 1990)

4.2.17 *Dynamic processes in the soil*

1. Soil as storage unit, buffer and converter of pollutants. Impact on the soil is primarily due to the acid formers sulphur dioxide and nitrogen oxides (Figure 4.2.17, 2); to heavy metals, which can be enriched as persistent pollutants in the soil; and to xenobiotics, which are difficult to break down. Overall they influence the material cycles starting with the soil, and they affect the food chain in general (Chapter 1). The functionality of soil as a filter is obvious from the water supply (Section 4.2.18).
2. Buffering systems in the soil. In addition to their ion exchange properties (Section 4.2.11), soils also function as a buffer. If free lime is present, upon the addition of acid, the calcium hydrogen carbonate buffering system first comes into play (pH 6.8–8.0):

$$CO_3^{2+} + 2H^+ \rightarrow CO_2 + H_2O$$

If free lime is no longer available, then the exchanger functions are activated: alkaline–earth and alkaline ions (silicate buffer range) are replaced by H^+ ions from the soil solution. This system buffers between pH 6.8 and 4.5. At pH levels below this, the aluminium buffer system takes effect. In contrast to ion exchange, in this irreversible process, H^+ ions draw out aluminium ions from the octahedra of the clay minerals (Sections 4.2.8–4.2.11). The aluminium ions are present in hydrated form at pH 3–4; at higher pH levels (induced by lime treatment for example), polymers with less damaging activity can occur on fine root systems (formation of aluminium–hydroxo–aquo complexes). At even lower pH levels, iron oxide hydrates can have a buffering effect:

$$FeOOH + 3H^+ + 4H_2O \rightarrow [Fe(H_2O)_6]^{3+}$$

In addition to the natural acidification of soils due to microbial processes (including humification), the introduction of acid from emissions, as acid rain, has led to an increased reduction in the buffering capacity of soils in the last few decades. The result is that nutrients bound to clay minerals have been washed out and that heavy metals with toxic activity such as Pb, Cd and Cu (from immissions) have entered the soil solution. Furthermore, at lower pH levels the root fungi, which promote the uptake of nutrients by higher plants such as trees due to symbiosis, are damaged. Therefore, the pH reduction in soils is associated with the death of forests (Chapter 2 and Section 4.3).

4.2.18 *Weathering and gas exchange*

Gas exchange with the atmosphere (Section 4.2.6) takes place in the three major horizons A to C (Section 4.1.7). In the A horizon there is intense weathering, enrichment and precipitation of inorganic salts, and enrichment and infiltration of organic materials. Materials and transformation products from the A horizon enter the B horizon as a result of precipitation. Oxidation of organic material takes place here, Fe(III) and Mn(IV) oxide hydrates precipitating out. In the C horizon, a small amount of weathering of the bedrock below it takes place. A solution (solubility) equilibrium is achieved here (as a soil equilibrium solution in this horizon). The dissolved organic C content is minimal here; the O_2 partial pressure is constant.

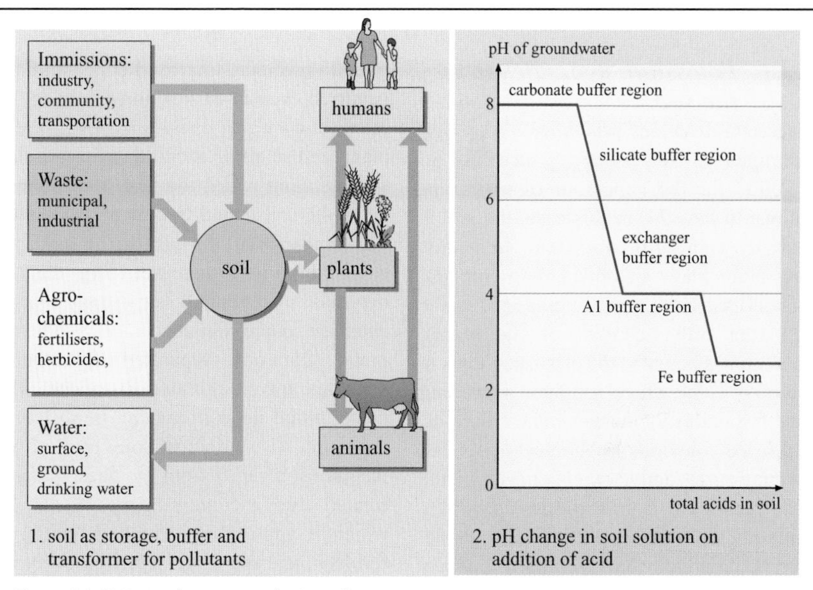

Figure 4.2.17 Dynamic processes in the soil

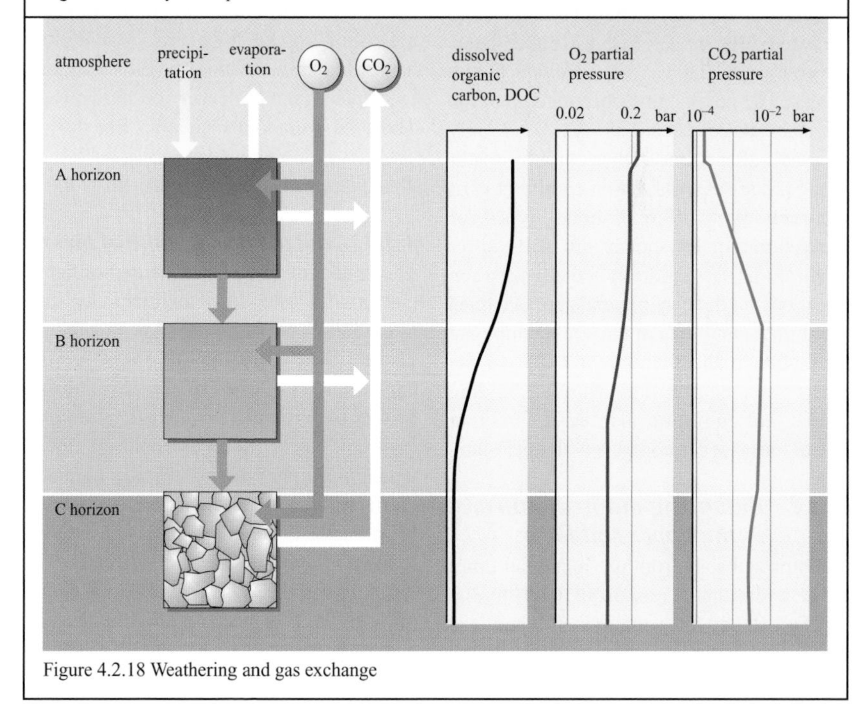

Figure 4.2.18 Weathering and gas exchange

4.3 Metals and acid rain

4.3.1 *Potassium dynamics in the soil*

Potassium dynamics, which have been well examined in soil science, serve as an example for the behaviour of metals or cations in the soil; in this case, the soil is viewed as an agro-ecosystem. In plants, potassium plays an outstanding role: K^+ ions influence the osmotic processes in the cell (the water content of the plant), potassium is bound to the plasma colloids in the plant cells, where it effects a swelling state favourable for the metabolic processes, and it has functions in photosynthesis and respiration (it activates enzymes). In the soil, potassium is bound largely to clay minerals, both inside and on the outside. The potassium that is more loosely adsorbed to the outer surfaces is readily available to the plants. The potassium bound between different clay layers is called interlayer potash and is less available to the plants. The potassium incorporated into the crystal lattice (Sections 4.2.8) is only available after a time in the course of weathering processes. Adsorption equilibria exist between the different bond types of potassium. In low-potassium soils, after potassium fertilisation the sorption locations on the clay minerals are occupied until potassium ions in the soil solution are available to the plants again due to the adsorption equilibrium. Similar binding ratios and equilibria also exist for other metal ions that can compete with potassium.

4.3.2 *Balancing out heavy metals in the upper soil layer*

Unimpacted soils from pre-industrial times have cadmium levels from 0.01 to 0.1 mg kg^{-1} (today the average is 0.5) and lead levels from 0.1 to 1 mg kg^{-1} (today the average is 30) of soil, as geogenic determined variations. Increased concentrations occurred even in earlier times near settlements and mining operations. Coal power plants and smelting operations are emission sources from which the heavy metals can reach the soil bound to dust particles after being transported through the air as deposition, both dry and wet. Dry dust is deposited more quickly and usually near an emission source; wet deposition with heavy metal fractions separated from dust particles travels greater distances. The heavy metal depositions can be differentiated into five principal sources: J_A from the atmosphere; J_B from biocides; J_S from treated sludge containing heavy metals, which is applied to agricultural areas as fertiliser; and as a result of fertilisation with mineral substances J_F and manure J_M. Heavy metals from the soil are delivered by means of uptake by plants J_P and their harvesting, by eluviation and erosion J_E and by displacement to deeper soil horizons J_V. Thus, the overall balance looks like this:

$$J = J_A + J_B + J_S + J_F + J_M - J_P - J_E - J_V$$

4.3.3 *Soil pH and uptake of metals*

The uptake of micronutrients and of heavy metals with potentially toxic activity such as Cd and Pb is clearly dependent on the pH of the soil. Except for Mo, the solubility and therefore the plant availability decreases with increasing pH, shown in Figure 4.3.3 for (nitrogen-fixating) clover, which requires Mo as a trace element. Agricultural soils without $CaCO_3$ reserves (Section 4.2.17) regularly have lime applied to them in order to maintain a pH of 6.5 in the soil solution. Many useful plants do not grow at pH < 5.5 because of Al toxicity. Fe and Mn deficiencies can occur when the pH is too high.

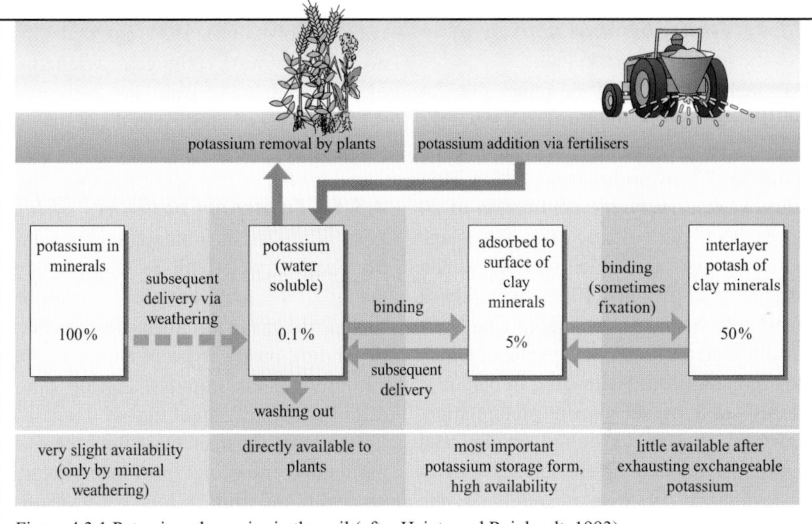

Figure 4.3.1 Potassium dynamics in the soil (after Heintz and Reinhardt, 1993)

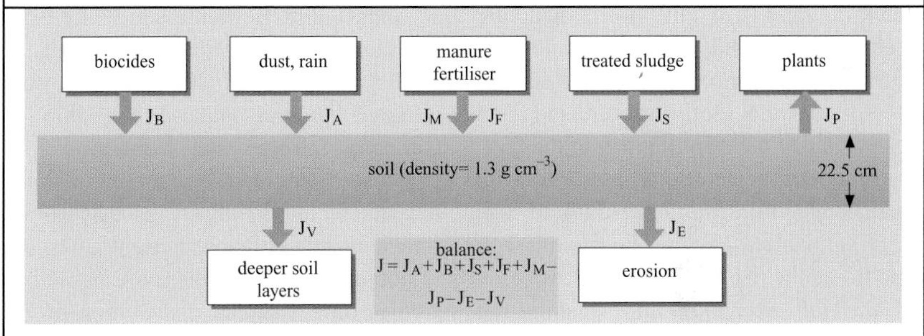

Figure 4.3.2 Balancing out heavy metals in the upper soil layer

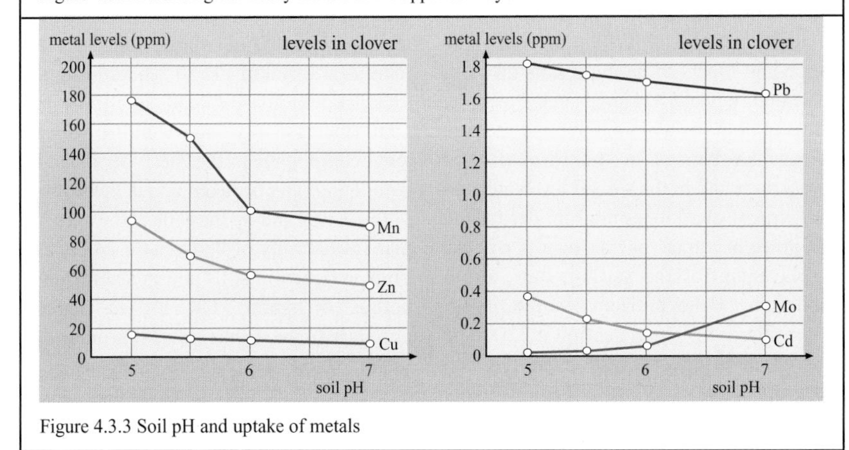

Figure 4.3.3 Soil pH and uptake of metals

4.3.4 *Transport and activity of melting salts*

The increased use of de-icing salt for combating slippery road surfaces since the early 1970s required an annual application of up to 1.3 million tonnes. With 50% being washed away, the application of 20 kg of salt per metre on a four-lane motorway in a single winter season resulted in a salt impact along the side of the road on both sides averaging 1 kg NaCl per qm. Part of the de-icing salt, of which the Cl^- is especially damaging to plants, is washed away by subsequent precipitation. However, the Na^+ is bound to clay and humus components of the soil and causes an exchange, a displacement of the elements that are important for the plant's nutrition and for the soil structure (Sections 4.1.8 and 4.1.11). This deteriorates physical properties of the soil such as pore space and clod structure (Section 4.1.6). Due to the displacement of the Ca^{2+}, dispersion of the clay particles occurs with increasing saturation by Na^+, and soil aggregates break down. The melting salt also raises the pH levels, up to 11, because of the ion exchange and washing-out processes. One of the results is a change in the spectrum of species and in the activity of soil organisms. Above the soil, the following effects of de-icing salts on plant life can be detected: changes in the osmotic ratios in both the soil solution and the root cells, disturbances in the formation of mycorrhiza and in the soil air levels with an impairment of root respiration, deficiency of nutrient ions as a result of the displacement of the sodium ions, and a reduction in the nutrient availability of other micronutrients (Section 4.3.3). The use of de-icing salt has been reduced considerably since 1985, so that Cl^- is no longer almost completely washed out of scattered areas from the soils along the roadsides and stored in deeper soil layers.

4.3.5 *Transport pathways of lead*

Lead is present as a natural component of the Earth's crust with an average of 16 mg kg^{-1}, whereby between 1 (plutonic rock) and 145 mg kg^{-1} (bauxites) can occur, depending on the type of stone. Primary anthropogenous impacts develop due to mining and the smelting of lead ores. Furthermore the main emission sources are particulate emissions from manufacturing processes and from thermal processes such as coal burning (2–40 mg kg^{-1} Pb in coal); secondary emissions include waste piles and landfills. Since the beginning of the 1950s an estimated 200 000 t Pb were released in the Federal Republic of Germany from the operation of combustion motors. The mobility of Pb is low due to low solubility, so that natural waters contain only low concentrations (up to about 10 μg L^{-1}). There are numerous studies on the global distribution of Pb, which indicate the dependence on the prevailing wind directions. Studies of the ice in Greenland show two distinct increases around 1750 (beginning of increased industrial lead smelting) and 1940 (increase in vehicular traffic with Pb alkyl compounds in the fuel). In plants, Pb compounds can be taken up via the stomata of the leaves and by the roots. However, the plant availability of Pb from the soil is low. Pb is enriched in soils and in sediments of bodies of water. The air in densely populated areas has levels around 0.7 μm m^{-3} (0.04 in areas with clean air).

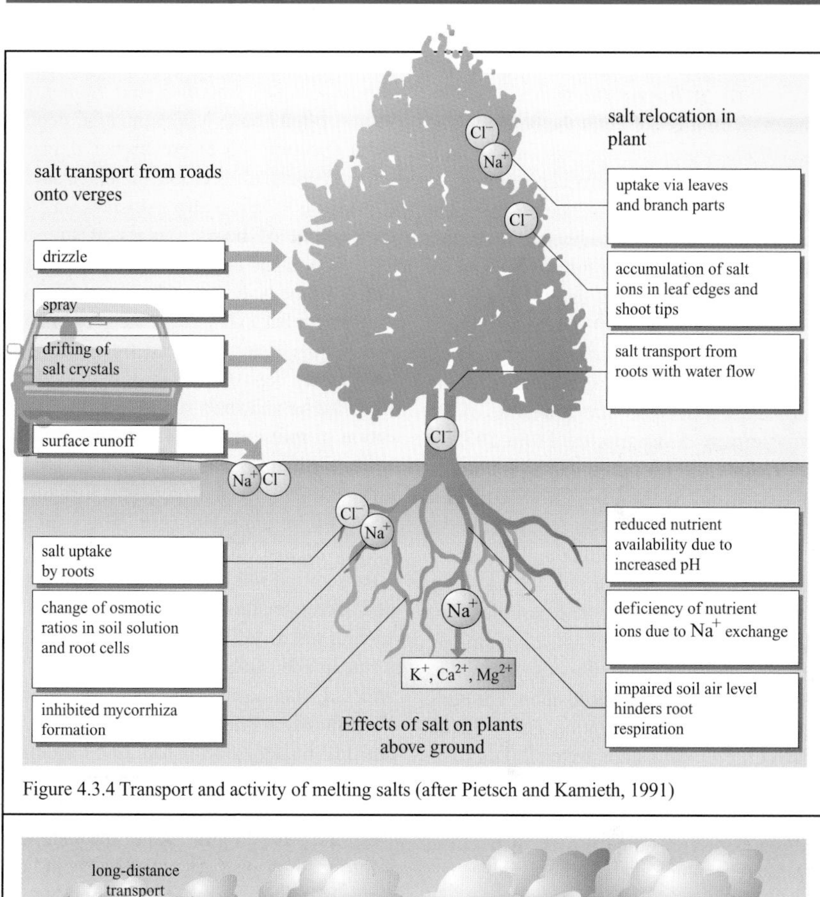

Figure 4.3.4 Transport and activity of melting salts (after Pietsch and Kamieth, 1991)

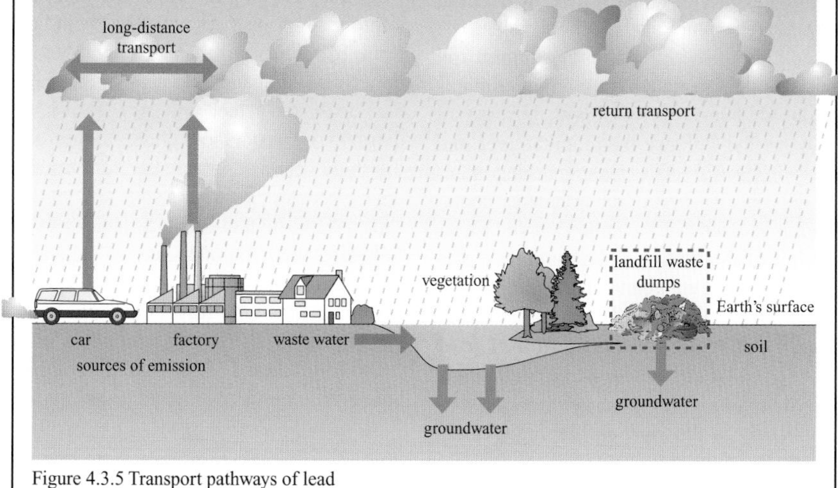

Figure 4.3.5 Transport pathways of lead

4.3.6 Formation of lead emission in airborne particles

The main emitters of Pb at the beginning of the 1990s were the combustion of fuel, metallurgical industry, furnaces and cement production. The impact on the environment due to Pb has become notably less in the last 20 years. The levels in airborne particles decreased from >1 $\mu g\ m^{-3}$ in 1974 to approximately 0.1 around 1990. The reduction of Pb in gasoline following the leaded gasoline law of 1976 in Germany in particular had an immediate effect on the lowering of Pb emissions in dust particles. From 1975 to 1976 alone the Pb precipitation impact in Bavaria declined from 0.27 to 0.10 $m^{-2}\ d^{-1}$. The Technical Instructions for Air ('TI for Air') contain a guide value and threshold limit as an emmission value of 2.0 $\mu g\ m^{-3}$ as a component of the airborne particles. The 'TI for Air' contain a limit of 0.25 $mg\ m^{-2}\ d^{-1}$ for Pb and inorganic Pb compounds as components of the dust deposition. The daily uptake of Pb via inhalation (with 20 m^3 outside air per day) amounts to <2.0 $\mu g\ d^{-1}$ in rural areas, 4–10 in densely populated areas, and 20–400 in areas near the emitter. In comparison, the figures are 10 $\mu g\ d^{-1}$ from 20 cigarettes and 100–150 from foods. The reduction of the Pb emmission impact can also be demonstrated via biomonitoring as reduced Pb levels in blood, with decreases of 60% in the Cologne metropolitan area and 25% in clean air areas. Biomonitoring in this context is an aspect of environmental monitoring as a total of the collection and monitoring of qualitative and quantitative data relevant to the environment and health.

4.3.7 Frequency distribution of lead and cadmium in garden soils

Garden soils differ from agricultural soils especially in the more favourable sorption properties due to enriched humus in the upper soil (A_P horizon) and higher pH values, often above 7. The material depositions (Section 4.3.2) are higher than the removal due to transfer into the plants and to harvesting. This also results in the enrichment of heavy metals in the soil horizon near the upper surface. Sources and causes of the heavy metal levels and depositions in garden soils include the geogenically and pedogenically induced fractions (as a base value), deposition of dust particles, both from the air and from streets, and the application of refuse materials and foreign soils, whereby the previous history of the garden and its previous environmental influences also play an important role. Heavy metals such as Pb and Cd can also enter the soil via fertilisers and soil conditioners, construction and other materials, pathway materials, foreign soil material and refuse materials. As a rule, garden soils have higher levels of Pb and Cd than farmlands: in North-Rhine-Westphalia, average Pb levels between 49 and 113 $mg\ kg^{-1}$ (farms had 29 on average) and Cd levels between 0.46 and 1.20 $mg\ kg^{-1}$ (0.40 in farms) were determined. The diagrams in Figure 4.3.7 provide an impression of the distribution of Pb and Cd levels by frequency for a large city, whereby the effects of abandoned depositions from the coal and steel industry (left-hand diagrams) are obvious. Guide values and threshold values of 100 $mg\ kg^{-1}$ for Pb and 1 $mg\ kg^{-1}$ for Cd for usable soils and garden soils have been in effect since January 1988 following the 'Minimal study of cultivated soil to estimate the risk of abandoned storage and other sites with respect to agricultural and gardening' (NRW). Based on this, 64% of the Pb levels and 83% of the Cd levels are below these limits (right-hand diagrams).

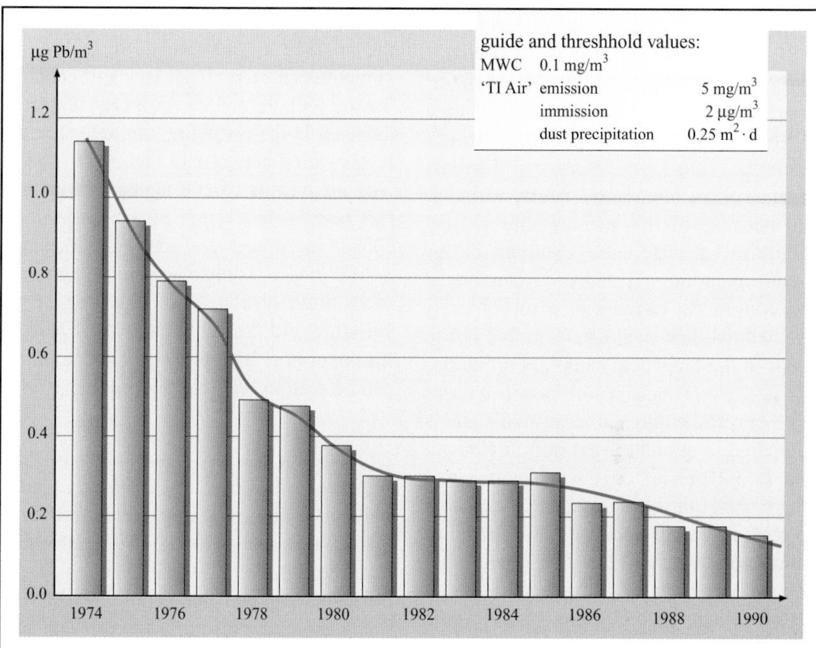

Figure 4.3.6 Formation of lead emission in airborne particles

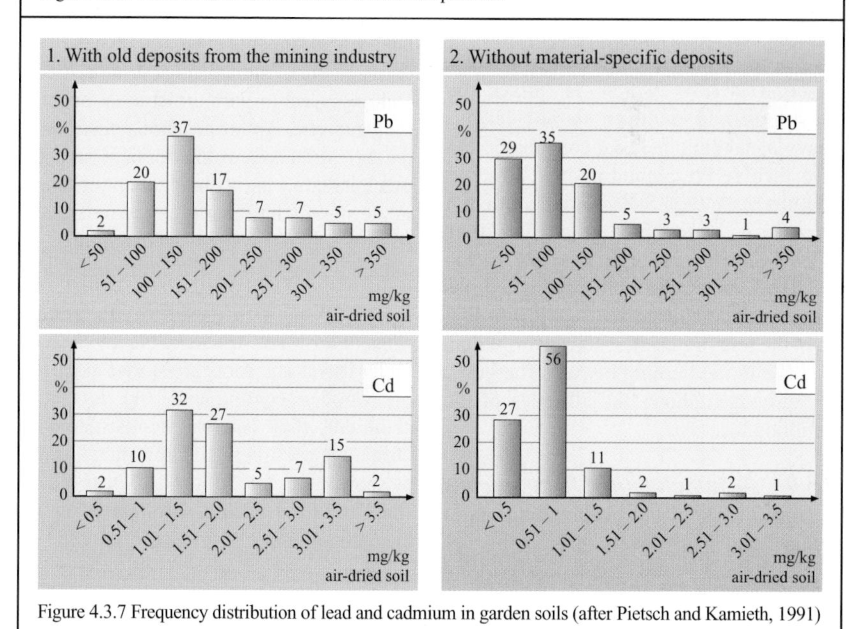

Figure 4.3.7 Frequency distribution of lead and cadmium in garden soils (after Pietsch and Kamieth, 1991)

4.3.8 *Standardised lead levels in the A_P soil horizon near a non-ferrous heavy metal emitter*

In order to be able to determine the source of Pb contamination, one must often use reference values or reference elements because of the often large natural variation and because of the various sources. By definition, a reference element corresponds to an internal standard in analytical procedures. An element or a substance is selected that has been demonstrated not to occur in the analytical sample. For studies of heavy metal loads in soils near foundries or other metal emitters, the emitted metal is compared to (standardised on) an element that is not present in the emissions and where levels in the soil change as little as possible as the distance from the source of emmission increases. In this manner, often one source among the heavy metal pathways shown in Figure 4.3.2 can be determined to be dominant. Figure 4.3.8 shows the test results (Fiedler and Rösler, 1988) for Pb in the A_P soil horizon (fraction <0.3 μm, mineral upper soil horizon (Section 4.1.7), which is mixed by ploughing). In addition to non-ferrous heavy metal emitters, coal combustion can also involve Pb contamination, where in contrast to emitters, vanadium is emitted. A comparison of the three graphs in Figure 4.3.8 shows that at a distance of 13 km from the emitter, the increased Pb concentration in the soil is due to its emission from the foundry itself, whereas further out it is due to the burning of coal. The graph at the bottom shows the altitudes of the sampling locations.

4.3.9 *Accumulation of heavy metals*

1. Lead in soils and plants along major motorways. Trace levels of elements are greatly increased in the soil layer close to the surface in municipal areas due to numerous emitters. High levels of contamination with Pb > Zn > As > Cu > Cd were detected in the Hamburg city area. The Pb levels varied between 13 and 3074 (1983), with an average of 208 mg kg^{-1} (from 486 soil samples at a depth of 0–5 cm). Some 73% of the values were above a guide value of 100 mg kg^{-1}; only 5% were in the range of natural levels. Both the degree of variation and the accumulation of lead in the uppermost layer of the soil and in plants can be seen in Figure 4.3.9 (Kloke, 1974).

2. Introduction of mercury in the sediment of a lake. Anthropogenously induced changes in the elemental trace distribution can generally be determined using soil or plant analysis (Section 4.3.6). Subsequent monitoring is possible if heavy metal contamination is conserved and arranged chronologically in soils or in plants. Local studies can be performed by analysing sediments in lakes with a defined drainage area, annual rings in older trees, and ice layers in permafrost areas such as the Antarctic. Anthropogenous changes in the terrestrial area of the drainage area are precipitated by the sedimentation processes in the slightly buffered ecosystem of a lake (shown here using Hg as an example).

3. Cadmium concentrations in annual rings of a fir tree. Traces of elements taken up by leaves and roots are stored in part in the wooden coating and, like layers of sediment, they can be reconstructed in chronological order using the annual rings, as shown here for Cd in a 180-year-old fir tree from the Fichtel Mountains at an elevation of approximately 650 m.

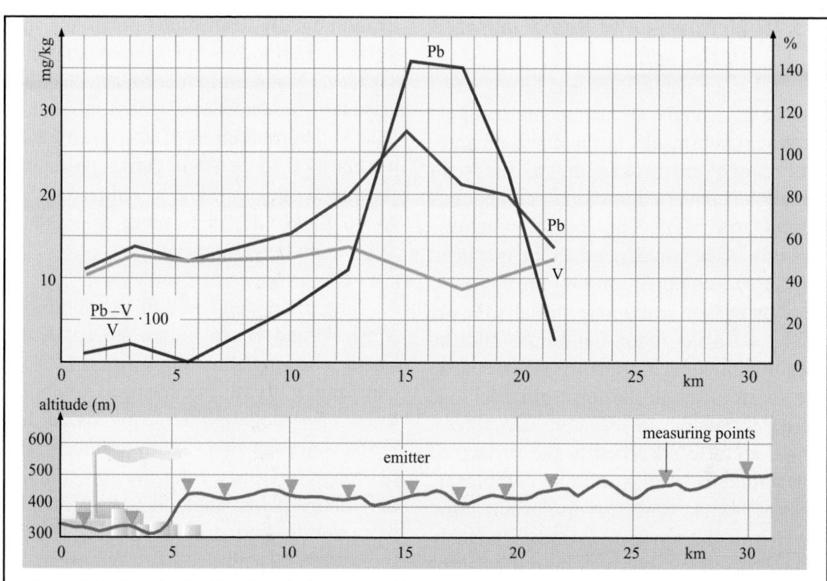

Figure 4.3.8 Standardised lead levels in the A_p soil horizon near a non-ferrous heavy metal emitter

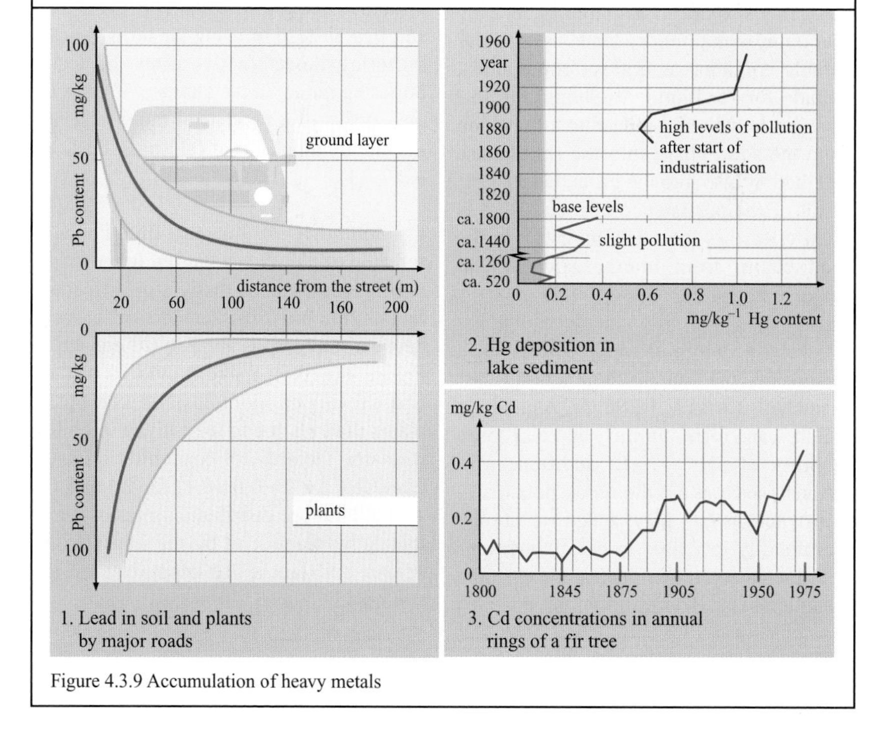

Figure 4.3.9 Accumulation of heavy metals

4.3.10 *Behaviour of heavy metals in the soil*

1. Influence parameters and heavy metal reactions. Metals are some of the most persistent substances in the environment – they cannot be broken down; they can merely be converted into other compounds (metal species). The basic reactions possible in the soil depend on the one hand on the soil composition and on the other hand on the soil reaction (especially the pH of the soil), the redox conditions and the reaction kinetics. Metals can be present in the soil solution in the form of metal ions or of inorganic or organic complexes, which are also adsorbed to the surface of clay minerals or clay–humus complexes. Sediments are the most important bond types in soils: ionogenic; interchangeably bound to clay minerals; bound adsorptively, e.g. to the surface of iron oxides and manganous oxides; as slightly soluble inorganic compounds; and as slightly soluble organic complex compounds, bound for example to humic acids. Changes in the soil can lead to a transition from the solid phase into the fluid phase and thereby also into the groundwater. One result of the persistence of metals that is often critical is their accumulation in the food chain, from microorganisms and plants to humans.

2. Relative mobility of elements as a factor of pH and redox conditions. Elements such as V, U, Se, Si, As and Cr, which can form anions, have a high mobility at pH > 7. In addition, with elements such as Cr the redox potential – shown here as E_h – plays a significant role: the mobility increases in basic conditions with increasing redox potential due to the oxidation of Cr(III) to Cr(VI), as CrO_4^{2-}. At experimentally determined potentials (not to be understood as normal potentials) of <0.33 V, O_2 can no longer be detected, nor can NO_3^- be detected at potentials below 0.22 V. The reduction of Fe(III) to Fe(II) begins at 0.15 V. At negative potentials starting at –0.05 V, SO_4^{2-} is reduced to S^{2-}. At –0.12 V, CH_4 is formed; at –0.18 V, SO_4^{2-} is no longer present.

3. The pH-dependence of the solubility of Si, Al and Fe. In Figure 4.3.10, 3, the graph showing the solubility of Si is compared with the characteristic solubility graphs for the amphoteric metals Al and Fe. Silicates are also slightly soluble in the pH range 0–4; they enter the soil solution only above pH 6. Thus, Si is considerably more solidly bound in clay minerals than the metals Al and Fe. All three elements with their compounds also form buffer systems (Section 4.2.17). However, with silicates, instead of using the term 'silicate buffering range' (active after the CO_3^{2-} buffering range, in the absence of $CaCO_3$), one speaks of a process of the weathering of primary silicates (of feldspars, Sections 4.2.7–11).

4. Solubility of Zn, Cd, and Pb as a function of pH. The solubility in the acidic region increases greatly for all three metals, the most for Zn. The differences between the graphs for Zn, Cd and Pb in Figure 4.3.10, 4, make it evident that Zn and Cd can be more easily taken up by plants than Pb due to their higher mobility (transfer factors: *F* concentration ratio plant/soil for Pb 0.01-0.1, for Zn and Cd 1–10). Thus the distribution rates of heavy metals by plants and by the transfer into deeper soil layers are dependent on the soil's pH.

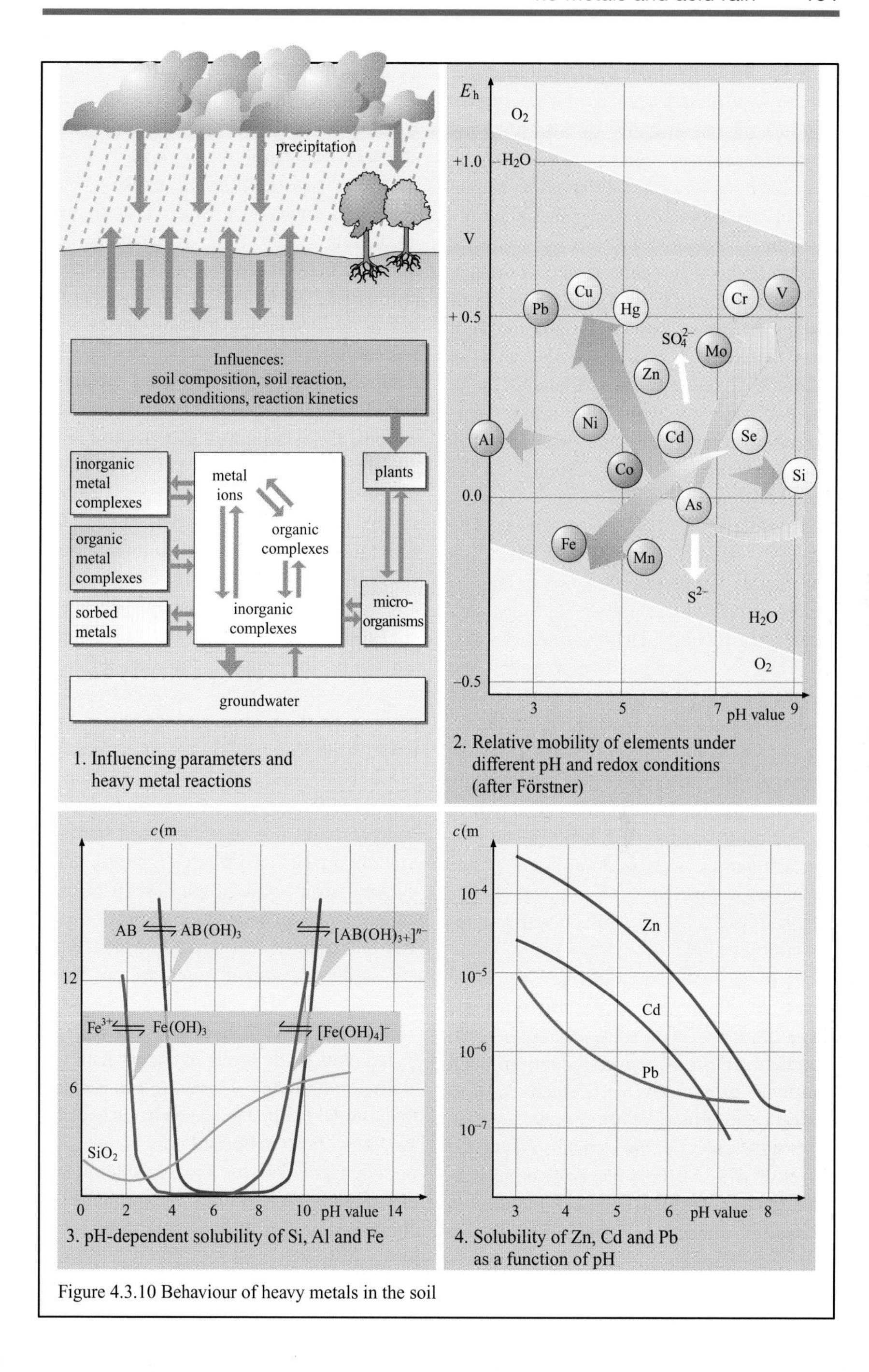

1. Influencing parameters and heavy metal reactions

2. Relative mobility of elements under different pH and redox conditions (after Förstner)

3. pH-dependent solubility of Si, Al and Fe

4. Solubility of Zn, Cd and Pb as a function of pH

Figure 4.3.10 Behaviour of heavy metals in the soil

4.3.11 Mobilisation of heavy metals with various extraction agents

Studies on the mobility of heavy metals from soils and of sediments are based on the possible bond types. We differentiate among the following bond types, depending on the complex composition of soils from mineral and amorphous components and of organic substances from the natural degradation of organisms. Those most easily substituted are ions bounds to clay minerals such as kaolinite, illites or montmorillonite. Due to ion exchange processes, they allow themselves to convert to the liquid phase. Components absorptively bound to the surface of iron oxides and manganese oxides are a little more difficult to release. The slightly soluble precipitates such as carbonates, sulphides and hydroxides of heavy metals form another group with decreasing mobility. Heavy metals can also be present as coprecipitates as a result of the coprecipitation properties of slightly soluble hydroxides such as that of Fe. If bound chemically to substrate components such as humic acids, heavy metals also form slightly soluble organic complexes. The bond type of an occlusion occurs when heavy metal ions are retained by encapsulation in crystalline precipitates. Finally, metal ions can also be incorporated into mineral crystal lattices (isomorphically).

In order to differentiate between total levels of elements in soils, soil scientists have conducted long-term leaching experiments to determine the availability of plant nutrients such as Mn, Fe, Zn and Cu or to evaluate the toxicity of heavy metal levels in soils with respect to the elements Pb and Cd (Section 4.3.10). This has been done with electrolyte solutions such as a magnesium chloride solution, which detects exchangeable and adsorptively bound components. Studies on the release of Pb from soils have shown that only small amounts (between 1% and 10%) are released with electrolyte solutions and that even complex-formers such as ethylene diamine tetraacetic acid (EDTA) or diethylene triamine pentaacetic acid (DTPA) convert only about 25% of the Pb into the extraction solution. Standard extraction series – as sequential extractions – were developed to determine binding strengths of metals in sediments and slurries: interchangeable cations are determined using ammonium acetate solution of pH 7 (solid matter:solvent ratio 1:20; shake 2 h) and carbonate components are determined after subsequent extraction with a sodium acetate solution of pH 5 (5 h). In these studies, kinetic effects must be taken into account as well as the chemical and physical conditions, which is the reason for the different extraction times. Metal fractions bound in easily reducible phases (manganese oxides, amorphous iron oxides) are determined at pH 2 (HNO_3 solution) with the addition of the reducing agent hydroxylamine hydrochloride. Metals are released from the moderately reduced phases (slightly crystallised iron oxide hydrates) using an ammonium oxalate solution at pH 3 (shake 24 h). In order to destroy organic bond types and to dissolve metal sulphides, one treats the soil (after performing the previous elution steps) with hydrogen peroxide at 85°C and then one extracts with an ammonium acetate solution (24 h). Finally, the residual fraction (silicate phases like clay minerals) must be opened up with acids (as per Förstner). The first two extraction stages are especially relevant to environmental chemistry, as they include a pH change.

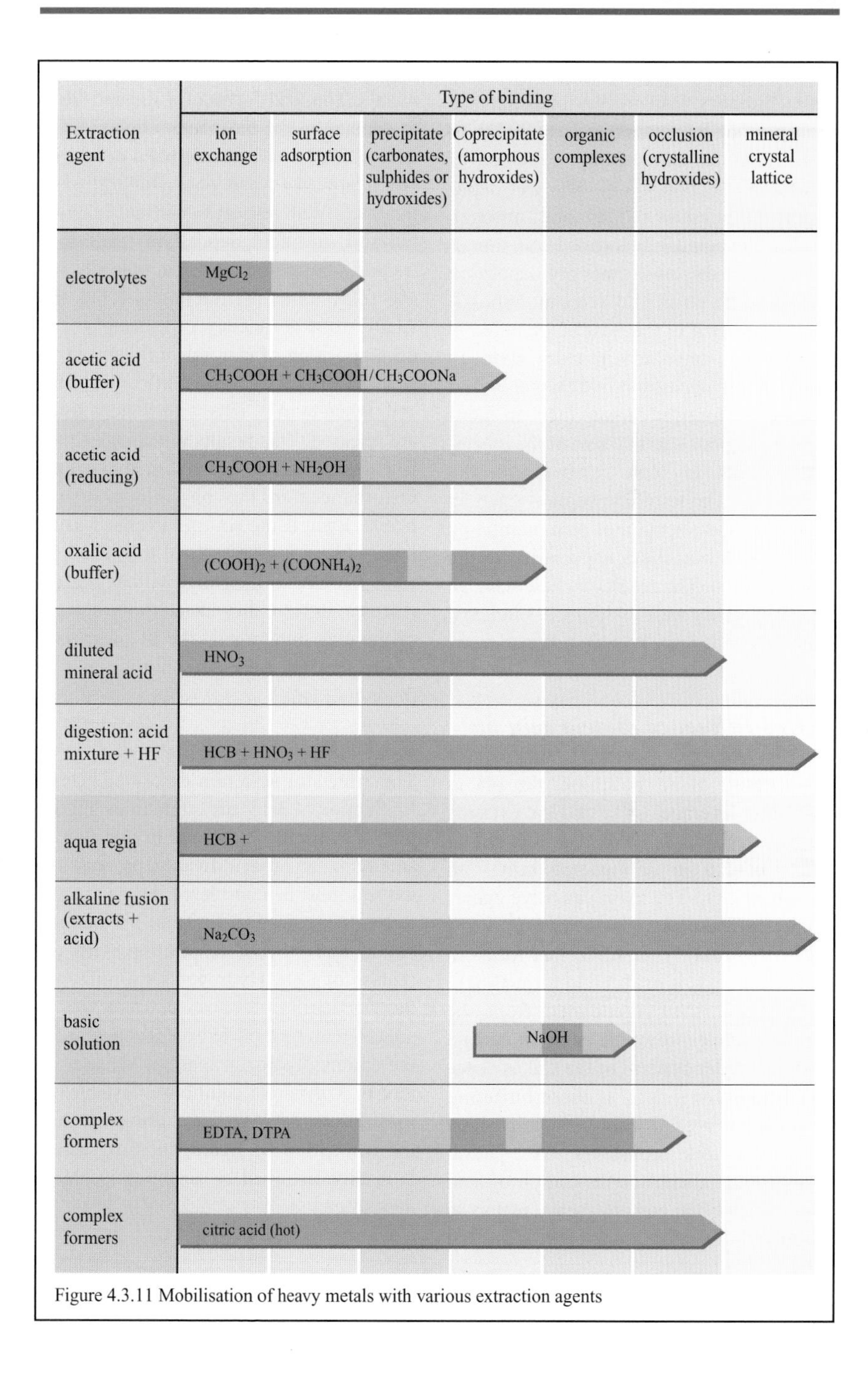

Figure 4.3.11 Mobilisation of heavy metals with various extraction agents

4.3.12 Deposition of acid rain

Tall chimneys lead to the long-distance transport of pollutants and therefore to deposition far away from the place of emission (Section 4.3.8). The wet deposition of sulphur oxides and nitrogen oxides (Section 2.3.2) is called acid rain or acid fog. At the same time, dry deposition must also be taken into account: approximately one-half of the total SO_2 emitted in Germany is deposited in a dry state. In addition, the deposition of SO_2 is 5–6-fold higher in forested areas than in non-forested regions – as that part of the precipitation that is first retained by the vegetation. The term interception refers in general to the retention of precipitation by leaves and branches, especially in the crown region. The results include damage to the trees as well as loss of water and nutrients with respect to the vegetation on the ground.

4.3.13 Influence of acid rain on the ecosystem of the forests

As a result of air pollution, the pH in rainwater averages 4.1, and in extreme cases it registers 2.3; compare this with a value of 5.6 in unimpacted rain with dissolved CO_2. The main causative agents are traffic and power plants that burn fossil fuels (oil, coal). The H_2SO_3 that forms in water from SO_2 is oxidated to form H_2SO_4. H_2SO_4 and HNO_3 (generated from the nitrogen oxides in water) come in contact with the buffer systems of the soil as deposition (Section 4.2.1). If these buffering systems are overloaded, over-acidification occurs. If one considers the natural processes in a soil, the deposition of acid rain, which often contains heavy metals in dissolved form, is associated with acid deposition and an accumulation of heavy metals. The acidification of the soil leads to damage in the biosphere. Organisms in the soil are killed; the microbial decomposition processes that occur naturally are inhibited. With respect to nutrient levels in the soil, the acid deposition leads to a washing-out of minerals such as Ca and Mg (clay minerals, Section 4.1.11) and finally to a deficiency of nutrients. The decomposition of clay minerals also takes place as a result of over-acidification of the soil, whereby Al^{3+} ions with toxic activity are released. This results in damage to the fine root system at first and then to deeper disturbances in the physiology of the plants. Even if the soil pH is raised using lime additives (above 4.3), the Al^{3+} ions cannot return to their original sites in the silicates; instead, they attach to the surface of the clay minerals. If the pH decreases again, they go back into solution as free (hydrated) ions.

4.3.14 Processes in the soil

The effects of acid rain described briefly in Section 4.3.13 are shown in Figure 4.3.14 again as equilibria. In addition to acid rain, natural acidification processes in the soil must be considered. These are due to the breakdown of biomass via humification and to root respiration with the release of CO_2. Hydrogen ions influence the weathering of stone. Nutrient cations are taken up by roots in excess compared to the anions; hydrogen ions are released in return. A dynamic equilibrium establishes itself in a soil between the generation and the utilisation of hydrogen ions, which is readily disturbed by acidic atmospheric deposition.

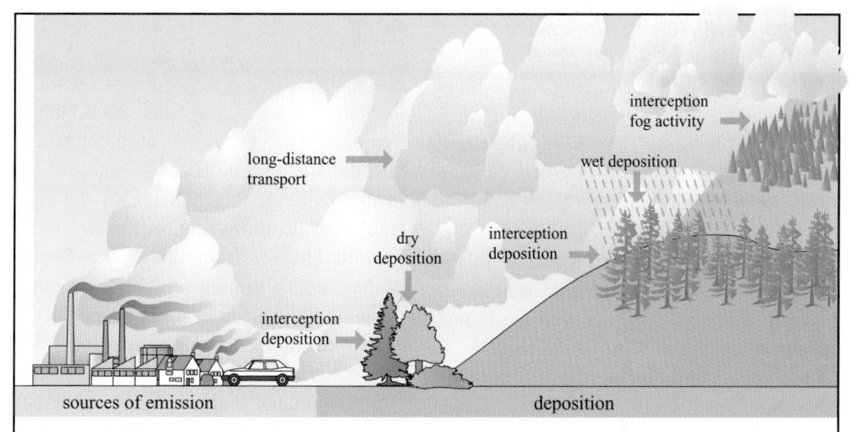

Figure 4.3.12 Deposition of acid rain

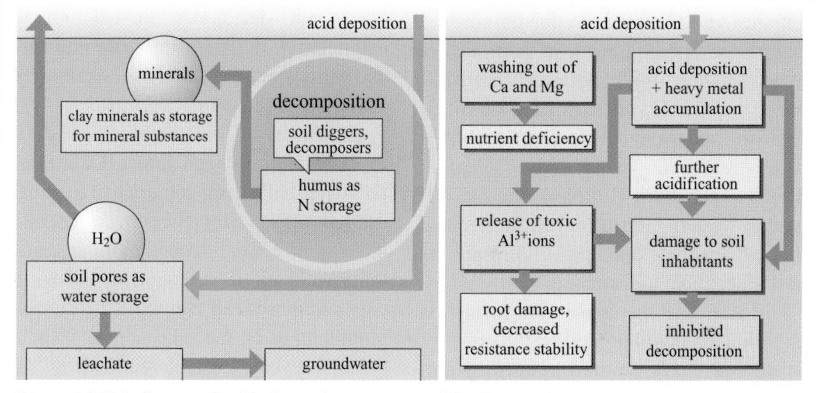

Figure 4.3.13 Influence of acid rain on the ecosystem of the forests

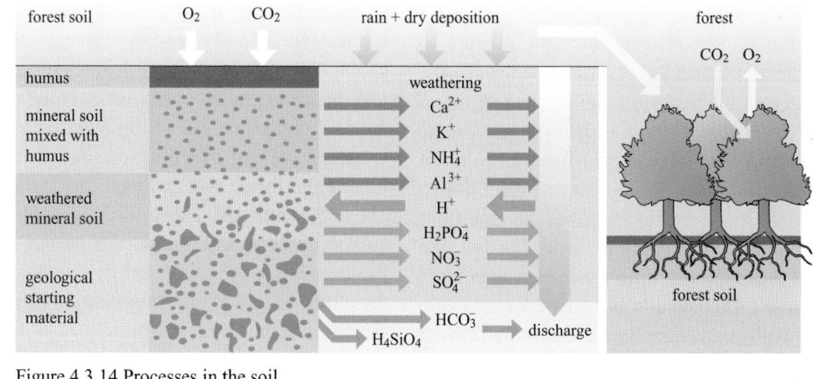

Figure 4.3.14 Processes in the soil

4.3.15 *Acid rain as the cause of new-type forest decline*

A new type of deforestation occurring over a large area, called *Waldsterben* (forest death) in Germany, has been identified since 1983 using the amount of damage from the previous year as a base. Conifers lose a large proportion of their needles, from the bottom to the top and from the inside to the outside, and deciduous trees show an early coloration of their leaves and in some cases drop their leaves as early as July. Increasing incidents of necrotic diseases (brown spots) have been observed. Vertical growth is stagnating, especially with the conifers; resin seeps out on branches and trunk in the needle-covered crown area. Spruce trees lose their needles in association with hanging branches (called the 'tinsel syndrome'), the tops of fir trees die and side-shoots form (called a stork's nest). In the early 1990s, 73% of all the trees in eastern Germany and 59% of those in western Germany were determined to be affected. Fir and spruce trees are the most susceptible to damage.

Research into the causes has identified the following possible factors: the effect of anthropogenously emitted airborne pollutants as a whole (Chapter 2); the acidification of forest soils, especially as a result of the acid deposition due to acid rain; weathering-related causes such as dryness, cold and snowbreak; climate-related causes; infestation; and electro-smog (the effects of electromagnetic waves from high-voltage lines and of microwaves from sections of radio link-ups). The centre of Figure 4.3.15 shows the increasing types of damage from bottom to top, ending in death. The left side of the diagram shows weathering-related causes. The negative effects of the emission of pollutants range from the photo-oxidising agents (Section 2.4) to acidic gases in general and to acid rain as a whole. Acid rain and acid fog attack the waxy layer of the leaves and can also cause scorching in tree bark. The continuous acid deposition into the soil (Sections 4.3.13 and 4.3.14) leads to irreversible damage of the self-regulating neutralisation (buffering effect) of soil. The result of this is finally the destruction of the entire ecological equilibrium: the biocommunity of the soil is destroyed, nutrients are washed out, and heavy metals are dissolved out of the soil structure and taken up by plants, where they can cause damage to enzyme systems and can enter the food chain. If there is a concurrent deficiency in calcium and magnesium, the released aluminium ions that come from the clay minerals injure the fine roots of the plants and trees.

Incidences of widespread deforestation as immission-induced forest decline have also occurred in past centuries in regionally limited areas after the development of industry or near refineries. With increasing industrialisation, associated with the construction of tall chimneys that enabled pollutants to be more broadly distributed, the deforestation has spread out in large areas since about the start of the 1970s. Usually from a political viewpoint, this phenomenon is also termed 'new-type forest decline', and represents a disturbance of the entire relationship between soil, tree and air, or an illness which has befallen the entire ecosystem. One of the most serious consequences of this deforestation is the loss of the forest as a groundwater storage place. A damaged forest also binds less carbon dioxide, with a resultant impact on the greenhouse effect. We hardly even know the global effects of this deforestation on the climate and the environment.

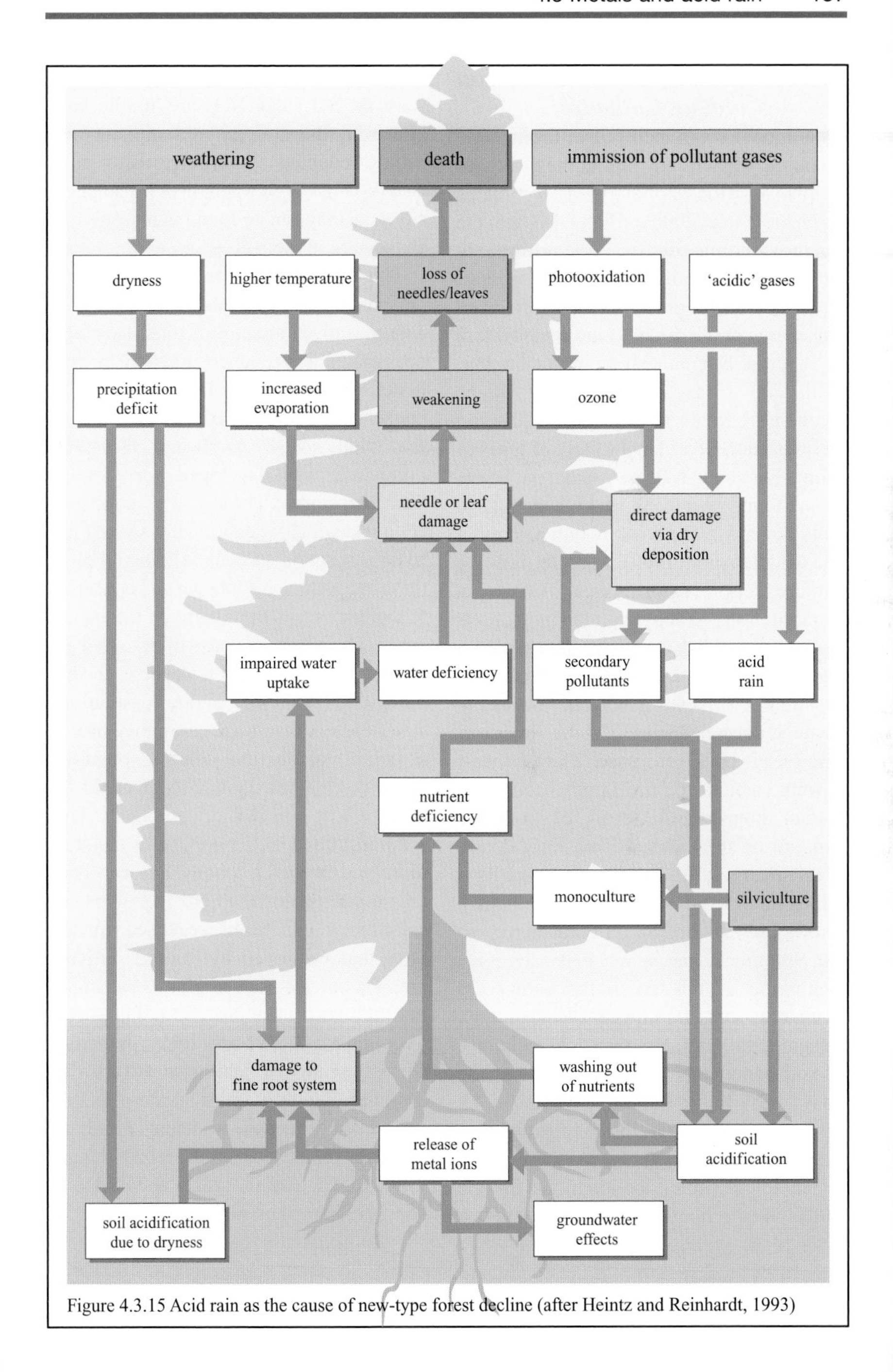

Figure 4.3.15 Acid rain as the cause of new-type forest decline (after Heintz and Reinhardt, 1993)

4.3.16 *Effects of acid emissions on Cologne Cathedral*

The annual costs incurred in the Federal Republic of Germany due to the effects of sulphur dioxide on material are estimated to be more than 2000 million DM (with the socio-economic costs amounting to 20 000–50 000 million DM). The master builder for Cologne Cathedral described the causes of the decay of the stonework in a 'Pictorial Documentation of Weathering' with the title 'Danger for Cologne Cathedral' thus. The different building periods are reflected in a variety of types of stone. For about 300 years, from the laying of the foundation stone on 15 August 1248 until the end of the construction activity in the Middle Ages (1560), trachyte from the Drachenfels region was used in construction. Trachytes are light grey, yellow to reddish extrusive rocks with a base of alkaline feldspars. Extensive restoration work was done in the nineteenth century, during which sandstone, limestone and basalt lava were used. The two west towers, which are the largest and most labour-intensive, in terms of both the amount of stone and the masonry detail, consist largely of the fine-grained light sandstone from Obernkirchen, which shows only slight traces of weathering so far. Soft limestones, which were used at the beginning of the twentieth century for sculpture work, are already disintegrating. When pelting rain hits them, they dissolve almost completely within a few years, and they must be removed or replaced by copies in poured stone, as has already been done on the south façade. In the selection of chemical protection methods for the stone in particular, protection against moisture and against aggressive substances such as SO_2 and NO_x are in the fore. However, the activity of (sulphate- and nitrate-reducing) microorganisms must also be considered when there is damage to stone, as they can be found in masonry just as they are in the soil.

The immission rates, i.e. the total amounts of pollutants picked up by a given area in a given amount of time, have been determined at Cologne Cathedral for long periods. The immission rates depend on the concentration in the air and especially on the wind speed. A special measuring procedure was developed for Cologne Cathedral studies. The test object consists of cellulose which is continually soaked in a basic absorption solution. The liquid phase drips into a storage container and is returned to the test object from there by means of a pump. The immission rates of SO_2, HCl and HF were determined at Cologne Cathedral. Since sulphates have a higher constitutional water content than carbonates, they take up a larger volume and tend to burst the structure. The water solubility of sulphate is also greater than that of carbonates. They are transported to the surface of the stone dissolved in water, where they can crystallise out (effloresce), so that gypseous crust forms. Finally, the weakened structure is further destroyed by rainfall. Silicone preparations are used to protect the stones, but they must meet specific requirements. They must not have any sticky properties, since they would otherwise adsorb dust particles, ending their water-repelling action. In addition to using protective measures for the stone in the refurbishing of Cologne Cathedral, stone restoration is being performed as well.

Figure 4.3.16 Effects of acid emissions on Cologne Cathederal (*source*: Cologne Cathedral site office)

4.4 Organic contaminants

4.4.1 Example of an impacted industrial area

The (contrived) example shown in Figure 4.4.1 considers various uses of an industrial region in the nineteenth and twentieth centuries. After the land was first used for the smelting of ore, it was filled with rubble and blast furnace slag. Heavy metals are the primary pollutants from this phase.

In the second phase, the area was used for a weapons and ammunition factory. From the debris and the backfill come heavy metals and organic substances from explosives. The explosive substances include nitro compounds as organic contaminants. These include organic nitric acid esters (nitroglycerine, nitroglycol, diglycol dinitrate, glycerin dinitrate and glycerin trinitrate, mannitol hexanitrate, nitrocellulose, etc.), aromatic nitro compounds (ammonium picrate, mononitrotoluene, dinitrotoluene, and trinitrotoluene, hexanitrodiphenylamine, etc.) and nitroamines. Nitroamines stem from the general formula $O_2N{-}NR_1R_2$ (nitro compounds of nitryl amide with $R_1 = R_2 = H$). Explosive materials that were produced and utilised include ethylenedinitramine (*N*,*N*-dinitroethylenediamine), tetryl (methyl picrylnitramine or *N*-methyl-*N*,2,4, 6-tetranitro- aniline), hexogen (1,3,5-trinitro 1,3,5-triazinan or cyclotrimethylenetrinitramine) and octogen (cyclotetramethyleneteranitramine). Most of these substances have a high toxic potential in soil as well. Aromatic nitro compounds, shown here using dinitrotoluene as an example, have the following general behaviour in the soil. Although they are slightly soluble in water (except for phenols), one can expect an average tendency toward bio- and geoaccumulation. In the soil, microbiological material transformations can occur which lead to corresponding amino compounds with higher water solubility than that of the starting compounds. An average persistence can be expected in the pedosphere and the hydrosphere. Because of the carcinogenic risk, the Environmental Protection Agency in the USA has provided threshold levels for drinking water. For example, at a concentration of 0.7 mg L^{-1}, the carcinogenic risk is considered to be 1%.

The gasworks located on this property during World War II – as the third phase of use – was destroyed by bombs. In this way, pollutants in the form of polycyclic aromatic hydrocarbons (PAHs), phenols, cyanides and heavy metals get into the debris and the backfill.

After 1948 this industrial area was used by a metal-working factory. Pollutants from this production process, as the fourth phase of use, can be expected to include chlorinated hydrocarbons in particular, which are found in the soil and which can also enter the groundwater. Materials used as solvents include those belonging to the group of readily volatile chlorinated hydrocarbons. Some of the chlorinated hydrocarbons can be used by bacteria as a carbon and energy source (in aerobic and anaerobic or methanogenous conditions and dehalogenation). The breakdown of chlorinated hydrocarbons depends on the pH, the oxygen concentration, the availability of hydrogen donors and macronutrients, and the absence of attendant substances that are toxic to microbes. The contrived industrial area represents a typical example of a hazardous waste site – as an abandoned site, i.e. a factory site with former facilities for the production or processing of environmentally unsafe materials.

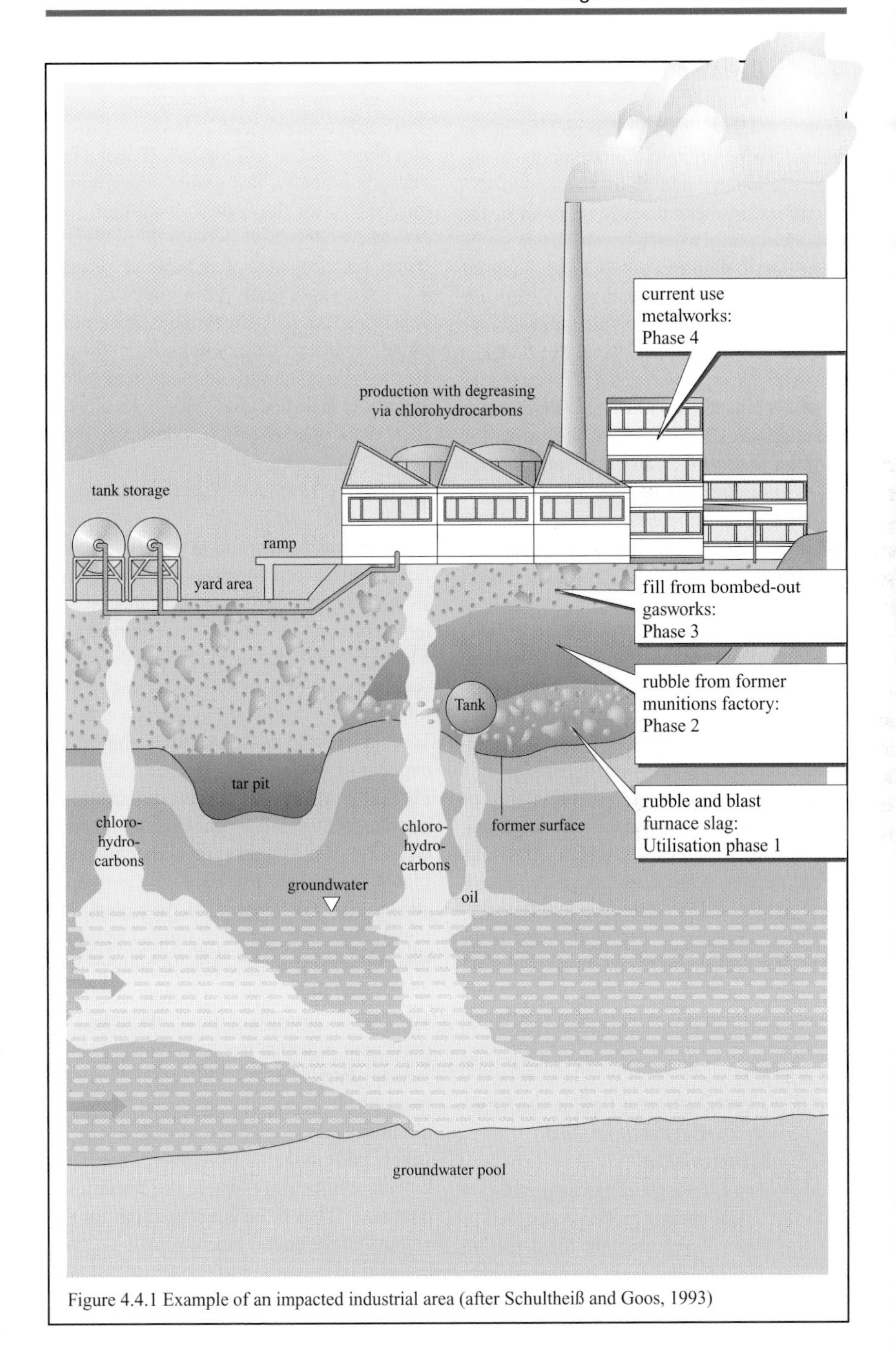

Figure 4.4.1 Example of an impacted industrial area (after Schultheiß and Goos, 1993)

4.4.2 *Behaviour of pollutants in the soil*

Figure 4.4.2 shows a global picture of the behaviour of organic contaminants in the soil. Pollutants bound to suspended dirt particles are mechanically retained in the soil by filtering, whereby even particles with very small diameters (<0.2 μm) can be filtered out of the percolating water in soils with a high proportion of fine pores. On the one hand, dissolved pollutants can be taken up by plants; on the other hand they can lead to the contamination of drinking water (Sections 3.3.2 and 4.4.3) via eluviation through the groundwater. The components of a pollutant that are of particular ecological or ecochemical relevance are those that are present in dissolved form or can be easily converted into the solution phase. The behaviour of organic substances in the soil is largely influenced by the activity of soil microorganisms. These in turn are decisive in determining the transformer function of a soil. Altogether the microbial transformations of organic pollutants effect a conversion into substances of a different chemical composition and of different states of aggregation. In so doing, metabolites with a higher toxicity than that of the starting substances can be generated. At the soil surface, photochemical transformations can play a role as well. A redistribution of pollutants in the pedosphere occurs via erosion and shore drifting of contaminated upper soil material by means of wind and water.

4.4.3 *Distribution of chlorinated hydrocarbons in soil atmosphere*

Studies of soil atmosphere can be performed with the aid of special probes connected to gas test tubes. A requirement for detecting soil contamination is a sufficiently high vapour pressure of the pollutants so that detectable concentrations are present in the gas phase of the soil spaces. If there is a correspondingly favourable distribution coefficient for the water–air system, it is possible to detect easily volatile pollutants using contaminated groundwater. Figure 4.4.3 shows an example for results from exploring the soil atmosphere for chlorinated, volatile hydrocarbons on a former factory site . The three-dimensional representation makes it possible to locate the focal point or source of the contamination.

4.4.4 *Behaviour of pesticides in the soil*

The general behaviour of biocides (as a generic term for chemicals to fight against living organisms) or pesticides (used synonymously) is shown in Figure 4.4.4. The right-hand side of the figure shows the requirements of a biocide – namely rapid degradability in addition to specific activity. Until a biocide is degraded it should be capable of killing pests (viruses, bacteria, algae, fungi, mites, insects, snails, and worms, specific substances existing for each type) or of being available to particular plants (herbicides for weeds). The picture shows the conversion of biocides and metabolites as a factor of time. The entire biocide amount is bioavailable immediately after application. Within a few days there is a reversible adsorption to soil particles, the maximum of which has been reached about 14 days later in this case; only 10% is still bioavailable in the soil solution. A biomineralisation takes place in the dissolved phase, which leads to the mobilisation of adsorbed biocide constituents, where the breakdown continues. Thus there is a maximum for the amount of the bound biocide.

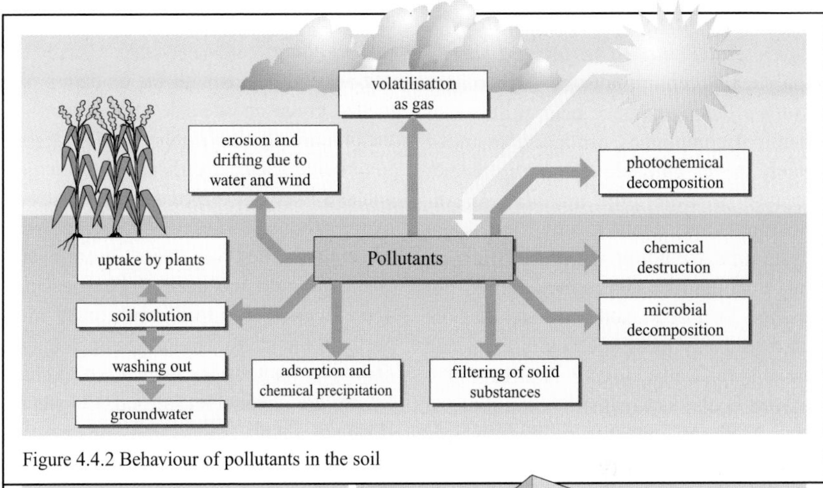

Figure 4.4.2 Behaviour of pollutants in the soil

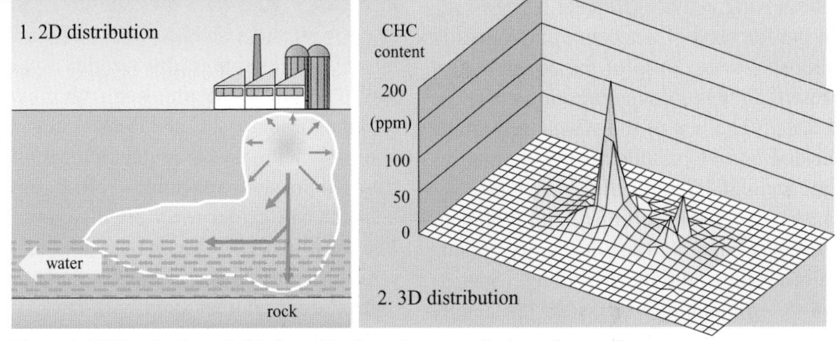

Figure 4.4.3 Distribution of chlorinated hydrocarbons in soil atmosphere

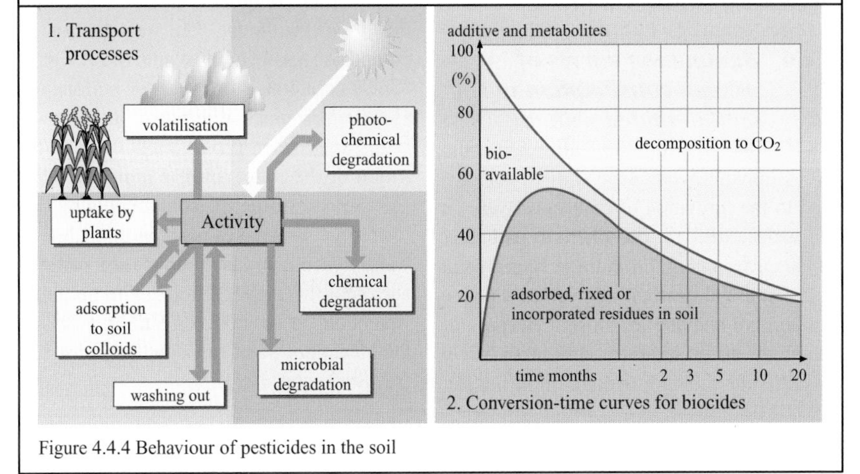

Figure 4.4.4 Behaviour of pesticides in the soil

4.4.5 *Amounts of measurable materials in impacted soil*

Analytical environmental chemistry assumes that there have been millions of chemical compounds produced, many of which have entered the environment. Because of these large numbers, an all-encompassing analysis of individual substances, including their identification and quantitative determination, is not possible. First of all, soil and water analysis can detect only those substances for which one is specifically looking. A broad-range analysis is also not realistic because of the cost. Methods other than the physico-chemical analytical methods are available for studying activity (Chapter 5). Dividing up the fractions of detectable substances in soils shows that 90% of the contaminants present in low concentrations cannot be detected. The rest of the substances can be divided into 9% readily volatile substances (Section 4.4.3), especially halogenated or chlorinated hydrocarbons, and 1% less volatile substances. The latter group include chlorophenols, PCBs, pesticides (Section 4.4.4), aromatic hydrocarbons (especially benzene, toluene and xylenes as BTX aromatic compounds) and PAHs.

4.4.6 *Summation curves of oxygen consumption in soil contaminated with oil*

Oil spills (from the aliphatic and cycloaliphatic hydrocarbons of petroleum) lead to the formation of emulsion layers on the surface of the hydrosphere in particular in the environment. From these layers, solid and liquid hydrocarbon aggregates separate out and are retained in the soil. Because of the slight water solubility and, with heavy oils, the low mobility overall, oil in the soil can be broken down by microbes before it reaches the groundwater table. Usually the degradation begins with the oxidation of a methyl group on one side of the chain – in the order of primary alcohols, aldehydes, up to carbonic acids. As early as 1906, microorganisms were discovered which break down hydrocarbons. These microorganisms, sometimes called ‘oil eaters’, occur everywhere on Earth and can even be detected in deep layers, down to a depth of 88 m in northern Germany. Straight-chained hydrocarbons with short to average-sized chain lengths are rapidly metabolised (Chapter 5). However, with increasing degrees of branching, the persistence of hydrocarbons increases. Figure 4.4.6 uses substrate respiration to show the breakdown of oil for a 1% by weight (relative to dry weight, dw) oil spill. The soil S contaminated with oil was mixed with compost C in different ratios, the compost being low-pollutant biocompost from domestic vegetable refuse from kitchens and gardens, collected separately. The most important factors for the decomposition are the interactions of the humus matrix and the microflora with the pollutants, and also the fertilisation effects, i.e. the addition of nutrients and trace elements due to the compost. The introduction of compost in a soil:compost ratio of 2:1 significantly accelerates the breakdown of the oil. However, in practice in remediation, lower ratios are chosen (up to 9:1). The optimisation of the water level is important. The metabolism of the oil components by soil microorganisms takes place in thin films of water which completely or at least partially surround the solid particles. The optimum water content lies between 50% and 80% of the maximum water capacity (Section 4.2.4) (see Stegmann, 1993).

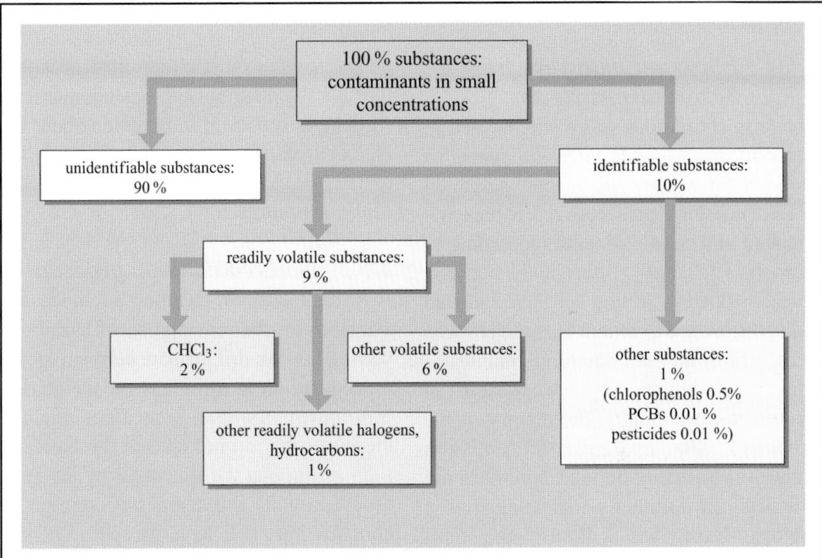

Figure 4.4.5 Amounts of measurable materials in impacted soil

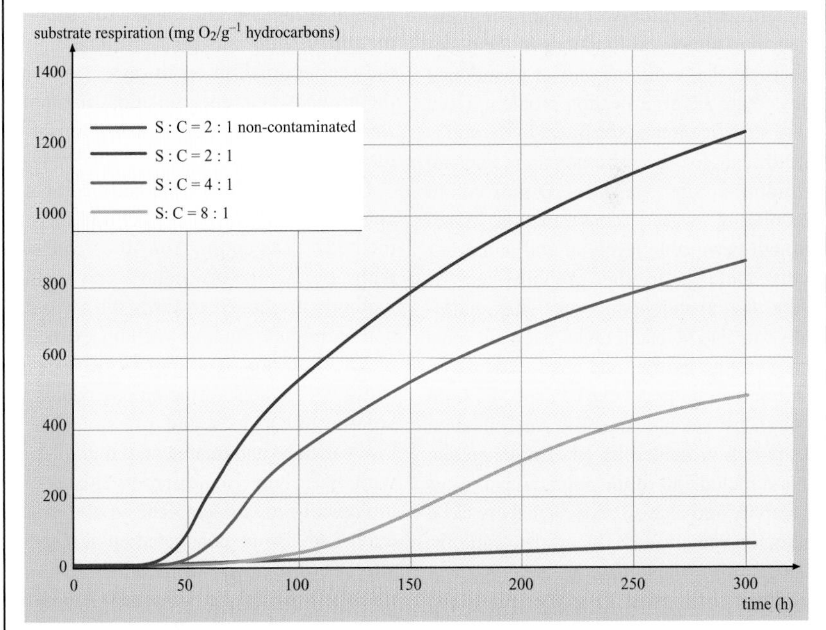

Figure 4.4.6 Summation curves of oxygen consumption in soil contaminated with oil (S = soil, C = compost; relative to dry weight) (after Stegmann, 1993)

4.5 Methods of soil remediation

4.5.1 Emission pathways from a hazardous waste site

The term 'hazardous waste site' is used to describe a definable area or piece of property for which there is some suspicion or evidence that it presents a risk to the environment and people's health (according to Römpp, 1993). We differentiate between abandoned sites (Section 4.4.1), abandoned storage sites and potential hazardous waste sites, which include hazardous munitions sites. To investigate hazardous waste sites, one must determine (i) the history of the property's use and (ii) the emission pathways. Narrowly defined, emissions are primarily air pollutants emanating from a factory. Here, we will use the term 'emission' in a considerably expanded sense (see Section 6.1.4 for estimating the risk). Contaminants in the soil impair the regulatory and biological functions in the pedosphere so that contaminants at abandoned sites create a soil protection problem. Even more significant than the threat to the soil is the threat to the groundwater that an abandoned site presents. Dissolved in percolating water, contaminants travel through permeable layers of soil and eventually end up in the groundwater. At overgrown abandoned storage sites, pollutants can also be taken up by the plant roots and therefore by the food chain. Additional environmental impact can be expected from gas exhaust and blowing dust. An evaluation of the risk presented by an abandoned site should include all of the potential pathways of activity and also all aspects that are to be protected (people's health, water, soil, air, plants and animals, material items such as buildings or lines of supply or removal). Due to the large number of possible contaminants, we limit the evaluation of the potential risk to just a few substances that threaten the environment. Selection criteria (if there is sufficient information about the type of contamination) include the mobility, toxicity, bioaccumulation and degradability of the substances.

4.5.2 Soil remediation procedures

A wide range of clean-up methods are available for the remediation of hazardous waste sites. We differentiate between *in situ* procedures on location and on-site/off-site procedures. On-site procedures include measures that do not change the location, such as limiting the use, securing the area, monitoring it, and temporary storage of contaminants that have escaped. Off-site measures include the clearing-out and relocation of waste landfills as a whole. *In situ* measures might consist of security measures with the goal of interrupting the contamination pathways (lowering the groundwater, encapsulation, immobilisation) or more extensive decontamination methods.

Decontamination methods include chemical–physical, biological and thermal methods. Depending on the emission pathway, *in situ* procedures can include methods orientated towards the impacted compartments, such as vacuum extraction of the soil air (and purification using activated charcoal, stripping procedures for volatile substances and precipitation of heavy metals) and treatment of the soil itself with the help of microorganisms for biological remediation. The *in situ* procedures can also be conducted on-site or off-site. The removal of residues requires renewed measures (e.g. thermal) and monitoring, as is the case when using protective and limiting measures.

precipitation
reaction gases
carrying away solid matter via erosion
leachate emerging at surface
hazardous waste site
surface runoff
leachates with pollutents
permeable soil layer
groundwater
direction of flow
contaminated groundwater
soil layer with little permeability

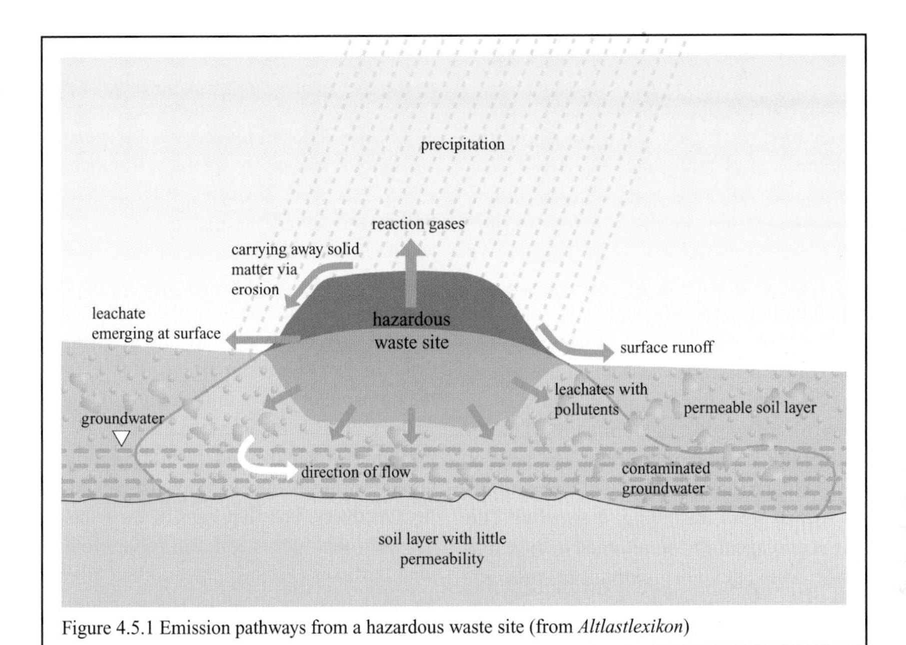

Figure 4.5.1 Emission pathways from a hazardous waste site (from *Altlastlexikon*)

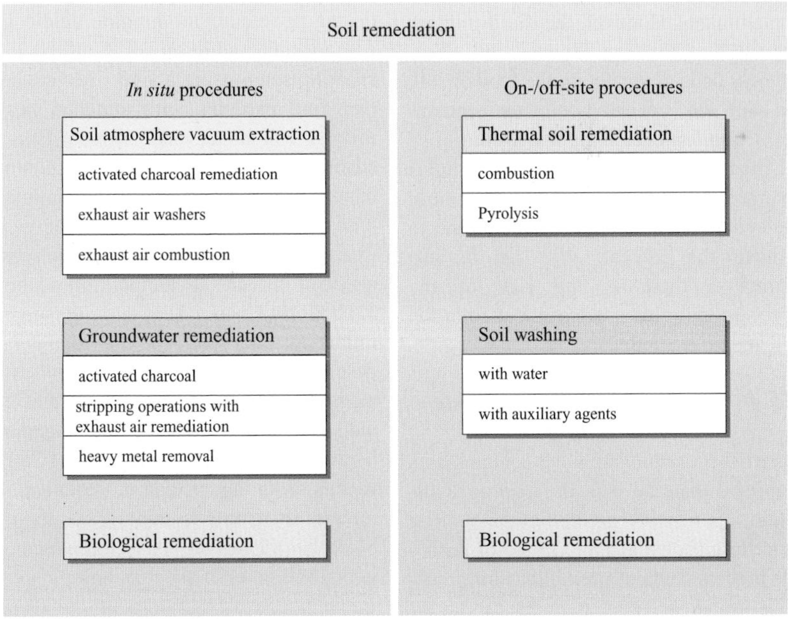

Soil remediation
In situ procedures
Soil atmosphere vacuum extraction
activated charcoal remediation
exhaust air washers
exhaust air combustion
Groundwater remediation
activated charcoal
stripping operations with exhaust air remediation
heavy metal removal
Biological remediation
On-/off-site procedures
Thermal soil remediation
combustion
Pyrolysis
Soil washing
with water
with auxiliary agents
Biological remediation

Figure 4.5.2 Soil remediation procedures

4.5.3 Counterflow extraction with solvents

As an example of a soil wash as a chemical procedure, in Figure 4.5.3 we show the principle of cleaning up a PCB-contaminated soil. After the soil is broken down, particles <2 mm in size are fed into the counterflow extraction setup. Because of the counterflow principle and the use of small particles, the amount of solvent A can be kept low. The solvent is separated out thermally; it is retrieved using a condenser C. The major portion of the solvent is separated from the contaminants B in an evaporator distillation process. The solvent and contaminant residue must be removed as special solid waste (e.g. thermally). The extraction agent can be returned to the cycle after recovery. This extraction process, which can be conducted on-site or off-site and is called soil scrubbing, can be used for the removal of both organic and inorganic contaminants. However, the disadvantages are the fact that the expense and effectiveness depend greatly on the type of soil, the state and concentration of the contaminants; that left-over extraction agent can be retained in the soil; that the soil structure is changed; and that it is necessary to remove the contaminants from the solvent and to remove the solvent. With an *in situ* extraction there is the risk of the uncontrolled release of chemicals into the soil and groundwater.

4.5.4 Thermal hazardous waste site remediation

Thermal procedures remove contaminants from soil material with the help of distillation, pyrolysis or combustion. The procedure shown as an example of an on-site process combines operations for preparation with a furnace, which is in turn linked to a five-step exhaust air remediation process (Sections 2.5.1 and 2.5.2). At the end of this series of procedures, the cleaned soil is returned back to the place from which it was removed. A drying stage is added before the actual thermal treatment. The setup shown uses three rotary kiln drums, and can process 5 $t\,h^{-1}$ of soil. After the extracted soil has been broken down to a particle size of 30–60 mm, the material is dried in a drying drum at 200–400 °C, and the burn-out occurs at temperatures of up to 1200 °C. The dwell times are listed as 30–45 min in the drying and burning drum; a cooling drum is attached downstream with heat recovery. The five steps of the exhaust air treatment are as follows: separation of dust particles (via cyclone and bag filter); thermal post-combustion (destruction of organic substances still present in the exhaust gas – 2 s at more than 1200 °C); a drying procedure for binding acidic gas components (addition of $Ca(OH)_2$); a filtering setup using a fine filter cassette (for fine particles with attached heavy metals); and an activated charcoal filter to adsorb the smallest amounts of contaminants. If the soil material is contaminated with PCB as in Section 4.5.3, the gases exiting from the rotary tube must be subsequently burned at temperatures above 1200 °C. Soil material can be dealt with in a milder manner than in the rotary kiln described above by using a pyrolysis method. In this situation a rotary kiln heated indirectly from the outside to temperatures between 500 °C and a maximum of 750°C is used. With this method of treatment, the mineral structure is largely maintained. More than 99% of the PAHs can be removed with dwell times of about an hour.

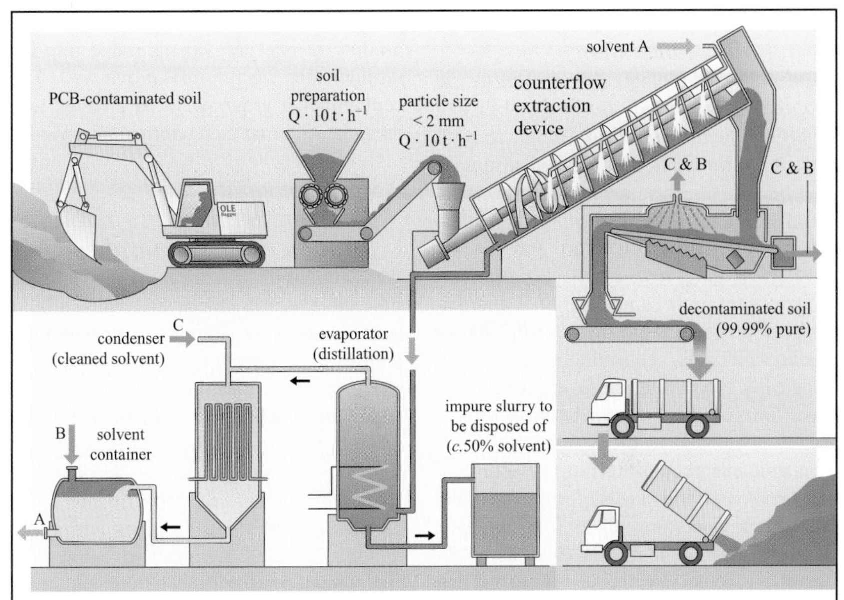

Figure 4.5.3 Counterflow extraction with solvents (after Bank, 1994)

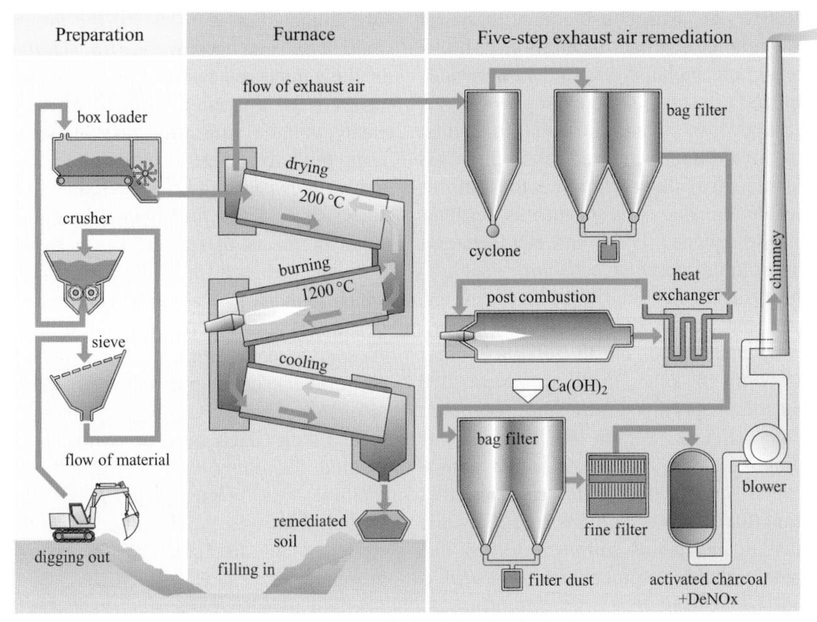

Figure 4.5.4 Thermal hazardous waste site remediation (after Bank 1994)

4.5.5 *Biological hazardous waste site remediation*

Procedures for biological hazardous waste site remediation make use of the ability of microorganisms to break down certain organic contaminants. The special advantage compared to the methods previously described lies in the mild *in situ* treatment, in which both the soil structure and the soil biology largely remain intact. Important prerequisites for the applicability of microbiological procedures are the degradability of the contaminants, concentrations in a physiologically favourable range, the absence of toxic (enzyme-inhibiting) substances such as heavy metals, and the solubility of the contaminants. Factors that must be optimised are the oxygen and nutrient levels, microbial activity, temperature, water level and pH. Disadvantages of biological procedures are that these factors must be monitored and that they fail with a significant increase of the biomass; it is only possible to monitor the effectiveness with considerable analytical expenditure; and that materials (e.g. heavy metals) can mobilise due to the microbially induced changes. An example for the breakdown of oil was presented in Section 4.4. In addition to refined oil, compounds that are easily degraded by microorganisms include aromatic solvents such as benzene, toluene, xylenes (BTX) and ethylbenzene, and soluble aromatic substances such as phenols and other substituted benzenes, e.g. from old gasworks (see Section 4.4.1). Compounds that are mostly difficult to break down, on the other hand, are chlorinated compounds, polycyclic aromatic compounds and pesticides. Figure 4.5.5 shows the combination of subterrestrial and above-ground procedures (Messer-Griesheim, Frankfurt). It is important to supply the microorganisms with sufficient oxygen. The object of the above-ground treatment stage is to accelerate the degradation process. The mixture of microorganisms is fed into the groundwater via the infiltrative well. The contaminated groundwater is directed into the bioreactor and then returned afterwards.

4.5.6 *Remediation of contaminated groundwater and of soil atmosphere*

With contamination from readily volatile substances such as chlorinated hydrocarbons (see Section 4.4), after they are localised, both the groundwater and the soil atmosphere must be cleansed. In the example shown in Figure 4.5.6, a remediation well down to the groundwater is directed to the source of contamination. The pipe contains filters for liquids and gases in the lower section. The two suction pipes within the remediation well are used to extract the contaminated water (left) or the soil atmosphere (right), which is blown through an activated charcoal filter to adsorb the CHCs. The funnel-shaped course of the groundwater level is induced by the suction activity of the submerged pump. The pumped-out water is fed through a stripper, where the easily volatile substances are blown out using fresh air and are then removed adsorptively, likewise with an activated charcoal filter. Air is blown directly into the groundwater by means of aeration lances that are introduced directly into the groundwater, which supports the gaseous suction removal of the soil atmosphere. The control lances are used to remove the sample in order to be able to check the progress of the remediation measures. The activated charcoal that is loaded with CHCs must be removed. One method for doing this (besides thermal removal) is to remove it using hot water vapour. The water–CHC mixture is then separated by distillation or by using a membrane procedure.

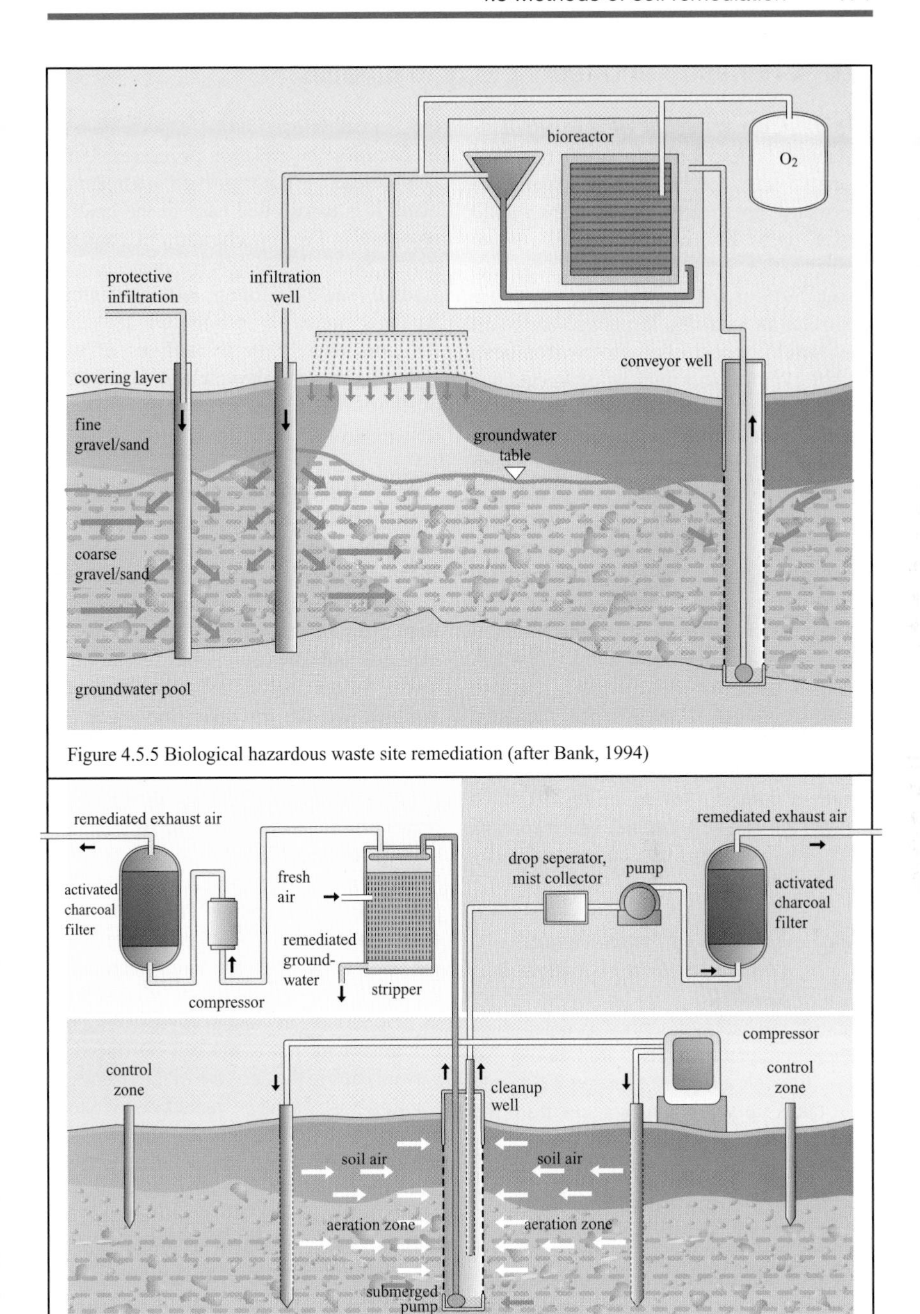

Figure 4.5.5 Biological hazardous waste site remediation (after Bank, 1994)

Figure 4.5.6 Remediation of contaminated groundwater and of soil atmosphere

4.6 Chemistry and technology in landfills

4.6.1 Construction of a landfill

A landfill is described as a locally restricted storage site for refuse. Currently in Germany, up to 70% of the urban solid waste, up to 90% of the inert solid waste (construction rubble, extracted soil) and about 50% of the special solid waste are deposited in landfills. To prevent emissions that would impact upon the environment, technical procedures must be conducted that prevent the transfer of pollutant-bearing landfill percolating water into the underground and into the groundwater (with landfill seals such as a base seal of clay and plastic sheeting). Percolating waters are directed into percolating water trenches via drainage pipes and, if necessary, are purified in a waste water treatment plant. The anaerobic fermentation of organic material generates a landfill gas (Section 2.6.7) which escapes from the landfill and leads to noxious odours if appropriate technical measures for redirection, detoxification or utilisation cannot be undertaken. For each tonne of domestic refuse, up to 250 m^3 of landfill gas can be generated, which consists of 55% by volume of CH_4 and up to 45% by volume of CO_2.

4.6.2 Sources of groundwater contamination risk near a landfill

Landfill leachate is described as water originating from precipitation or from refuse that flows through a landfill (Section 4.6.1) and that takes up soluble substances from the refuse in the process. An average precipitation of 750 mm per year yields about 5 m^3 of landfill leachate per hectare per day. The major components of landfill leachate are water-soluble nitrates, sulphates, chlorides, heavy metals and high amounts of organic substances (Section 6.1.5). Impact and composition depend on the type of refuse, the weathering and especially on biochemical degradation processes. If the landfill leachate is not purified in a treatment plant, it is usually fed back to the landfill, resulting in the concentration of contaminants. Since the long-term protection of landfill seals is often not guaranteed, landfills must be monitored for their tightness using regular analyses of well water. Figure 4.6.2 shows a landfill as well as other sources of hazards for groundwater such as agriculture.

4.6.3 Development of the waste volume and its composition

The refuse composition is determined by the analysis of average samples that are as representative as possible and that are compiled from many samples from different detection locations and collection points several times a year. Screen analysis is used to divide the samples into the fractions fine waste (<8 mm), medium waste (8–40 mm), coarse waste (40–120 mm) and sieving residue (>120 mm). Figure 4.6.3 shows the development of the specific volume of waste (1) and its composition (2) for the city of Stuttgart with 560 000 inhabitants (1995) as per Sattler and Emberger (1990). The two diagrams also show the division of refuse at the same time. The continuous decrease of the waste area density (waste mass/volume in 1950~200 kg m^{-3} and in 1990~100 kg m^{-3} per inhabitant per year) can be traced in general back to the decrease of the fine waste fraction (ash, sand – reduction of stove heating since 1950) and to the increase in the proportion of packaging material up to 1990. At the beginning of the 1990s, the amount of municipal waste in Germany totalled 40 million tonnes, of which 15 million tonnes was domestic refuse, 2 million tonnes was bulk waste and about 13 million tonnes was domestic-type commercial refuse.

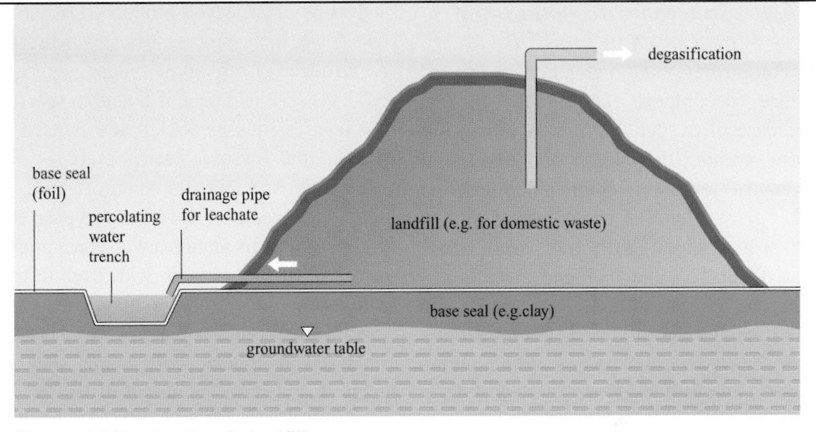

Figure 4.6.1 Construction of a landfill

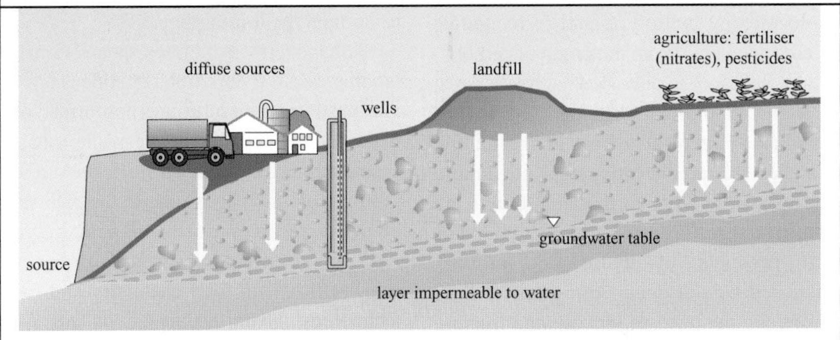

Figure 4.6.2 Sources of groundwater contamination risk near a landfill

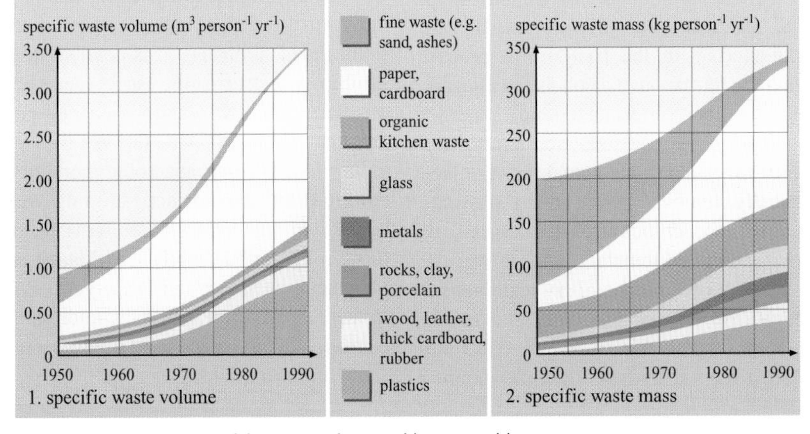

Figure 4.6.3 Development of the waste volume and its composition

4.6.4 Anaerobic decomposition processes in landfills

In composting, the organic components of refuse decompose aerobically in the presence of oxygen via microorganisms to form carbon dioxide, water and solid (compost) residue. In contrast, fermentation takes place in landfills under the exclusion of air. In four stages – hydrolysis, degradation, acid formation and methane formation – methane, carbon dioxide, trace gases and solid residue are produced. Putrefaction is used in waste water treatment and sludge treatment as well (Sections 2.5.7 and 3.5). In the first stage of the degradation process, there is an extracellular splitting of macromolecules by hydrolytic and fermentative microorganisms. From the group of carbohydrates, cellulose is only slowly broken down to form cellobiose, a reducing disaccharide. Xylobiose is generated from hemicellulose. Starches are rapidly broken down. Maltose and sucrose are formed. Pectins are hydrolysed to form galacturonic acid. The end products of hydrolytic splitting, catalysed by enzymes from the hydrolase group, are the hexoses glucose and fructose and the pentose xylose. In the second stage, the building block molecules (carbohydrates, fats, proteins and nucleic acids) are fermented, with the best-known process being the formation of alcohols from sugars. Fermentations are anaerobic, energy-yielding processes. Reduction equivalents (electrons or coenzyme-bound hydrogen) are typically removed from a carbohydrate cleavage product such as glucose, fructose or xylose, and are transferred to an organic acceptor (see a biochemistry textbook for details). All fermentation processes have in common the formation of pyruvic acid, which is formed during the decomposition of sugars as intermediate products. Oxidation takes place in the process, whereby hydrogen is removed from some sugar cleavage products. The hydrogen is bound to specific coenzymes such as NADP and is used to reduce the pyruvic acid itself or its breakdown and transformation products. In addition to methanol and ethanol, typical end products of fermentation are acetone, propyl alcohols, acetoin, organic acids from formic acid (with 1 C atom) up to butyric acid and succinic acid (with 4 C atoms). In the third stage, methanogenic compounds such as acetic acid, formic acid and methanol are produced by acetogenic bacteria; this provides the substrate for the methane bacteria in the fourth stage.

Methane fermentation is a special fermentation process. Even carbon dioxide is one of the substrates. The substrates are formed due to the activity of cellulytic organisms (phases 1 and 2) with subsequent fermentation (optimal pH 4–7) according to the pathways of the fermentation of lactic acid, butyric acid, or mixed acids, which also plays an important role in food chemistry and food technology. Mixed fermentations (pH 7–8) take place in sewage treatment plants, and also in the sediments of bodies of water (with the formation of marsh gas) and in the stomachs of ruminants. It is possible that a symbiosis between the strict anaerobic methane bacteria and the cellulose fermenters is based on the fact that the latter induce the strong anaerobic conditions required by the former. Overall, very complex mechanisms are the basis for the processes described briefly here (Maurer and Stegmann, in Sattler and Emberger, 1990). In the end, all of the processes result in the formation of biogas and solid residues, especially of mineral salts.

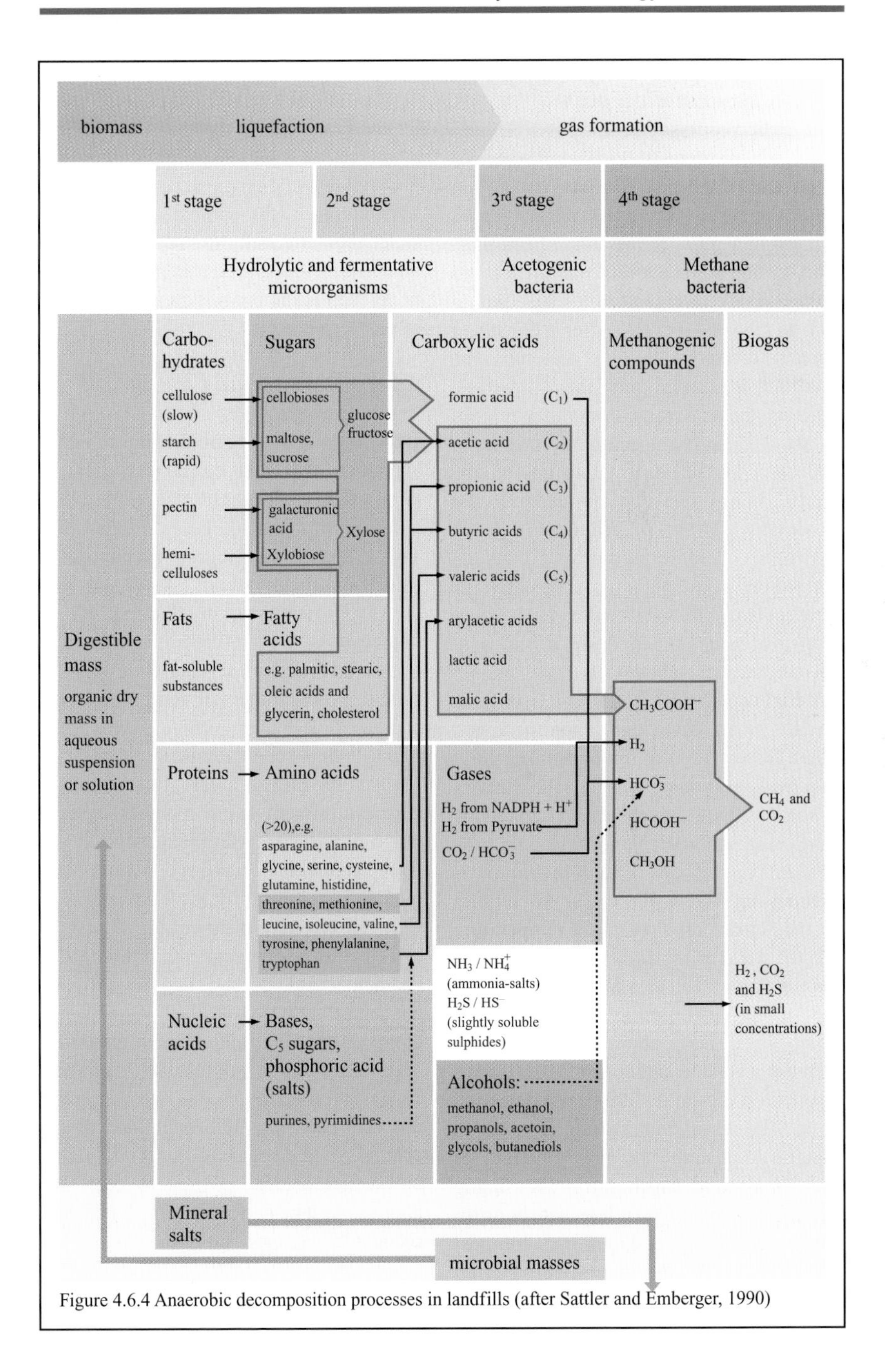

Figure 4.6.4 Anaerobic decomposition processes in landfills (after Sattler and Emberger, 1990)

4.6.5 *Flow process of a waste fermentation plant*

The basic processes described in Section 4.6.4, which several bacterial strains (acetogenic and methogenic bacteria) undergo in the different transformation stages, can be utilised in a waste fermentation plant thus. The starting materials used are kitchen and garden refuse as organic wet refuse with a dry weight of 40–50%. After mechanical sorting to remove the inert materials, the material is ground and then liquefied, whereby the percentage of dry matter sinks to about 13%. The anaerobic fermentation processes are set up in either one or two steps. In a one-step procedure, the processes shown in Figure 4.6.5 of hydrolysis, acid formation and methane formation take place concurrently or consecutively in a fermenter. In the two-step procedure, the hydrolysis and acid formation (see above) of the organic material occurs microbially in a hydrolysis reactor in the first stage. Then the liquid phase fermentation of the acids takes place in a methane reactor. For the solid fermentation residues, which are generated by thickening and water removal using a travelling screen press, a subsequent post-composting stage is added. The dwell times in the fermenters (operating temperature 35 °C) are 12–15 days (Rottweil Waste Fermentation Plant as a pilot plant – as per Sattler and Emberger, 1990) and the post-composting lasts about 8 weeks. The fermenter volume amounts to 45 m^3 with a population of about 4000 in the detection area. The term composting is used in general to describe the directed decay of municipal solid waste and sludge using bacteria and fungi. We differentiate between three main working stages in a compost system for municipal solid waste: the preparation of raw municipal solid waste for the raw compost material (the fermentation residue is present as the compost raw material in this case); decaying to form fresh compost (the optimal adjustment of water content and air supply for the microorganisms is necessary); and the post-decay step is the recovery and preparation of the fresh compost.

4.6.6 *Removal and use of landfill gas*

One kilogram of carbohydrates yields approximately 0.8 m^3 biogas (50% CH_4), 1 kg of proteins produces 0.7 m^3 (70% CH_4), and 1 kg of fat yields 1.2 m^3 (70% CH_4). The disposal of biogas, which has a calorific value of about 24 MJ m^{-3} with 60% CH_4, consists of flaming off or combustion. Possibilities for utilisation depend on the calorific value. Biogas can be used to generate thermal and process heat and electrical energy even on a large scale. Landfill gas is divided into a methane-rich product stream and a carbon dioxide-rich permeate stream using membrane separation procedures (with modules of hollow fibre membrane bundles). Natural gas quality is achieved by the elimination of CO_2. However, a purification must take place before an energetic use, in which H_2S and halogenated hydrocarbons in particular are separated out adsorptively on activated charcoal or using long-chained paraffins. The stages before this consist of drying, condensate separation, and separating out dust and particles. Before the landfill gas can be used as a fuel for combustion motors, measures to prevent corrosive damage must be performed which take into account the composition of the gas.

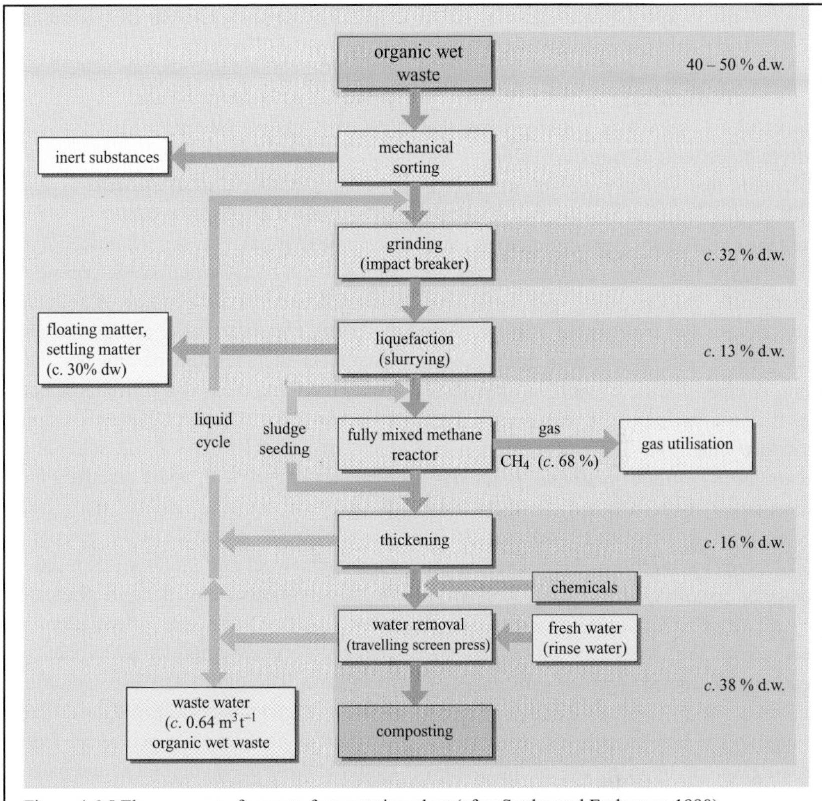

Figure 4.6.5 Flow process of a waste fermentation plant (after Sattler and Emberger, 1990)

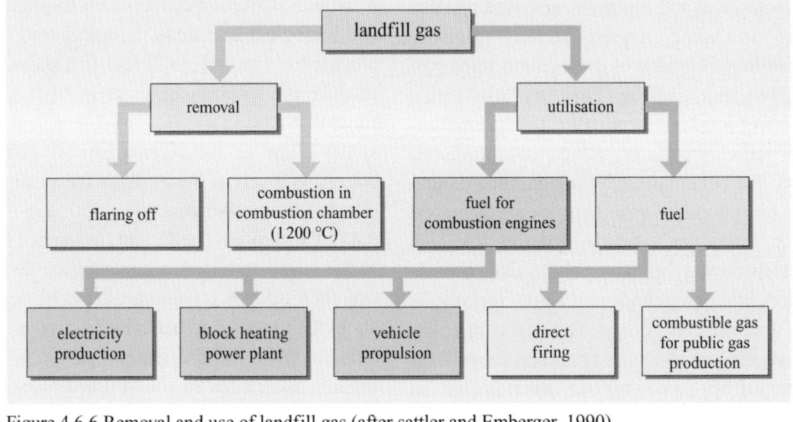

Figure 4.6.6 Removal and use of landfill gas (after sattler and Emberger, 1990)

5 Environmental Chemistry of Selected Xenobiotics and Heavy Metals

5.1 Pathways of pollutants, food chains and properties of materials

5.1.1 Pollutant pathways in ecosystems

Figure 5.1.1 shows a summary of the different sections of transfer pathways for pollutants that we have already discussed, with an emphasis on xenobiotics. Derived from the Greek *xenos* (foreign/foreigner) and *bios* (life), this term describes organic substances which are generated in ecosystems by something other than biological, i.e. enzyme-driven, transformations. A characteristic of these substances is that they are foreign to the biosphere in their structure and their biological properties. Xenobiotics include synthetic pesticides, chlorinated hydrocarbons, tensides and softeners made of plastic. Transformation and degradation reactions can change considerably the ecotoxicological potential (Chapter 6) of xenobiotics, either increasing or decreasing the toxicity. Pollutants are generally described as those substances or mixtures that induce disadvantageous changes when they transfer into ecosystems or are taken up by living organisms or even on objects. Of this large group, xenobiotics are those that are present in the environment as a result of anthropogenic activities and that are to some degree already globally distributed because of propagation and distribution mechanisms. They are also called environmental chemicals. Thus, environmental chemicals are substances introduced into the environment by human intervention that can occur in quantities or concentrations that endanger animals, plants, microorganisms and humans. They also include inorganic compounds and heavy metal salts of anthropogenic origin. The effects of pollutants in close-to-nature or natural ecosystems range from damaging individual species to serious damaging effects for the overall ecological equilibrium, such as acidification of the soil and eutrophication.

5.1.2 Basic principles of pollutant distribution and transformation

Chemodynamics and chemical and biochemical transformations determine the behaviour and the distribution of pollutants on Earth. The data profile of environmental chemicals includes the parameters listed under the term chemodynamics as selected properties. For hexachlorobenzene (vapour pressure 1.45×10^{-3} Pa (20°C), solubility in water about 5μm L^{-1}), under reactivity there is a note (Koch, 1991) that hexachlorobenzene is stable with respect to physicochemical and biochemical reactions (high persistence) and a rapid photolysis takes place with the formation of pentachlorobenzene and tetrachlorobenzene in hexane. The data relative to solubility, volatility (vapour pressure), and the different distribution coefficients (such as the Henry coefficient, sorption coefficient and bioconcentration factor) make it possible to make statements regarding the mobility and the distribution trend between non-biological and biological structures. In addition to the constant H from Henry's law (for gaseous substances), a substance-specific 1-octanol-water distribution coefficient (determined by Henry's law or the Nernst law of distribution) is given as log P. With these values (for hexachlorobenzene $H = 450$, log $P = 6.44$), the sorption coefficient (as log SC, 5.1 in this case), and the bioconcentration factor (log $BCF = 5.5$–6.1), statements can be made about the distribution behaviour between the compartments of the water biosystem or organic structures in the soil/sediment or between water and atmosphere.

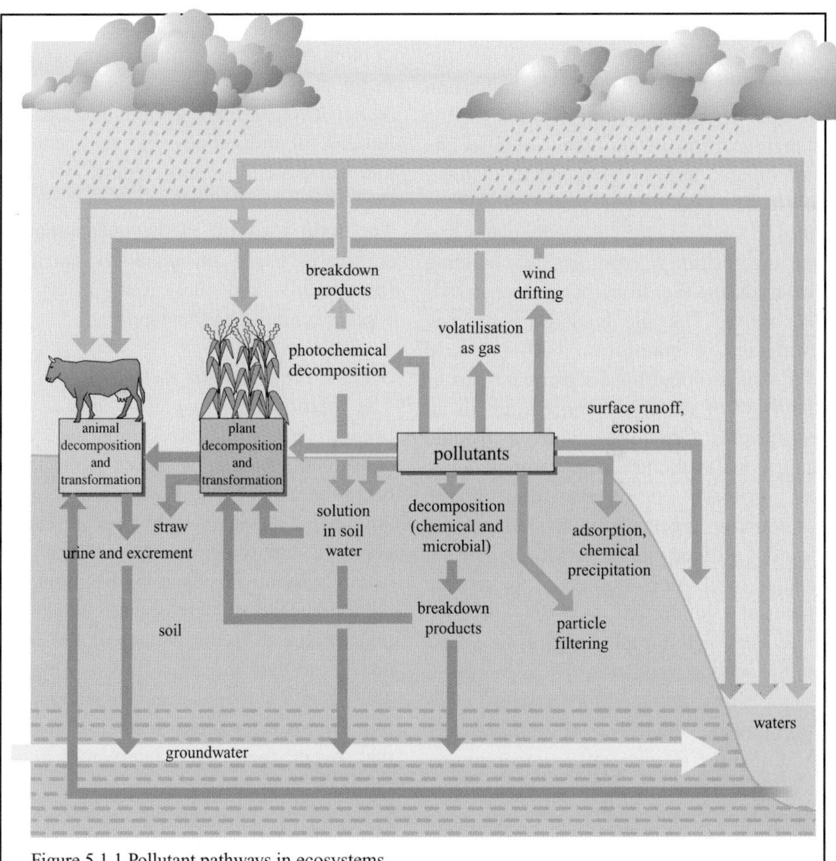

Figure 5.1.1 Pollutant pathways in ecosystems

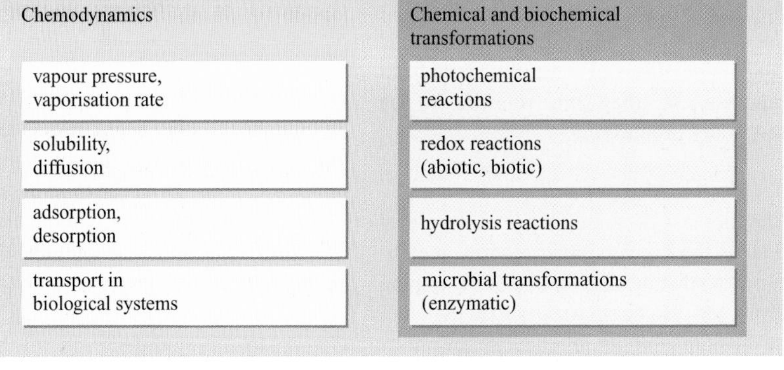

Figure 5.1.2 Basic principles of pollutant distribution and transformation

5.1.3 Food chain and energy flow

The determining physical basis for the formation of food chains is the utilisation of solar energy, which is required for the transformation of inorganic compounds into organic ones by green plants (Müller, 1991). Thus, autotrophic plants stand at the head of the food chain as higher plants that are in a position to construct their building materials and fuel from inorganic materials (see above). They are described as primary producers in connection with the food chain, as a model for a significant part of the flow of materials and energy in an ecosystem. Organisms with a trophic (nutritional) function form a trophic level. As primary producers of organic substances, green plants are the starting point of all food chains; they form the first trophic level. As consumers of the organic material generated by the primary producers, heterotrophic organisms follow on the second trophic level – as biophytes (plant eaters, predators or parasites) or as saprophytes, which live off dead biomass. At the third trophic level, decomposers break down dead biomass residues to form inorganic substances. The producers and consumers generate 'refuse' in the form of foliage, excretions, faeces and cadavers. The refuse is incorporated into another food chain of the decomposers, also called the detritus food chain. Organisms in this chain mineralise and at the same time form biomass, so they are referred to as secondary producers.

5.1.4 Food pyramid

The biomass decreases from level to level due to the conversion in energy metabolism, so that the distribution of the trophic levels can also be shown as a food pyramid. Because of the respirated high proportion of biomass taken up, it often decreases by a factor of 10 from a lower level to the next higher level. Overall, towards the top, the number of individuals, biomass, energy and reproduction rate decrease, whereas the body size and radius of action increase. The food pyramid in the ocean that is shown is based on plankton nutrition (producers) and the four levels of consumers building thereupon.

5.1.5 Food chain network

The relationships in places such as the oceans between the producers and consumers can also be shown in the form of a food chain network, with a differentiation based on plants, herbivores and carnivores. If slightly degradable pollutants are taken up by organisms at the beginning of the food chain, an increasing accumulation takes place over the course of the food chain, so that the final links – higher animals and humans – can take amounts in that are toxic for them. It is also obvious that several food chains interwoven like a braid or network merge together at the point of the end consumers, shown in Figure 5.1.5 as tuna fish and sharks in species-rich biocenoses such as those in the ocean. If pollutants (environmental chemicals or pesticides) end up in a biocenosis, often one link in the chain or one strand of the braid is broken, so the link just before it can develop unchecked. The number of subsequent links is reduced due to the deficiency of nutrients that develops. Thus, in order to evaluate the effects of environmental chemicals, it is necessary to have information about these kinds of ecological interactions as well.

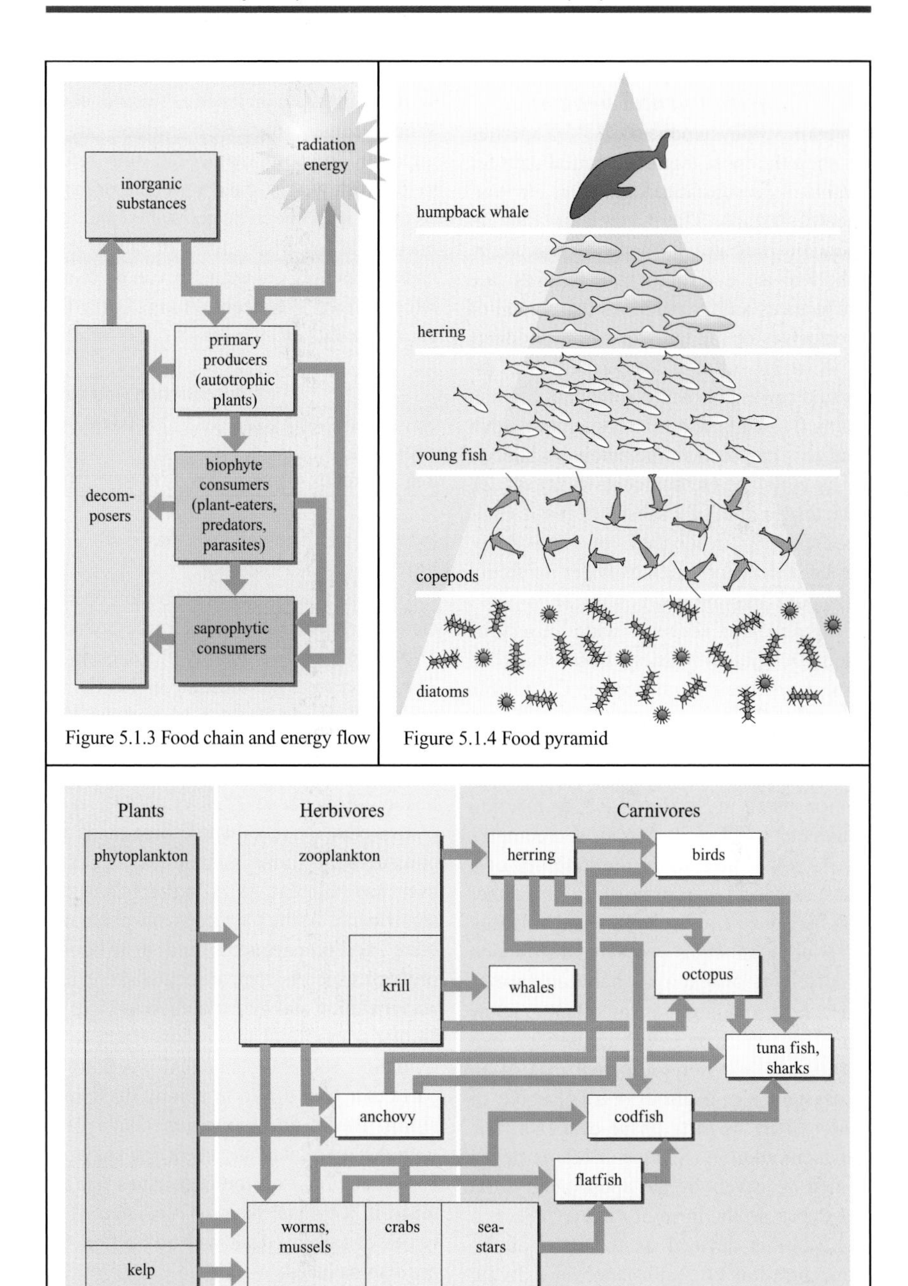

Figure 5.1.3 Food chain and energy flow

Figure 5.1.4 Food pyramid

Figure 5.1.5 Food chain network

5.1.6 *Materials properties and environmental behaviour*

Overall, physicochemical properties provide the most important initial data for evaluating a substance's potential environmental hazard. The Chemicals Law in Germany indicates that materials that can be labelled as environmental hazards are 'substances and preparations that in and of themselves or the transformation products of which are suited to change the state of the natural contents of water, soil or air, climate, animals, plants and microorganisms such that this can lead immediately or later to hazards for the environment' (Section 6.2). The distribution of a substance in the environment is influenced by substance-dependent factors such as water solubility or vapour pressure (Section 5.1.2).

Figure 5.1.6 shows a wide variety of physicochemical values, from materials properties such as structure, UV/visible spectrum, fixed points and vapour pressure curves (as a factor of the temperature), to the coefficients for adsorption or distribution listed in Section 5.1.2, to reaction parameters such as hydrolysis and complex formation. Clues about the mobility of a substance from one compartment to another can be gained from solubility and volatility (vapour pressure), as well as the distribution coefficients of materials between various phases. In so doing, it is important to know whether a substance will accumulate in a biological or a non-biological part of an ecosystem, e.g., in the sediment of a body of water or in the fatty tissue of a fish. The bioaccumulation (Section 5.1.2) is determined decisively by the liposolubility of a substance. In the form of the *BCF* (Section 5.1.2), it is defined as the ratio of the concentration of a substance in a living organism to the concentration in the surrounding medium (such as water or soil). When considering *BCF* values, it is important to know whether the value refers to the fresh weight, the dry weight, or (as is common with lipophilic pollutants) the fatty mass. The log *BCF* values of polychlorinated biphenyls, for example, can be calculated from concentrations in the compartments as follows:

	Concentration (mg kg^{-1})
Sea water (North Sea)	0.000,002
Sediment	0.02
Sea mammals	160

log *BCF* sea mammals/sea water = 8,
log *BCF* sediment/sea water = 4.

The 1-octanol–water distribution coefficient (Section 5.1.2), P_{OW}, is a measurement of the water or fat solubility of a substance. The higher the P_{OW} value, the higher the fat solubility of a substance is, and the less readily it will dissolve in water. By comparing the P_{OW} values of benzene and hexachlorobenzene (1.8 vs. 6.44), it is obvious that the chlorinated compounds are considerably more soluble in fat. The overview (Bliefert, 1994) differentiates the significance of the physicochemical parameters by compartments and distribution processes such as compartmentalisation and accumulation and by the transformations of abiotic or biotic decomposition. Hydrolysis, oxidation and photochemical reactions in particular play a role in abiotic decomposition. Biotic decomposition takes place with the aid of naturally occurring enzymes in association with the organism's metabolism in a compartment or ecosystem. The issues of toxicity dealt with in Section 6.2 are also included.

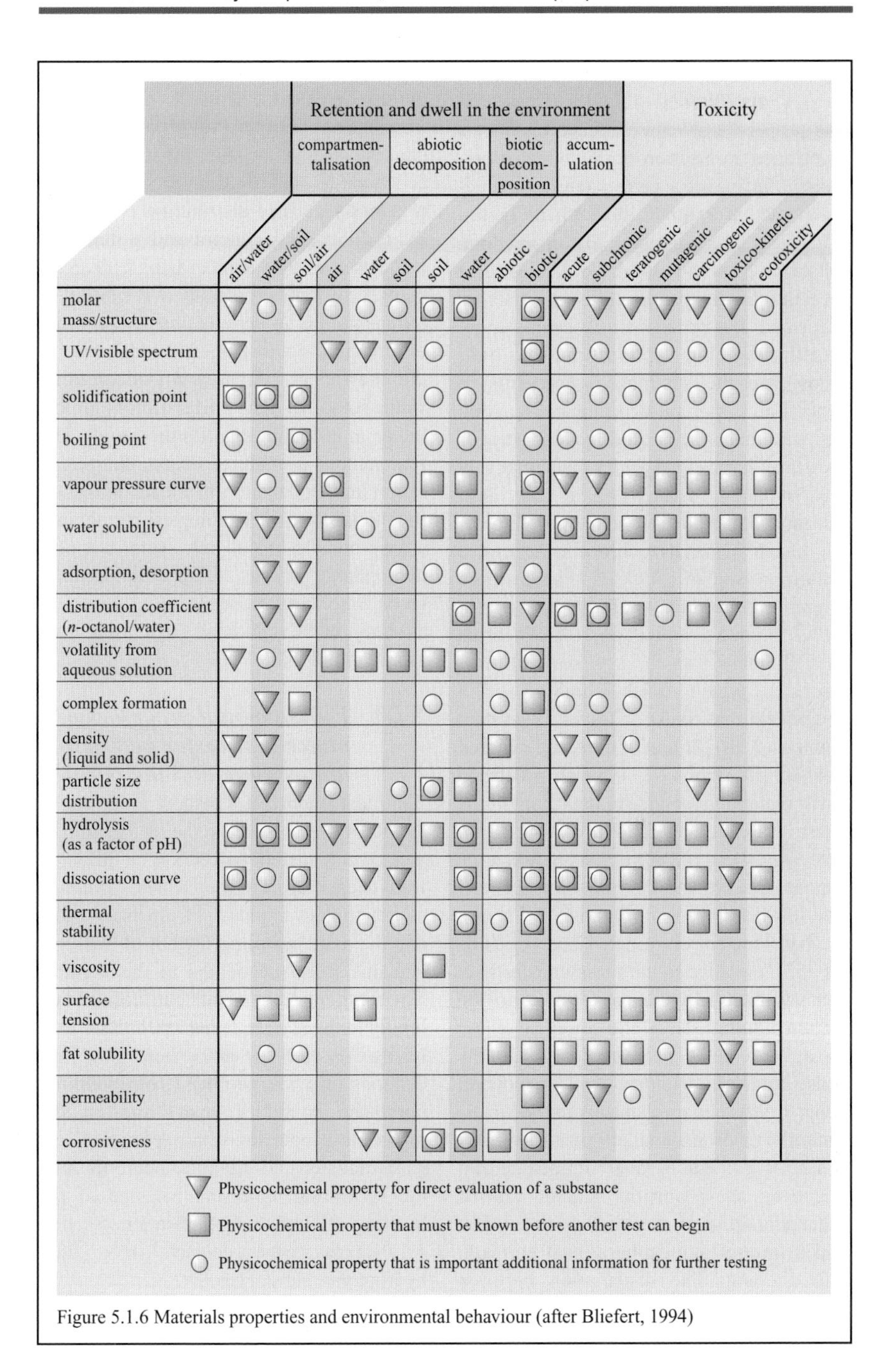

	Retention and dwell in the environment										Toxicity						
	compartmentalisation			abiotic decomposition			biotic decomposition		accumulation								
	air/water	water/soil	soil/air	air	water	soil	soil	water	abiotic	biotic	acute	subchronic	teratogenic	mutagenic	carcinogenic	toxico-kinetic	ecotoxicity
molar mass/structure	▽	○	▽	○	○	○	⊡	⊡		⊡	▽	▽	▽	▽	▽	▽	○
UV/visible spectrum	▽			▽	▽	▽	○			⊡	○	○	○	○	○	○	○
solidification point	⊡	⊡	⊡				○	○		○	○	○	○	○	○	○	○
boiling point	○	○	⊡				○	○		○	○	○	○	○	○	○	○
vapour pressure curve	▽	○	▽	⊡		○	□	□		⊡	▽	▽	□	□	□	□	□
water solubility	▽	▽	▽	□	○	○	□	□	□	□	⊡	⊡	□	□	□	□	□
adsorption, desorption		▽	▽			○	○	○	▽	○							
distribution coefficient (*n*-octanol/water)		▽	▽					⊡	□	▽	⊡	⊡	□	○	□	▽	□
volatility from aqueous solution	▽	○	▽	□	□	□	□	□	○	⊡							○
complex formation		▽	□				○		○	□	○	○	○				
density (liquid and solid)	▽	▽							□		▽	▽	○				
particle size distribution	▽	▽	▽	○		○	⊡	□	□		▽	▽			▽	□	
hydrolysis (as a factor of pH)	⊡	⊡	⊡	▽	▽	▽	□	⊡	□	⊡	⊡	⊡	□	□	□	▽	□
dissociation curve	⊡	○	⊡		▽	▽		⊡	□	⊡	⊡	⊡	□	□	□	▽	□
thermal stability				○	○	○	○	⊡	○	⊡	○	□	□	○	□	□	○
viscosity			▽				□										
surface tension	▽	□	□		□					□	□	□	□	□	□	□	□
fat solubility		○	○						□	□	□	□	□	○	□	▽	□
permeability										□	▽	▽	○		▽	▽	○
corrosiveness					▽	▽	⊡	⊡	□	⊡							

▽ Physicochemical property for direct evaluation of a substance

□ Physicochemical property that must be known before another test can begin

○ Physicochemical property that is important additional information for further testing

Figure 5.1.6 Materials properties and environmental behaviour (after Bliefert, 1994)

5.1.7 Ecochemical materials properties

The properties of environmental chemicals mentioned in Section 5.1.6, which decisively influence their environmental behaviour, are shown in depth in Figure 5.1.7 using three examples. The transfer of a substance from its solid or liquid aggregate state into the atmosphere (gaseous state) is described by processes of evaporation, volatilisation and vaporisation. The transition into the gas phase represents in general an important ecochemical reaction pathway that is enhanced by high vapour pressures. Even if pesticides have relatively low vapour pressures as a rule, still substantial amounts can be converted to the gas phase. Thus, hexachlorobenzene at a vapour pressure of 1.45×10^{-3} Pa registers a saturation concentration of 0.17 $\mu g\ L^{-1}$ air at 20 °C (Kümmel and Papp, 1990). Terms used in mass transfer are material flow ($mol\ s^{-1}$), which is the amount of a component that enters or leaves the system (compartment) per unit of time, and material flow density ($mol\ s^{-1}\ m^{-2}$), which is defined as the material flow relative to the unit of the phase transition area. With evaporation processes, there is a correlation between the vapour pressure and the mass transfer rate. The left-hand graph in Figure 5.1.7 shows the behaviour of chlorinated hydrocarbons (at 25 °C) relative to the mass transfer rate. Decisive factors for the mass transfer are the degree of turbulence of the liquid phase and the diffusion resistance that are active in the phase boundaries (in detail in Kümmel and Papp, 1990). The mechanisms in the vaporisation of environmental chemicals from the soil are considerably more complicated. In this case, the adsorptive interactions or sequential diffusion steps in the soil solution and in the soil atmosphere must be taken into account. The distribution between liquid phases is characterised by the distribution coefficient (middle graph). The distribution between water and organic phases is one of the most important ecochemical distribution processes. The previously defined distribution coefficient for the system 1-octanol and water (P_{OW} value) serves as a standard for hydrophobicity, i.e. for the probability or the extent of a transition to an organic phase. For structurally similar substances (middle graph), the coefficient increases with decreasing water solubility. The correlation is almost linear in a double logarithmic graph. The P_{OW} value is also suited for the characterisation and explanation of bioaccumulation (Section 5.1.6), since biological membranes are permeable to not only water but also hydrophobic particles. A higher bioaccumulation factor corresponds to high P_{OW} values, as shown in the right-hand graph using trout as an example.

5.1.8 Fundamental processes of material transfer between atmosphere and the ocean

The most important transport processes in the ocean with respect to the distribution of xenobiotics include advection or diffusion from deeper layers to the surface, formation of gas bubbles in places of supersaturation, and the daily travelling rhythm of microorganisms from the depths to the surface. Environmental chemicals can accumulate briefly in the surface film (0.001–1000 μm) before they enter the atmosphere. How long depends on wave activity (from breaking open the boundary layer) and on the materials properties. Solid particles and gases transfer into the atmosphere through the spray and splashing because of the bursting of bubbles. Volatilisation (Section 5.1.7) plays the decisive role with hydrophobic substances.

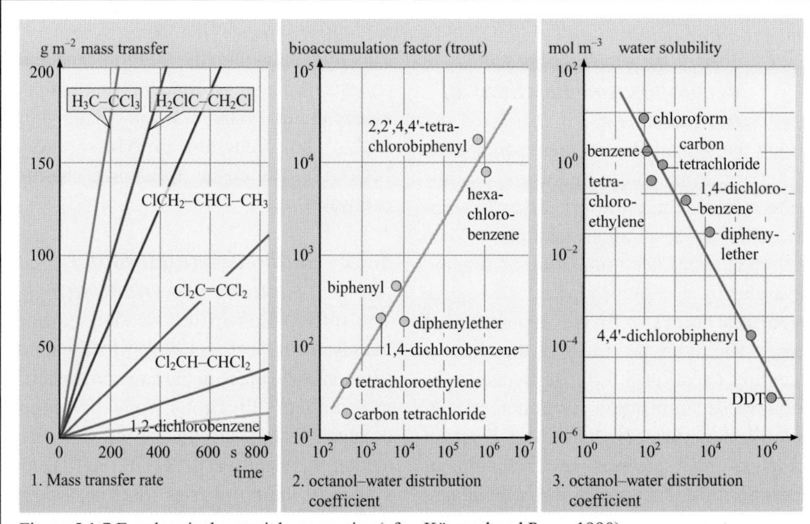

Figure 5.1.7 Ecochemical materials properties (after Kümmel and Papp, 1990)

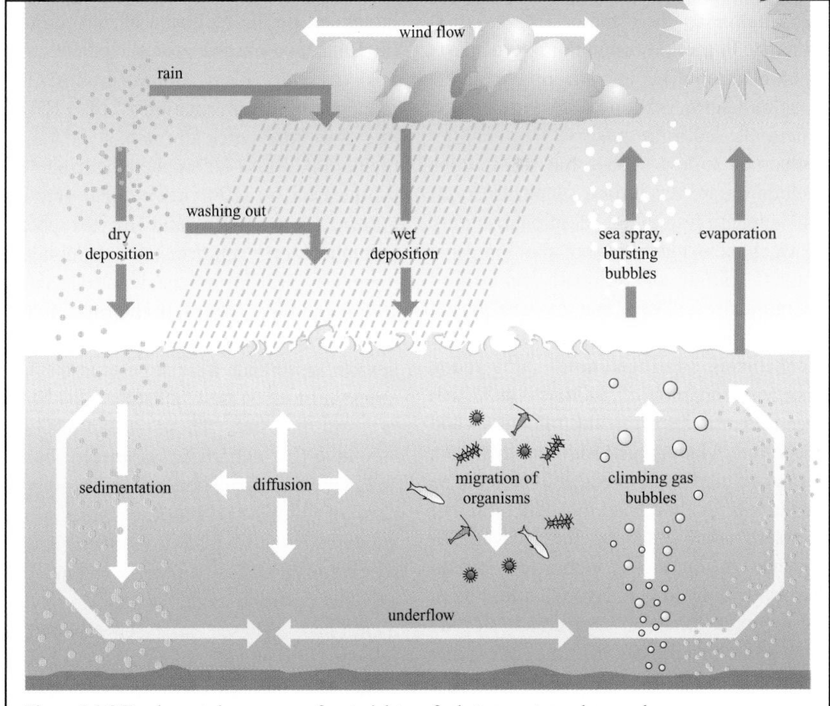

Figure 5.1.8 Fundamental processes of material transfer between atmosphere and ocean

5.2 General decomposition pathways

5.2.1 *Reaction enthalpies for decomposition reactions of biomass*

During the oxidation of organic compounds by heterotrophic microorganisms, assimilation and dissimilation processes (energy production via respiration) take place. Thus the biochemical decomposition of organic substrates by non-photosynthetically active organisms occurs with the generation of energy. The redox potentials of the electron acceptors (Section 3.3.8) are decisive for oxidative decomposition. The energetically controlled stepwise reduction of electron acceptors characterises all biochemical decomposition reactions in the hydrosphere and the pedosphere. Figure 5.2.1 shows the steps not with respect to the redox potential, but relative to their molar free reaction enthalpy. In aerobic respiration, oxygen acts as an electron acceptor; anaerobic respirations and fermentations with lower redox potentials and more positive free reaction enthalpies follow. Anaerobic respirations include nitrate respiration (denitrification) and sulphate respiration (desulphurisation), in which the oxidation potential of inorganic anions – nitrate and sulphate – is utilised energetically under slight oxygen partial pressure. From the standpoint of reaction mechanisms, fermentations are redox disproportionation of biomass in which organic substrates act as electron donors and acceptors. Methanogenesis and hydrogen formation are linked to bacterial systems, to methanogenic bacteria: the generation of methane takes place by the cleavage of acetates or by the reduction of carbon dioxide with hydrogen. Hydrogen formation in turn takes place in a precursor step by acetogenic bacteria, in an endergonic reaction with positive free reaction enthalpy, which is only made possible thermodynamically by linking with the exergonic methanogenesis. A negative free reaction enthalpy means the exergonic reaction occurs voluntarily. The reduction of MnO_2 and of $FeO(OH)$ can ensue following chemical pathways.

5.2.2 *Reductive (anaerobic) reactions of xenobiotics*

The information in this section is drawn mainly from Korte, (1992). Mostly abiotic reactions take place in the soil and especially in sediments. Electrons from the reduced organic substrate are transferred to environmental chemicals via the Fe(II)–Fe(III) redox system or via porphyrins (derived from the decomposition of biological material, such as haemoglobin) that are bound to proteins. Examples are the reduction of nitro groups (in pentachloronitrobenzene and parathion in this case) to form amino groups; the reductive dechlorination of DDT (dichlorodiphenyltrichloroethane = 1,1,1-trichloro-2,2-bis(4-chlorophenyl)ethane; synthesised in 1874, insecticidal activity known since 1939) and of toxaphene (camphechlor: chlorinated camphene, roughly $C_{10}H_{10}Cl_8$, insecticide introduced in 1948); and the complete dechlorination and aromatisation of lindane (γ-hexachlorocyclohexane; insecticide against pests in the soil and against forest pests inhabiting the bark) to form benzene. These reductions are viewed in part as transitions between abiotic and biotic transformations, since the proportions of abiotic and biotic causes are not known exactly. Oxidation reactions in the soil are catalysed by enzymes and also by metal oxides such as the Fe and Mn oxides mentioned. These catalysed oxidation reactions also include abiotic mineralisation, which takes place in sands, for example.

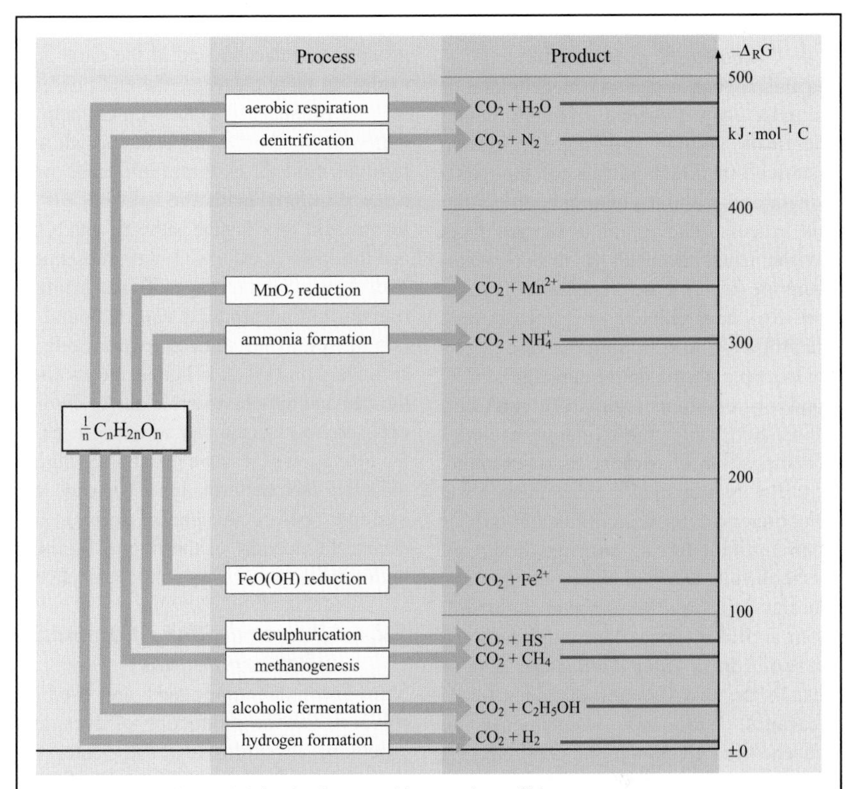

Figure 5.2.1 Reaction enthalpies for decomposition reactions of biomass

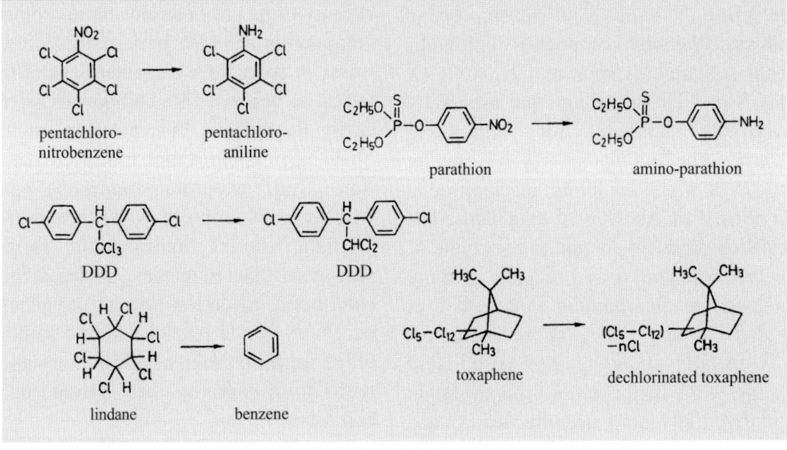

pentachloro-nitrobenzene → pentachloro-aniline

parathion → amino-parathion

DDD → DDD

lindane → benzene

toxaphene → dechlorinated toxaphene

Figure 5.2.2 Reductive (anaerobic) reactions of xenobiotics

5.2.3 *Bacterial decomposition of aromatic compounds*

Organisms such as bacteria, higher plants or animals can change environmental chemicals without utilising the carbon contained in the chemicals for their own growth and without yielding energy from this process. The products formed from environmental chemicals via this co-metabolism are in turn environmental chemicals, and also hazardous to the environment according to the definition (Section 5.1.6). For example, the oxidative cleavage of C–C bonds is ecotoxicologically significant, shown here using the oxidative bacterial decomposition of benzene as an example. With the assistance of the coenzyme NAD (nicotinamide adenine dinucleotide) of hydrogen-transferring enzymes in energy metabolism, a two-fold hydroxylation takes place at first. An *ortho*-diphenol is formed. Then a cleavage can occur either between the two hydroxy groups (*ortho* cleavage), or next to the two functional groups (*meta* cleavage). Side-chain substituents on benzene with 'electron pushing' or 'electron pulling' activity appear to be responsible for whether *ortho*- or *meta*-cleaving enzymes determine the subsequent decomposition pathway. Phenolic compounds with additional substituents such as CH_3, CH_3–O, CH_3–S or CH_3–SO groups are especially favourable substrates for *meta*-acting enzymes. The decomposition yields smaller molecules, which are further metabolised in the normal metabolism of organisms. The oxidation product is bound to coenzyme A. Acetyl-coenzyme A and succinic acid are produced as dicarboxylic acid with four C–atoms. Coenzyme A (abbreviated CoA or CoA-SH) takes up carboxylic acid groups (e.g., in the citric acid cycle, in oxidative fatty acid decomposition) and transfers them to other substrates (the acetyl groups to oxalacetic acid in the citric acid cycle). In *meta* cleavage, the end products that are produced – again with the help of the coenzyme NAD – are acetaldehyde and pyruvic acid (2-oxopropionic acid with three C–atoms) with the splitting off of formic acid, or using the pathway with NAD and the splitting off of CO_2 with subsequent hydrolysis and cleavage. These primary transformation products can be bound as conjugates to substances in the body in secondary processes. In human metabolism, the damage to bone marrow due to chronic exposure to benzene can be traced back to the formation of reactive metabolites during oxidative biotransformation of epoxide, for example (before the phenol forms), and during the opening of the ring to form the highly reactive *trans*,*trans*-muconaldehyde.

5.2.4 *Decomposition of aromatic nitro compounds*

Nitro-aromatic compounds are used as solvents, for the production of dyes, and especially in explosives. They enter the environment from emissions, whereby the waste water pathway plays the greatest role because of its low volatility. Nitro-aromatic compounds are also part of the leftover waste in soils from armaments manufacturing. In general, nitro compounds are not easily degradable, but microbial metabolism is possible under modified conditions. Under anaerobic conditions, again with the interaction of NAD (see above), aromatic amines (anilines) are formed step–wise via reductases. Under aerobic conditions, oxidation (via an oxygenase) occurs to form a phenol with the splitting off of the nitro group as nitrite. Substances with toxic activity are produced in both cases.

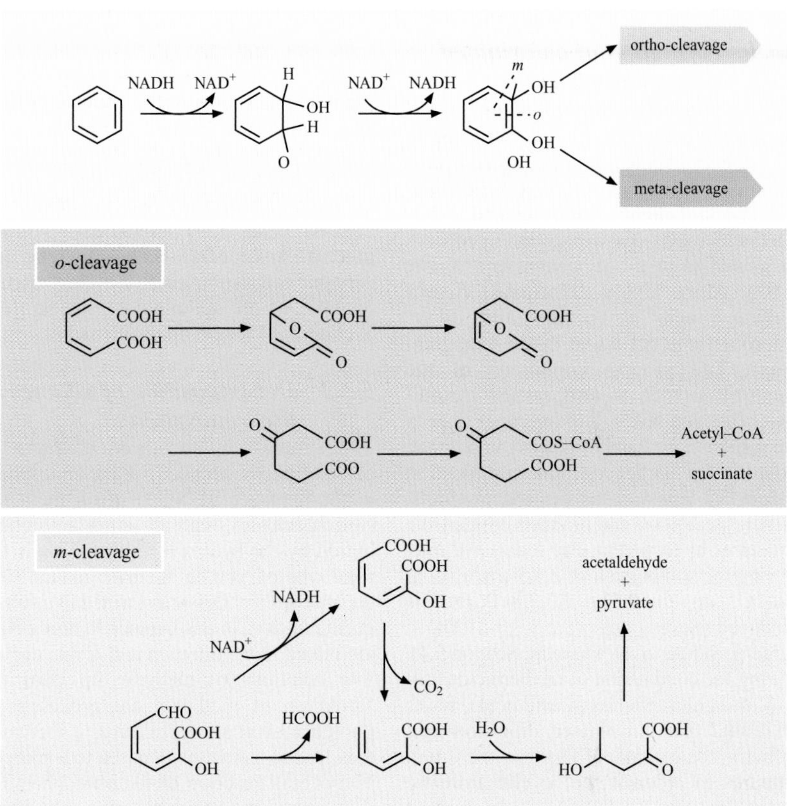

ortho-cleavage

meta-cleavage

o-cleavage

Acetyl–CoA + succinate

m-cleavage

acetaldehyde + pyruvate

Figure 5.2.3 Bacterial decomposition of aromatic compounds (after Korte, 1992)

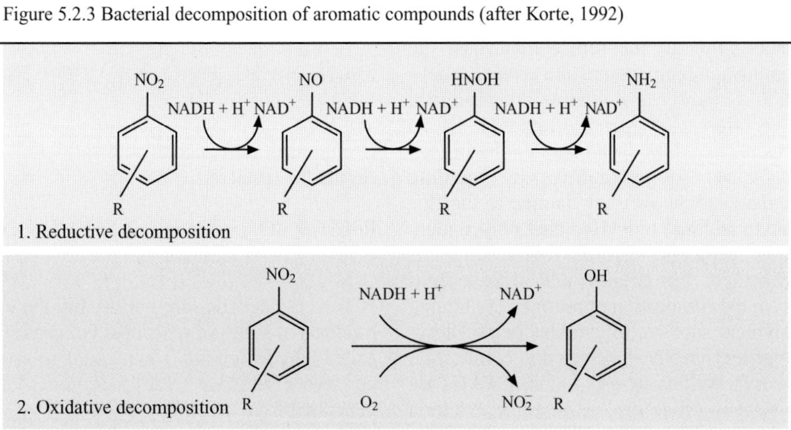

1. Reductive decomposition

2. Oxidative decomposition

Figure 5.2.4 Decomposition of aromatic nitro compounds

5.3 Hydrocarbons: PAHs and PCBs

5.3.1 *Formation of chlorinated compounds during combustion processes*

During the combustion of polyethylene in the presence of NaCl, the chlorinated hydrocarbons chlorobenzene, 1,3-dichlorobenzene, 1,2-dichlorobenzene and 1,2,3,5-tetrachlorobenzene were found in concentrations of >1 $\mu g\ g^{-1}$ of polyethylene (Korte, 1992). More highly chlorinated hydrocarbons such as penta- and hexachlorobenzene are found in the nanogram range. The process conditions of the combustion such as temperature maintenance, oxygen levels, and length of time in the combustion chamber determine both the extent of the partial reactions combined in Figure 5.3.1 as well as the end products. With respect to the end products emitted, the structures in particular play a decisive role. During the combustion of 2,4,5-triphenol at 600 °C, up to 0.5% 2,3,7,8-TCDD is produced (Seveso poison: 2,3,7,8-TCDD = tetrachlorodibenzo (1,4)dioxin; Section 5.4). During the combustion of the herbicide 2,4-D (2,4-dichlorophenoxyacetic acid), fewer toxic and less chlorinated dibenzodioxin isomers are produced. If there are small amounts of chloride, polycyclic aromatic hydrocarbons (PAHs) are also formed during combustion. In addition to the composition of the refuse materials, the combustion temperatures are critical. Above 1000 °C hardly any dioxins are produced even from 2,4,5-trichlorophenol. It is possible to reduce the amount of dioxin even at lower temperatures by adding triethanolamine or triethylamine to the flue gas. In addition to PAHs, other polycondensated aromatic hydrocarbons such as azaarenes, nitroarenes and cyanoarenes (from the combustion of nitrogen-containing polymers such as polyamides or polyacrylonitriles) have been detected in combustion gases as well as fly-ash and slag. PAHs are present in crude oil and in algae, bacteria and even higher plants. They are generally produced during incomplete combustion processes and during the pyrolysis of organic materials such as wood, coal, gasoline and oil, as well as during grilling and frying (see Section 5.3.3 and 5.3.4 regarding decomposition). Below 1000 °C, mostly tri- to tetra-nuclear PAHs occur; above this temperature (e.g. in combustion motors), penta- to hepta-nuclear PAHs are present. Anthracene, benzo[a]pyrene and perylene can also be nitrated in the presence of NO_x as in automobile exhaust (see Section 5.2.4 for decomposition).

5.3.2 *Decomposition of alkanes in the troposphere*

The simplest hydrocarbon, methane, is oxidised photochemically in the atmosphere in the presence of NO to form methanal (formaldehyde), with an important contribution by the hydroxyl radical. This is the ideal type of 'classic' methane oxidation in the atmosphere. Ozone is formed in a subsequent reaction. In this manner, hydrocarbons are linked to the nitrogen and ozone cycles. The reactions of methane up until the formation of methanal are processes in photochemical smog; relative to methane, they lead to a sink (as opposed to a source). The overall reaction of the photochemical decomposition of aliphatic hydrocarbons up to the stage of the aldehydes looks like this:

$$R\text{–}CH_2\text{–}H + 2O_2 + {}^2NO \rightarrow R\text{–}CHO + 2NO_2 + H_2O$$

If one takes into account the formation of ozone, the result is:

$$R\text{–}CH_3 + 4O_2\ (+\ NO/h\nu) \rightarrow R\text{–}CHO + 2O_3 + H_2O$$

However, aldehydes are not usually the end products of series of reactions. For example, acetaldehyde (ethanal) can react to form peroxyacetylnitrate (PAN); ketones are generated from secondary alkanes.

For further discussion see Section 2.2.

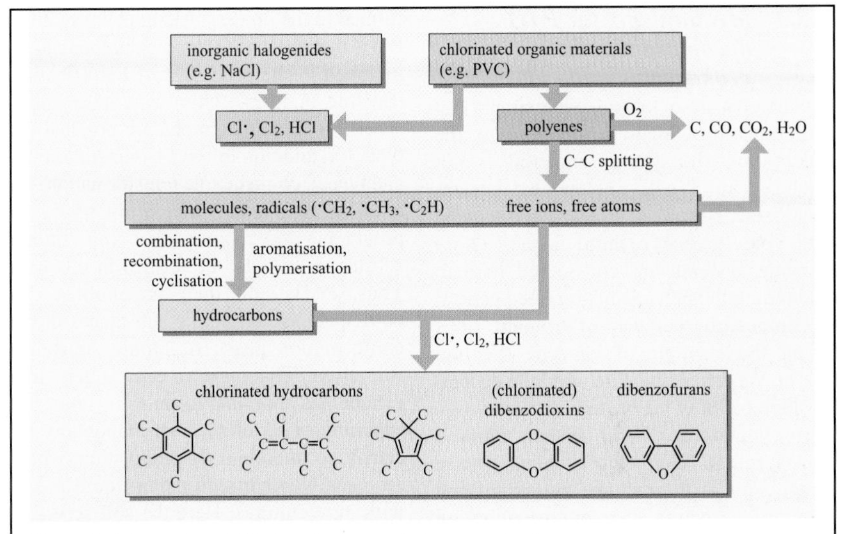

Figure 5.3.1 Formation of chlorinated compounds during combustion processes

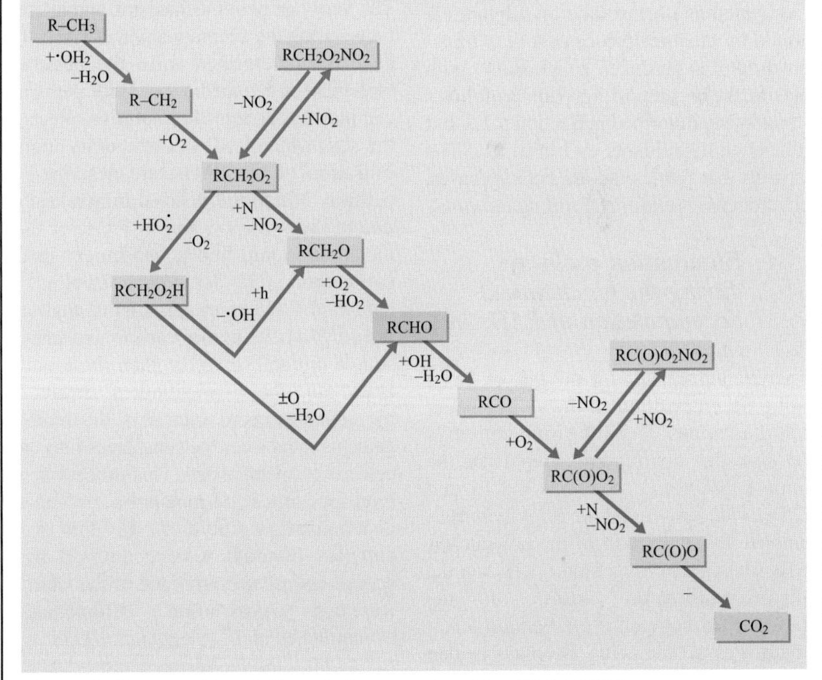

Figure 5.3.2 Decomposition of alkanes in the troposphere (after Bliefert, 1994)

5.3.3 Mechanisms for PAH decomposition in prokaryotes and eukaryotes

Prokaryotes are organisms with simple cell structures in which, in contrast to the eukaryotes, the genetic material is not enclosed in a cell nucleus by nuclear membrane. The prokaryotes include bacteria and 'blue-green' (cyano) algae. Other enzyme systems are considered here in addition to the previously described general decomposition pathway of aromatic rings using the example of benzene (Section 5.2.1). We differentiate between the decomposition pathway via bacteria and via fungi. The cytochrome P-450 system plays a crucial role in eukaryotes, as shown here in fungi. This is the enzyme system of cell respiration, which also oxidises foreign chemicals especially in the human body. By means of an epoxide, a substituted phenol (conjugated as glucuronide or sulphate) is formed by chemical processes, or a *trans*-dehydrodiol is produced enzymically. With bacteria, the decomposition pathway follows as previously described in Section 5.2.3, but with muconic acids as end products. The enzymes that participate are deoxygenases, dehydrogenases and ring-forming enzymes.

5.3.4 Elimination pathways during the biochemical decomposition of PAHs in the soil

Experimental models for the fate of PAHs during decomposition in the soil consider both the various decomposition pathways and also the surrounding area (like the humic substances – Sections 4.2.15 and 4.2.16). Pathway 1 in Figure 5.3.4 includes a complete mineralisation of the polynuclear PAHs (detailed in Stegmann, 1993) via the 2-hydroxycarboxylic acids of the $(n-1)$nuclear compound (in Section 5.3.3: 2-hydroxymuconic acid). Biomass is also formed in the process. Along pathway 2 the PAHs themselves or their metabolites can be included in the processes of humification (Section 4.2.12). This way, even non-specific radical oxidations are possible (as type 3 in addition to type 1, mineralisation, and type 2, co-metabolic transformation). In addition to phenolic oxidation products, polymers and conjugates with humic substances can also be formed in this way. Thus, PAHs and their metabolites can participate as non-humic substances in the conformation phase of humic substance formation. This is why only a few free metabolites are found in soils, in contrast to experiments performed in the laboratory for clarifying microbial decomposition, which are usually performed using liquid media with pure cultures. Here the soil serves to fix pollutants and also provides a 'detoxifying' function for the ecosystem of the soil. The very complex mechanisms and interactions during the decomposition of PAHs that have been described form the basis for biological soil remediation. Both pathways can play a role here. The sorptive effects in the soil, which lead to a temporary immobilisation of the PAHs, are described as pathway 3, and the delayed mineralisation due to the incorporation of PAHs or their metabolites into humic substances in the soil-carbon (C) depot are labelled as pathway 4. If microorganisms that can utilise PAHs as a sole carbon and energy source mineralise PAHs, then these pollutants or their metabolites must be bioavailable in pore water. First the microorganisms must steer the substances into their cells to be metabolised. This process is not required for a co-metabolic or radical decomposition. Studies on this very complex material were conducted by a special research commission of the German Research Association, 'Remediating contaminated soils' (Stegmann, 1993).

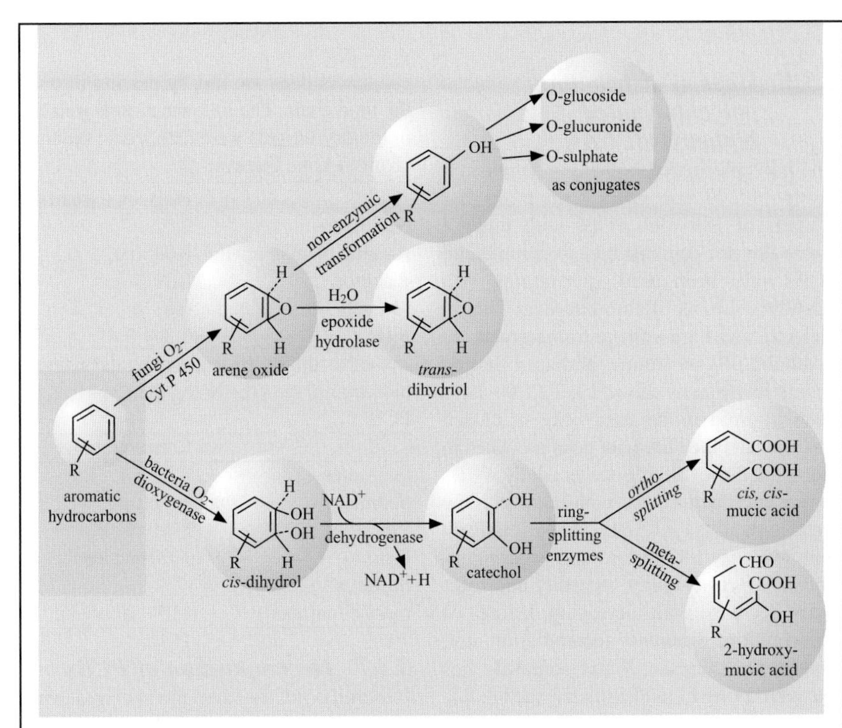

Figure 5.3.3 Mechanisms for PAH decomposition in prokaryotes and eukaryotes

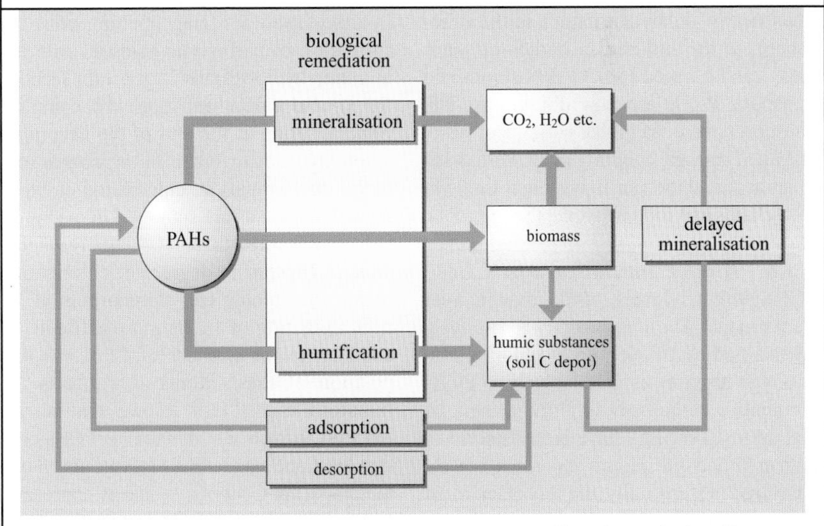

Figure 5.3.4 Elimination pathways during the biochemical decomposition of PAHs in the soil

5.3.5 *Ways of distributing polychlorinated biphenyls (PCBs)*

The polychlorinated biphenyls represent a group of 209 congeners with molecular weights of 189–499. PCBs with four Cl atoms are not combustible, so commercial PCBs have been used as mixtures with 30–60% Cl as flame-retardant liquid dielectrics in high-voltage transformers, as hydraulic oils in mining, and as softening agents in plastics. Since 1972 PCBs have been allowed to be used only in closed systems, and they have not been produced in Germany since 1983. Ecochemically, PCBs have a high persistence with relatively low water solubility, and thus have a relatively high tendency toward bio- and geoaccumulation. Water solubility, volatility and reactivity decrease with increasing degrees of chlorination, whereas accumulation and persistence increase. Water solubility lies between 0.1 $\mu g\ L^{-1}$ and 6 $mg\ L^{-1}$; the log P_{OW} increases from 4.56 for monochlorobiphenyls to 9.6 for decachlorobiphenyls. Their nearly global distribution in the atmosphere, in the soil and in bodies of water today can be traced back to depositions and transport; PCBs are now ubiquitous. The transport occurs via global currents of water and air. The high lipophily of PCBs has led to an accumulation in living organisms via food chains and food networks (Section 5.1).

5.3.6 *Bioaccumulation of PCBs*

PCBs with Cl levels >60% induce liver cancer in rats. The acute toxicity is decisively determined by the degree of chlorination, although toxicity is relatively low. PCBs accumulate in the body fat of organisms, so that up to 10 $mg\ kg^{-1}$ have been detected in human fatty tissue. Highly chlorinated PCBs are stored preferentially, the less chlorinated being are preferentially excreted via breast milk. The relative levels of low-chlorinated congeners decrease just by passing through the food chain. The following levels (relative to the dry weight) were detected in the food chain of Lake Geneva:

	PCB level (ppm)
Sediment	0.02
Aquatic plants	0.04–0.07
Plankton	0.39
Mussels	0.6
Fish	3.2–4.0

In the North Sea, marine mammals present a bioaccumulation factor log *BCF* = 7 to 8 for PCBs:

	Concentration *c*
Sea water	2 $ng\ L^{-1}$
Sediment	5–16 $\mu g\ L^{-1}$
Plankton	8–10 $mg\ kg^{-1}$
Fish	1–37 $mg\ kg^{-1}$
Seabirds/ marine mammals	110–160 $mg\ kg^{-1}$

5.3.7 *Decomposition of PCBs*

Generally PCBs are metabolised very slowly by microorganisms and higher organisms. The most important steps are hydroxylation and ring opening with the formation of carboxylic acids as with the aromatic hydrocarbons in general (Sections 5.2.3), and a dechlorination with complete mineralisation at the end of the decomposition chain. Altogether, in the course of a biotic dechlorination, chlorinated environmental chemicals are subjected to oxidative, reductive, hydrolytic and conjugative mechanisms. Increasing degrees of chlorination make the biological transformation or decomposition of PCBs more difficult. In addition to the number of Cl atoms, the position of these atoms also affects the transformations. It is known that warm-blooded animals can metabolise PCBs only if at least one ring is free of Cl atoms on two neighbouring C atoms. *p*- and *m*-conformations of unsubstituted C atoms make transformation easier.

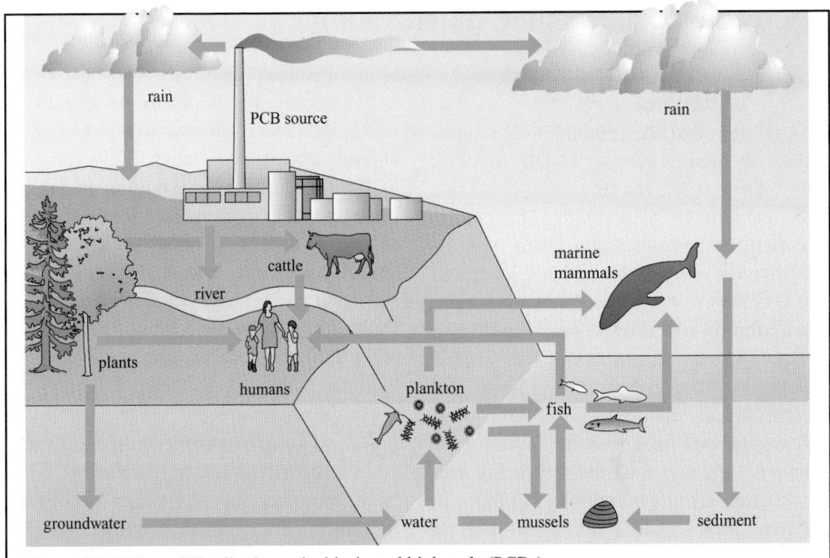

Figure 5.3.5 Ways of distributing polychlorinated biphenyls (PCBs)

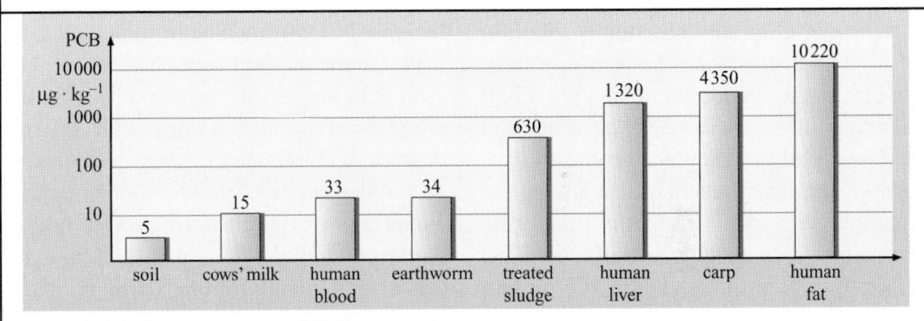

Figure 5.3.6 Bioaccumulation of PCBs

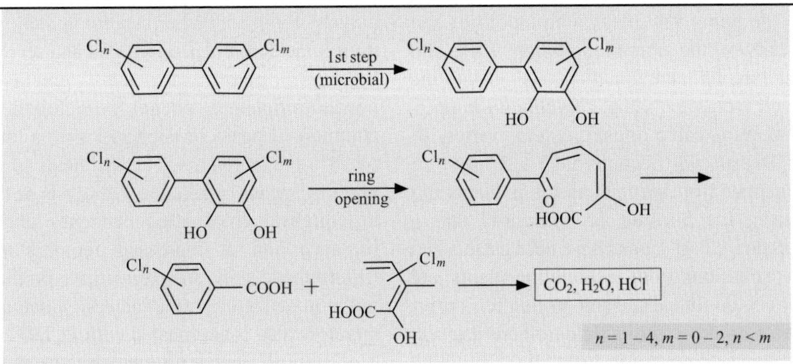

Cl_n — biphenyl — Cl_m → 1st step (microbial) → Cl_n — biphenyl — Cl_m (HO, OH)

Cl_n — biphenyl — Cl_m (HO, OH) → ring opening → Cl_n — ... O, HOOC, OH →

Cl_n — COOH + Cl_m, HOOC, OH → CO_2, H_2O, HCl

$n = 1 - 4$, $m = 0 - 2$, $n < m$

Figure 5.3.7 Decomposition of PCBs

5.4 Dibenzodioxins and dibenzofurans

5.4.1 Sources and pathways of dioxins

W. Sandermann described 2,3,7,8-tetrachlorodibenzo(1,4)dioxin, TCDD, in 1957. The public first became aware of this substance in 1976, after the spectacular accident in Seveso, Italy, when 1–5 kg TCDD were released during the synthesis of the herbicide 2,4,5-trichlorophenoxyacetic acid after overheating. Poisoning led to chloracne, to disturbances in nerve functions, aching in muscles and joints, and psychological disorders. Polychlorinated dibenzodioxins and dibenzofurans (see Sections 5.4.2 and 5.4.3 for structures) were never manufactured industrially, but they can be generated in different ways. During thermal processes at >200 °C, they are formed during incomplete combustion in the presence of chlorine compounds (Section 2.5.4). These substances have always been present in the environment in very small trace amounts due to fire clearing and lightning strikes. They are produced as a result of secondary reactions, e.g. during the manufacture of aromatic chlorine compounds such as the now outlawed wood preservative pentachlorophenol PCP or that of PCB. Sources are branches of industry such as paper factories (during the bleaching of the paper raw mass with chlorine) and metalworking operations where PVC and chlorine-containing cutting oils get into the smelt (scrap recycling, recycling of copper-containing cable refuse). A wide variety of TCDD species (Sections 5.4.3 and 5.4.4) are generated from waste incineration plants and during the burning of household refuse (Section 2.5.4). Limits have been established for emissions from incineration plants and for treated sludge and soil on children's playgrounds and on sports fields, where slag was applied as copper slag material (*'Kieselrot'*) from the smelting of copper ore. Even distillation residues from dry cleaners (with chlorobenzene for example) are sources of dioxin. In general, all industrial processes of chlorine chemistry (Section 2.2.9) can lead to the formation of PCDD and PCDF. In waste landfills it is possible for these environmental toxins to occur as a result of smouldering fires. The formation of these substances has even been demonstrated in Beilstein's test (the detection of halogens in organic compounds with the help of copper by flame coloration; Section 5.4.4).

5.4.2 Contaminants in commercial chlorophenol products

Both commercial 2,4,6-trichlorophenol (intermediate for organic syntheses, termite extermination agent) and pentachlorophenol can contain polychlorinated dibenzo-*p*-dioxins (PCDD) and dibenzofurans (PCDF). With the dioxins and furans, we are dealing with chlorine derivatives of cyclic aromatic ethers from two phenyl rings (with differing degrees of chlorination), which are linked to each other via two *ortho*-positioned oxygen atoms or one oxygen atom and a C–C bridge. Corresponding polybrominated compounds are also generated from flame-retardant substances containing bromine. The particular PCDFs are easily produced from polychlorodiphenylether. Approximately 10 ppm octachlorodibenzo-*p*-dioxin and several hundred ppm of the precursor nonachlorophenoxyphenol were found in commercial pentachlorophenol with a total of 13% contaminants. The synthesis takes place by means of alkaline hydrolysis of the next-highest chlorinated benzenes under pressure and at increased temperature. Chlorophenols are an intermediate product in the manufacture of herbicides, which are therefore also contaminated with PCDD (an ecochemically better alternative is catalytic phenol chlorination on an Fe contact).

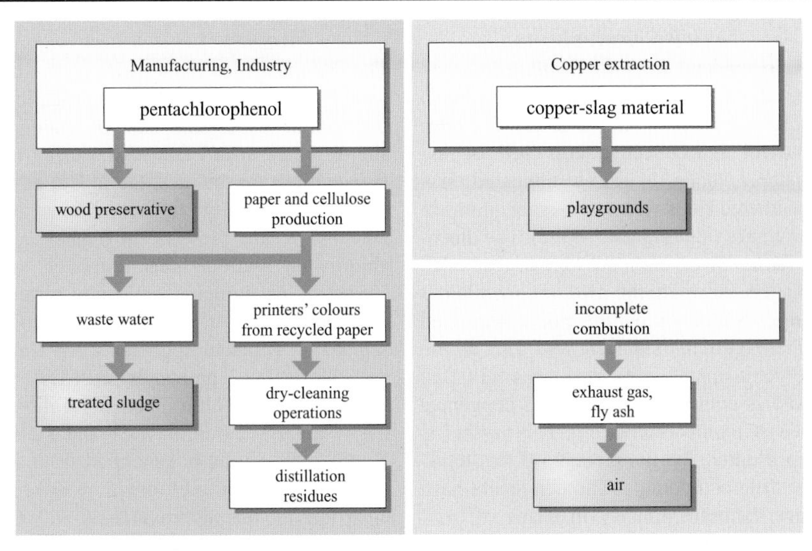

Figure 5.4.1 Sources and pathways of dioxins

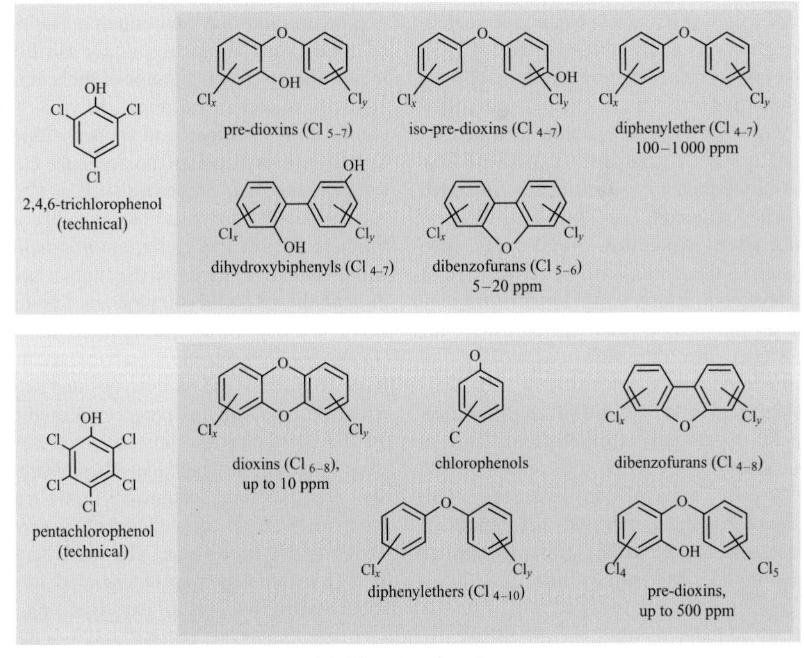

Figure 5.4.2 Contaminants in commercial chlorophenol products

5.4.3 *Structural formulas and toxicity equivalents*

The compound 2,3,7,8-tetrachloro-dibenzo (1,4)dioxin (Figure 5.4.3, 1) is also known as Seveso poison (Section 5.4.1). The term 'dioxins' is a collective term used for ring systems made of three annealed six-membered rings with two oxygen atoms in the centre ring (Figure 5.4.3, 2). The dibenzofurans have a carbon bridge instead of an oxygen between the two six-membered rings. Some 75 of the polychlorinated dibenzodioxins (PCDD) and 135 of the polychlorinated dibenzofurans (PCDF) exist as compounds labelled as congeners (see PCB), of which the largest number is represented by the group of the tetrachloride compounds. The congeners have very different activity in terms of toxicology. The compound 2,3,7,8-TCDD is extremely toxic to animals such as guinea pigs and mink. That is why this compound is used as a reference substance for calculating toxicity equivalents (Figure 5.4.3, 3). The factors cited must be multiplied by the concentrations of each individual substance in order to be able to compare its toxicity with that of 2,3,7,8-TCDD. Toxicity equivalence factors are contained in the appendix to the 17th Federal Immission Protection Regulation, also known as the BImSchV. The summation of the concentrations weighted in this way yields a TCDD equivalent (TE) for mixtures. In addition, according to Eisenbrand and Metzler (1994), 'The toxicological profile of TCDD in animal experiments is highly multi-faceted and is characterised by stark differences in the sensitivity of various species. ...The reasons for these extreme differences in sensitivity among different animal species are not understood at this time.'

5.4.4 *Formation and intake by humans*

The previously described pathways for the formation of dioxins and furans (Sections 5.4.1 and 5.4.2) are shown again in simplified form in Figure 5.4.4 using the Beilstein test (in the presence of Cu) as an example. Typically the conditions are 200–400 °C, the presence of oxygen, and chlorinated hydrocarbons (aliphatic and aromatic) or inorganic chlorine-splitting compounds. Two starting substances were selected – pentachlorophenol (PCP) and polychlorinated biphenyls (PCBs) – whereby a tetrachlorodibenzo-1,4-dioxin might be formed from the PCP and a tetrachlorodibenzofuran is generated from the PCB. The mechanism of formation makes it clear that the hot (approximately 300 °C), carbon-containing fly ash from incomplete combustion processes represents a continual danger for the formation of TCDD and its distribution into the environment. The risk of the dioxin combination of fly ash from incineration plants as a factor of the temperature can be seen in the graph, shown for the combustion of refuse and treated sludge. Due to the distribution of dioxins in the environment and their properties such as slight water solubility and good bioaccumulation, they are part of the global environmental contaminants. They enter the human body through the air (via respiration and the skin) and more importantly through foodstuffs (Figure 5.4.4, 3). Via the digestive tract, they reach liver, fatty and skin tissues and accumulate there. Humans present the highest TCDD concentrations in fatty tissues, in contrast to rats, in which they accumulate in the liver. Therefore, conclusions drawn from the effects on rats and their behaviour in humans have little value. The half-life for TCDD in humans was determined to be about 8 years, but only 17 days in the rat.

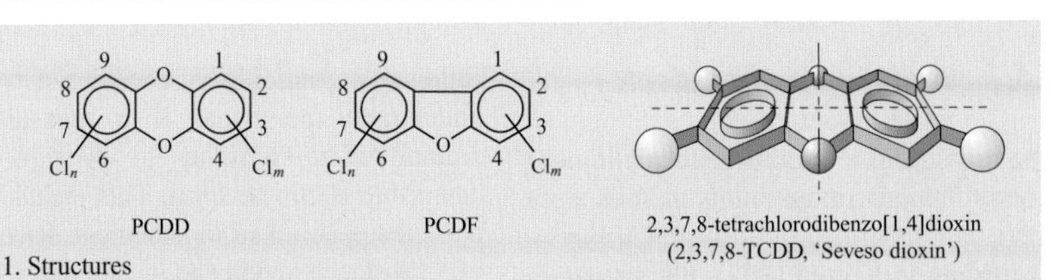

1. Structures

Number of chlorine atoms	Number of CDD isomers	Number of CDF isomers
1	2	4
2	10	16
3	14	28
4	22	38
5	14	28
6	10	16
7	2	4
8	1	1
sum	75	135

2. Isomers

Compound	TEF
2,3,7,8-*T*CDD	1
1,2,3,7,8-*Pe*CDD	0.5
1,2,3,4,7,8-*Hx*CDD	0.1
1,2,3,4,6,7,8-*Hp*XDD	0.01
*O*CDD	0.001
2,3,7,8-*T*CDF	0.1
2,3,4,7,8-*Pe*CDF	0.5
1,2,3,4,7,8-*Hx*CDF	0.1
1,2,3,4,6,7,8-*Hp*CDF	0.01
*O*CDF	0.001

T Tetra, *Pe* Penta, *Hx* Hexa, *Hp* Hepta, *O* Octa

3. Toxicity equivalence factors (TEF)

Figure 5.4.3 Structural formulas and toxicity equivalents

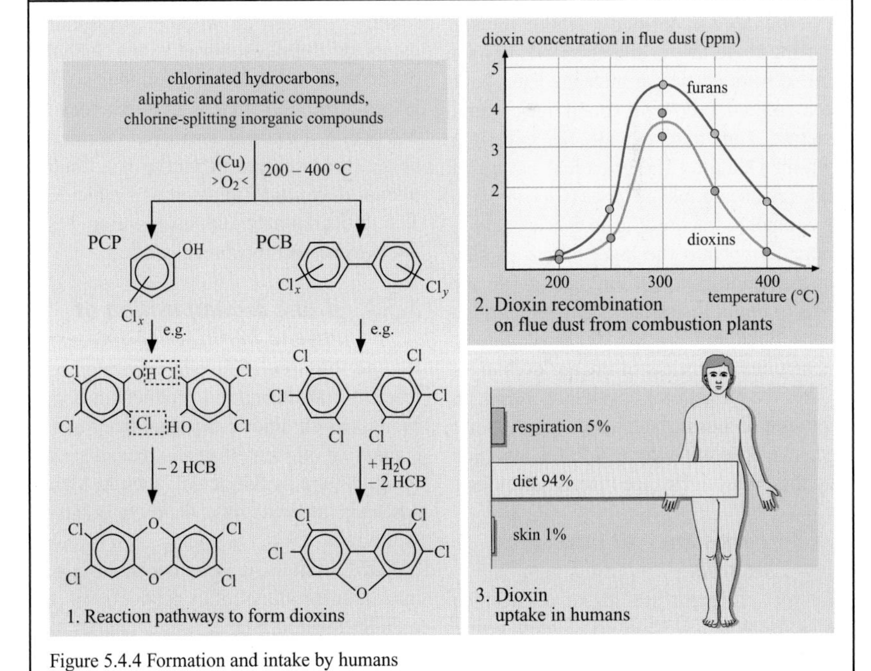

Figure 5.4.4 Formation and intake by humans

5.5 Pesticides and tensides

5.5.1 *Behaviour of pesticides in the soil*

The biotic and abiotic transformations of environmental contaminants in soils were previously described in detail in Chapter 4. According to Korte (1992), the long-term behaviour and the long-term distribution and transformation mechanisms in the soil are not sufficiently understood qualitatively or quantitatively for any organic foreign substance. This statement is more or less valid for pesticides as well, for which processes and reactions will be drawn out once again, whereby kinetics must be taken into consideration in order to be able to perform reliable calculations. To evaluate the behaviour of pesticides in soil, the formation of non-extractable or bound residues is especially important for long-term statements. Non-extractable residues of pesticides are defined as those species (starting materials or metabolites) that are formed from pesticides after the latter have been applied following sound agricultural practice and that cannot be extracted without changing the chemical nature of these residues. Linkage types (Chapter 4) to be considered according to Korte include intercalation into the layer lattice of clay minerals; non-covalent intercalation in the voids of humic substances; hydrogen bridge bonds; bonding using van der Waals forces; interactions by means of charge exchanges (redox reactions) and covalent bonds with monomeric humic substance precursors; and, in the context of humification, insertion into the humic substance macromolecules.

5.5.2 *Persistence of pesticides in soils*

The very different periods of persistence range from a few days to several years. IUPAC defined the term 'persistence' as follows: a pesticide is considered to be undesirably persistent if a measurable amount of it continues to exist in any detectable chemical form. This means that persistence is linked to analytical detection of a substance and not to its activity, which would be more meaningful ecologically. Thus, pollutants such as PCB or DDT have not only a high degree of stability, but also a high tendency to accumulate, whereby activity thresholds (Chapter 6) can be exceeded. In addition to DDT (dichlorodiphenyltrichloroethane), other chlorinated hydrocarbons such as γ-hexachlorohexane and dieldrin (as insecticides) are highly persistent. Derivatives of phenoxyacetic acid (used as herbicides) such as Picloram, MCPA (methylchlorophenoxyacetic acid) or 2,4-D (2,4-dichlorophenoxyacetic acid) decompose much more readily. The group organic phosphorus esters or thiophosphorus esters includes diazinon, disulfoton and parathion (used as insecticides). Simazine is an N-heterocyclic compound, a triazine derivative; diuron is a urea derivative; and trifluralin is a dinitrophenol derivative (all used as herbicides). The thiocarbamate Barban (Section 5.5.3) decomposes the most readily of all.

5.5.3 *Biotic decomposition of ethene bisthiocarbamates*

Ethene thiourea (ETU) is generated as a primary transformation product (in both a biotic and an abiotic manner) from metal-containing ethylene bisthiocarbamates used in agriculture as fungicides, such as Maneb with Mn against grey scale on potatoes. Although ETU is toxic for warm-blooded animals, it is transformed in part into natural metabolic products in the regular metabolism of cows (Parlar and Angerhöfer, 1991).

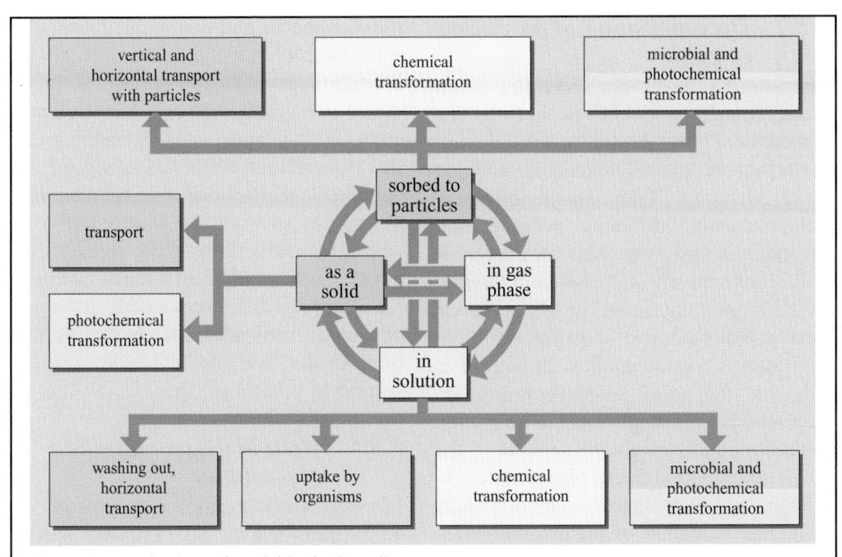

Figure 5.5.1 Behaviour of pesticides in the soil

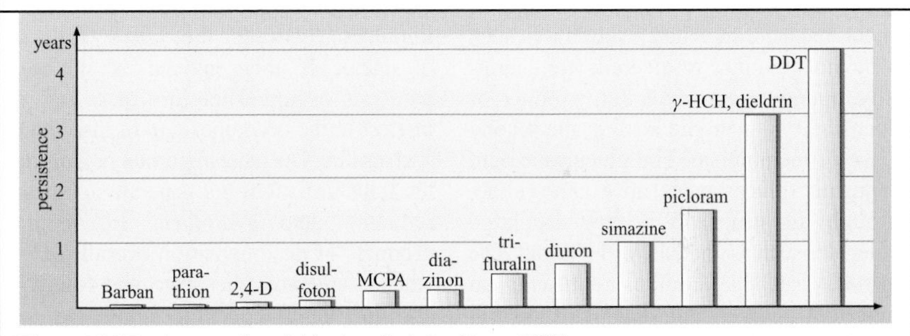

Figure 5.5.2 Persistence of pesticides in soils (after Korte, 1992)

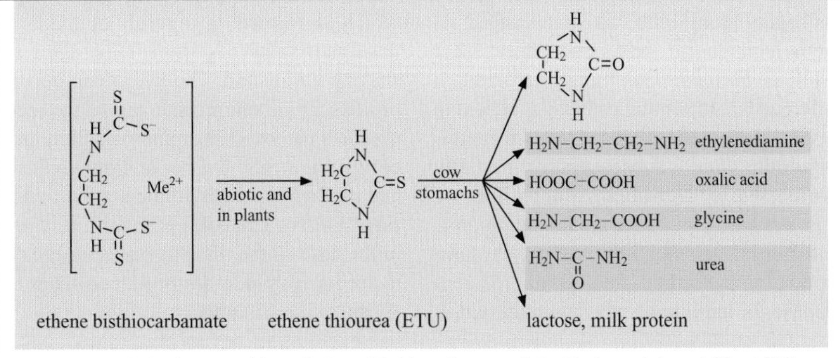

Figure 5.5.3 Biotic decomposition of ethene bisthiocarbamates (after Parlar and Angerhöfer, 1991)

5.5.4 Decomposition of parathion in cattle

The information in this section is drawn mainly from the work of Korte, in Parlar and Angerhöfer (1991). The mean lethal dose of the organophosphorus insecticide parathion is 25–50 mg kg^{-1} body weight for cattle. Fundamentally, different primary and secondary transformations can occur not only consecutively but also concurrently within a given organism. The oxidation (via microsomal oxidases) of parathion in the liver leads to a splitting off of the sulphur as sulphate; the actual effective metabolite paraoxon is generated which, when further transformed (hydrolysed), forms the cleavage products (due to phosphatases) *p*-nitrophenol and diethylphosphoric acid (after the oxidation of diethyldithiophosphoric acid). They are excreted as glucuronide either directly or bound to glucuronic acid. In general, substances such as phenols, amines or alcohols are chemically bound to the oxidation product of glucose, i.e. glucuronic acid, in the second phase of metabolism. The glucuronic acid conjugates (glucuronides) are excreted either through the urine or, if they are large molecules such as phenols, with the gall. The formation of glucuronides represents an important pathway for detoxification via excretion. Transferases can replace the ethyl groups of the diethylphosphoric acid by one hydrogen atom each, so that instead of diethylphosphoric acid, phosphoric acid itself is formed. On the other hand, a different decomposition pathway is argued in Pansen: parathion is reduced to form amino-parathion and subsequently partially hydrolysed to form diethylthiophosphoric acid. Another part of the amino-parathion is transported to the liver, where amino-paraoxon is formed due to microsomal oxidases. Cleavage of this substance yields *p*-aminophenol and diethylthiophosphoric acid. Aminophenol is transformed (likewise a detoxification reaction in stage II metabolism) into glucuronide or a sulphonic acid ester and is then excreted. Something else that is significant about the activity of this pesticide is the fact that different mammals hydrolyse parathion to greatly differing degrees (rabbits do so more rapidly than mice by a factor of about 40) and that there are significant differences among the organs of a single mammalian species (transformations in the liver are higher than in the kidneys by a factor of 10).

5.5.5 Microbial decomposition of parathion

Phosphoric acid esters and thiophosphoric acid esters inhibit the enzyme acetylcholinesterase, which serves an important function in the transmission of nerve impulses (between the nerve cell and muscles). If this enzyme is inhibited, paralysis occurs. Therefore there is great interest in the decomposition of insecticides such as this. The most important processes in the transformations are generally oxidation, reduction and hydrolysis. Intermediate products of decomposition overall include ethanol, inorganic phosphate, and 4-nitro- or 4-aminophenol (Section 5.5.4). As was already mentioned, the paraoxon (phosphoric acid diethyl-4-nitrophenyl ester), which is formed as a result of oxidative desulphurisation, is more toxic than the starting compound. The hydrolytic decomposition of parathion itself and of the redox products paraoxon or amino-parathion leads to two fragments: 4-nitro- or 4-aminophenol and diethyl- or diethylthiophosphoric acid ester. Finally, ethanol, phosphoric acid and, in the case of the thiophosphoric acid ester, hydrogen sulphide are produced from the phosphoric acid esters.

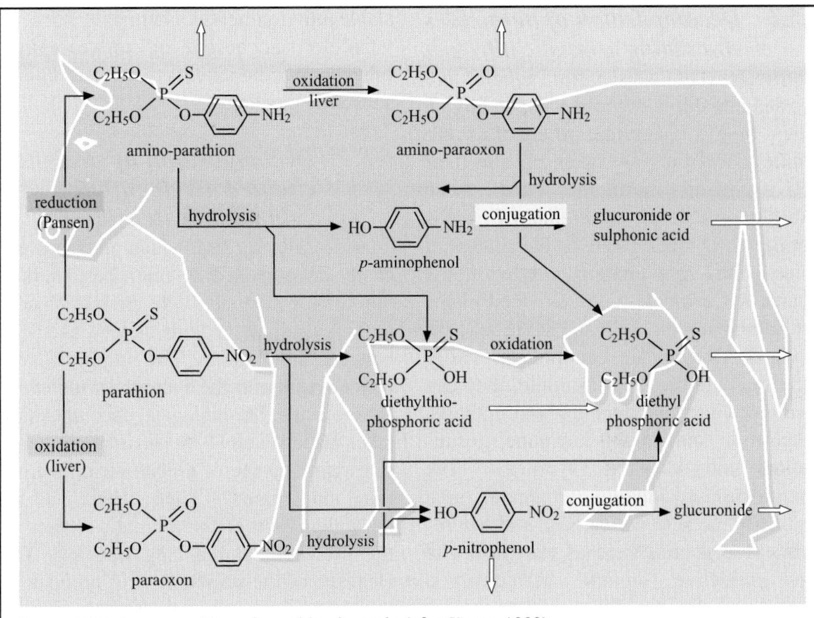

Figure 5.5.4 Decomposition of parathion in cattle (after Korte, 1992)

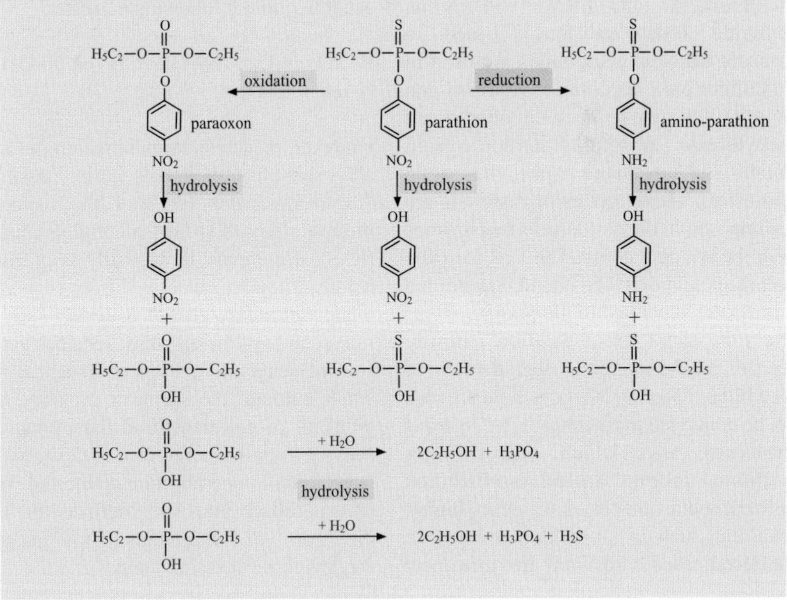

paraoxon ← oxidation — parathion — reduction → amino-parathion

paraoxon —hydrolysis→ p-nitrophenol + $(C_2H_5O)_2P(=O)OH$

parathion —hydrolysis→ p-nitrophenol + $(C_2H_5O)_2P(=S)OH$

amino-parathion —hydrolysis→ p-aminophenol + $(C_2H_5O)_2P(=S)OH$

hydrolysis:

$$H_5C_2{-}O{-}P(=O)(OH){-}O{-}C_2H_5 \xrightarrow{+H_2O} 2C_2H_5OH + H_3PO_4$$

$$H_5C_2{-}O{-}P(=O)(OH){-}O{-}C_2H_5 \xrightarrow{+H_2O} 2C_2H_5OH + H_3PO_4 + H_2S$$

Figure 5.5.5 Microbial decomposition of parathion

5.5.6 *Decomposition of herbicides containing aromatic nitrogen*

Herbicides, or weed-killer agents, are divided into different groups, depending on when they are applied, where they are applied, how they are taken up and their activity. Soil herbicides are those substances that are taken up by weeds through their roots. Weeds in the context of weed control are those (dicotyledonous) plants that compete with cultivated plants for water, nutrients, light and living space as a whole. Herbicides can present vastly differing activities. For example, they can be divided into inhibitors of photosynthesis, respiration and growth hormone, germination and carotene synthesis. The chemicals used include carbamates, urea and carbonic acid derivatives, and heterocyclic compounds such as triazines. The first selective organic herbicide is considered to be 2-methyl-4,6-dinitrophenol, which was used as early as 1892 as an insecticide and after 1934 as an herbicide. Today, substances based on aromatic nitrogen compounds are used in agriculture on a large scale (Kümmel and Papp, 1990); these include substituted phenylureas, substituted carbonic acid anilides and substituted *N*-phenylcarbamic acid esters. These groups of materials are decomposed in the soil largely by enzymes from the amidase group. The first step is a hydrolysis of the CO–NH bond. Substituted anilines are generated in the primary stage. The subsequent decomposition pathway depends on the microbial degradability of the anilines (Section 5.2.4). Simple anilines can be converted to catechol (1,2-dihydroxybenzene), after which ring splitting reactions follow. Chlorine-substituted anilines, on the other hand, which are highly persistent, such as 3,4-dichloroaniline, are not decomposed oxidatively but form new stable environmental chemicals such as tetrachloroazobenzene or can react with other soil components to form humic substance precursors.

5.5.7 *Decomposition of straight-chained alkylbenzene sulphonates (tensides)*

Tensides (detergents), which are relatively easily decomposed by microbes, include non-ionic detergents (with straight chains) and alkylsulphonic acids as well as alkylbenzene sulphonates. The sulphuric acid residue represents the hydrophilic moiety of the molecule. The biological decomposition takes place via an ω- or β-oxidation with a stepwise cleavage of acetate residues from the side-chains. Then there is a desulphonation of the aromatic moiety of the molecule and a ring opening. The cleavage of the sulphonic acid group with the aid of the enzyme monooxygenase in the presence of oxygen takes place according to the following equation:

$$R\text{–}SO_3^- + O_2 + NADH + H^+ \rightarrow R\text{–}OH + NAD^+ + HSO_4^-$$

However, reductive desulphonation to form alkyl-substituted benzenes is also possible. Alkylsulphonates (without the benzene group), also called fatty alcohol sulphates (FAS), decompose biologically even more readily. They can be derived from vegetable oils of secondary growth materials such as rape seed or of the oil palm. Tensides in the soil influence soil parameters such as the water content, the sorption capacity, the mobility of pollutants and the biological activity overall. Therefore it is desirable to have a rapid mineralisation connected with incorporation into the biomass or the direction of fragments (via acetyl coenzyme A) to assimilation.

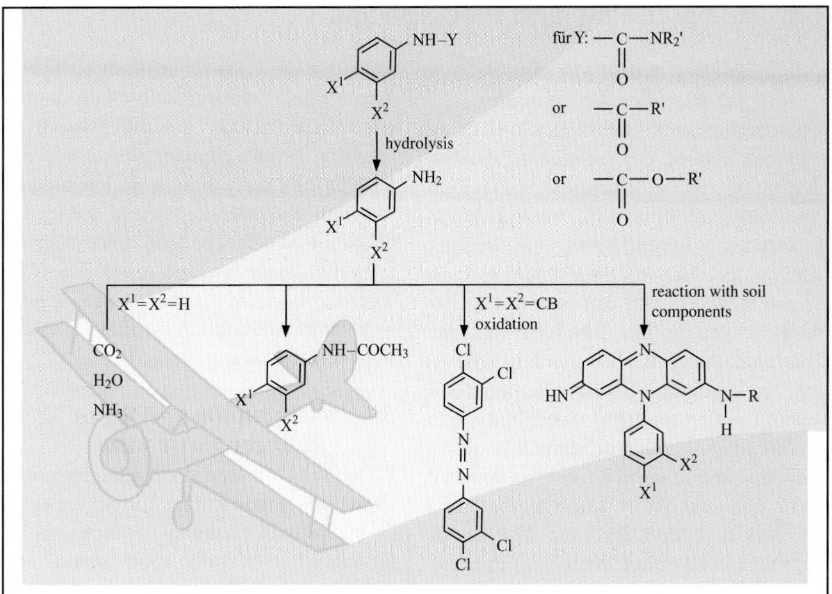

Figure 5.5.6 Decomposition of herbicides containing aromatic nitrogen

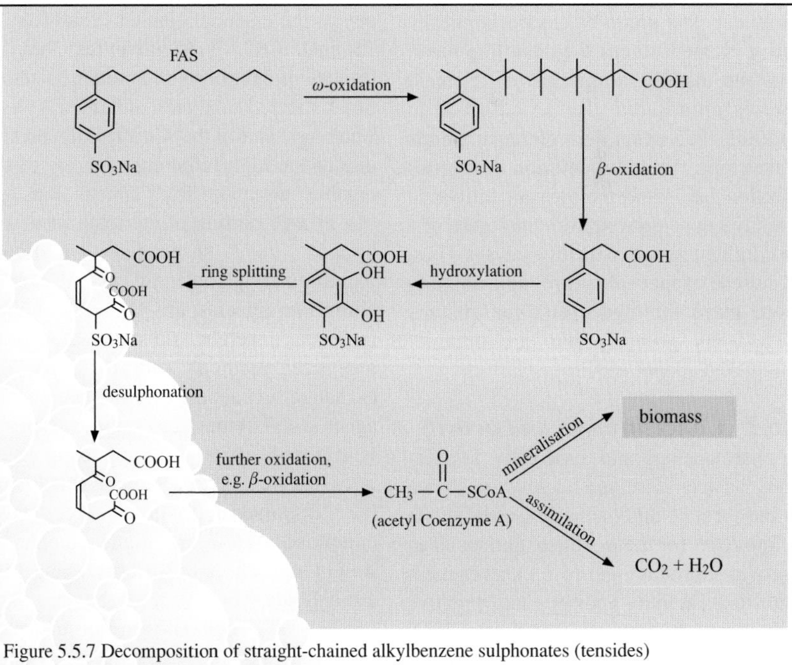

FAS

ω-oxidation

COOH

SO_3Na

β-oxidation

ring splitting

hydroxylation

OH

desulphonation

further oxidation, e.g. β-oxidation

$CH_3—C(=O)—SCoA$

(acetyl Coenzyme A)

mineralisation

assimilation

biomass

$CO_2 + H_2O$

Figure 5.5.7 Decomposition of straight-chained alkylbenzene sulphonates (tensides)

5.6 Heavy metals and their species

5.6.1 *Heavy metal species in natural bodies of water*

The physical and chemical states of an element, called the element species, determine its ecochemical behaviour. Solubility, mobilisation, sedimentation properties, bioavailability and toxicity depend decisively on the element species. In natural waters, one differentiates analytically between the dissolved and the filterable fractions. The element species that are in solution can be differentiated – again, from an analytical standpoint – into a fraction of electrochemically active element traces (which can be detected using polarography or voltammetry) and a chemically bound fraction. The stable organic and inorganic complexes can only be detected electrochemically after treatment, such as an oxidising UV treatment. The group of electrochemically active element traces includes free metal ions and instable complexes. All element species groups can also be adsorbed to colloids. For example, reference points concerning the sedimentation properties (Section 5.6.3) and the bioavailability and toxicity can be derived from the results of a discriminating heavy metal analysis. Thus, hydrated copper ions are considerably more toxic for algae (and for enzyme systems in general) than are organic-complexed copper species.

5.6.2 *Concentration and activity*

In pharmacology and toxicology, correlations between dose and activity are shown in the form of dose–activity graphs. Even at low doses (or in low concentrations in an organ as shown in Figure 5.6.2), an element with toxic activity (poison) has negative effects which impair relative growth. A non-essential element is tolerated by an organism in a low concentration range. Toxic effects appear only with higher concentrations. On the other hand, an essential (vital) element shows a curve with an optimum: both a deficiency and too high concentrations have a negative impact on growth. The term 'trace element' is used in connect with essential activities, whereas the term 'element trace' is used as a neutral description for small amounts or concentrations.

5.6.3 *Chromium species in tannery waste water*

Cr(III) salts are used in the manufacture of chromium-tanned leather. Chromium binds to the protein skeleton of animal skins (collagen) to form a solid, pliant, and water-, heat- and bacteria-resistant leather. At an oxidation stage of +3, chromium is one of the essential elements; however, as chromate CrO_4^{2-} (+6), it has toxic activity (mutations have been observed in bacteria). Sophisticated analyses of tannery waste water have shown that Cr(VI) is not present due to the high reducing capacity of the organic substances from animal skins and that a large portion of the total amount is bound to suspended particles. Approximately half of the dissolved chromium species are bound to higher molecular weight substances. Various analytical methods (ultrafiltration, ion exchange, extraction with the complexing agent acetylacetone, using ion exchangers with a wide variety of supporting structures) can be used to determine that 46% of the dissolved fractions are stable complexes, 16% are higher molecular weight substances (>30 000 D), and 15% are colloids. Thus, 95% of the chromium can be precipitated out with flocculants.

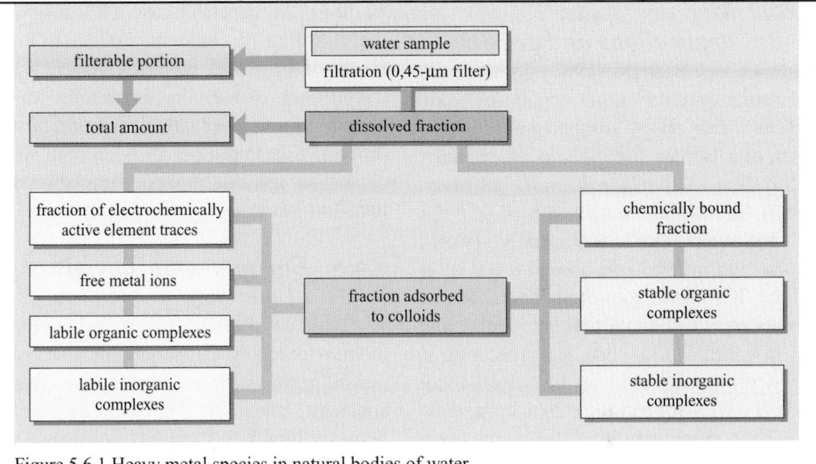

Figure 5.6.1 Heavy metal species in natural bodies of water

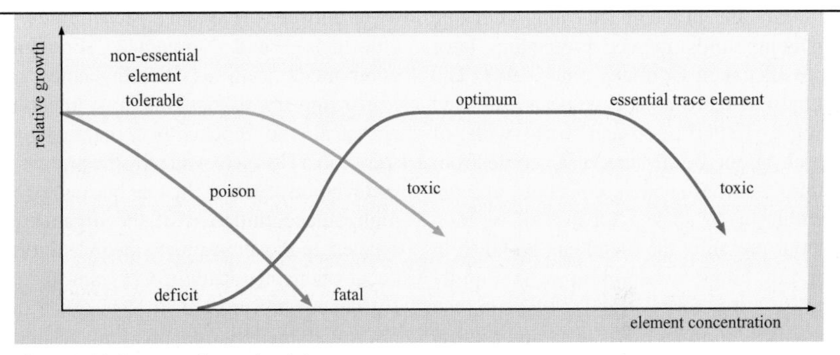

Figure 5.6.2 Concentration and activity

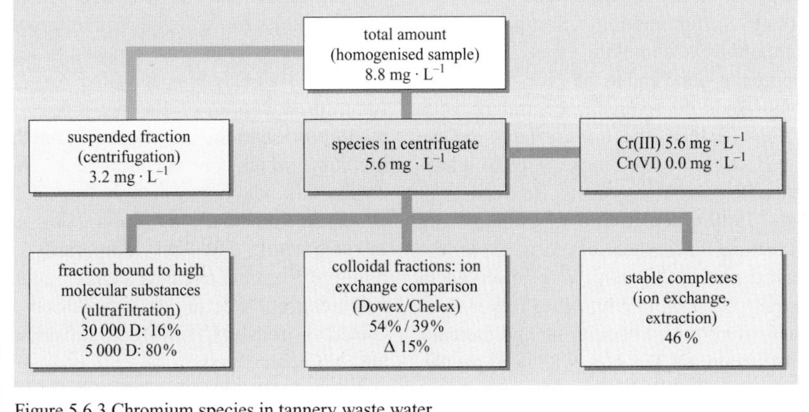

Figure 5.6.3 Chromium species in tannery waste water

5.6.4 Mercury 'spider': applications and activity

Kemper and Bertram (1987) depicted the exposure options, target organs of toxic activity, and depot compartments in the form of a 'spider' using Hg as an example. According to *'Lexikon Umwelt'* (Römpp, 1993), in 1985 in Germany a total of 635 t of Hg were sold industrially (world production in 1987 was around 6 000 t), of which 309 t were exported. The scope of emissions is listed as 4 t Hg into the air and 1 t in water. World–wide, Hg processing is divided between 30% for the chloroalkali industry, 21% for batteries, 20% in the electrotechnical industry, 15% for paints (only in the USA), and 14% for other goods such as thermometers and dental fillings. Organic Hg compounds as seed protectants have been illegal in Germany since 1988. The global Hg depositions have been estimated at up to 10 000 t per year world–wide, of which about 2 500 t per year come from natural sources such as volcanoes and the weathering of rocks. Because of its high vapour pressure, the metal can be taken in via the lungs, whereupon it is rapidly oxidised to form Hg(II) ions in the organism. Inorganic Hg salts and methyl mercury (Section 5.6.5) enter the body via the gastro-intestinal tract, where they can induce inflammation. Inorganic Hg compounds accumulate especially in the suprarenal gland and in the liver; organic Hg compounds do so in the central nervous system (brain). The toxic effects can be traced back to the blockage of sulphur-containing enzyme systems. Between 1953 and 1960, sensitivity disorders and increasing incidence of damage to the central nervous system were observed in fishing families in Minamata Bay, Japan. These were traced back to methyl mercury intoxication. Of 121 highly toxified people, 46 died of Minamata disease. The cause was the ingestion of fish with extremely high levels of dimethyl mercury, which were traced back to mercury-containing waste waters from a plastics manufacturing process. Concentrations of 6–25 ppm were present in fish due to bioaccumulation via plankton intake.

5.6.5 Environmental chemistry of mercury

To evaluate mercury from an environmental–toxicological standpoint, one must consider the various bond types (metal species): metallic Hg (differentiating between liquid and vapour); inorganic Hg salts with extremely differing water solubilities between HgS and the nitrates or chlorides; and organic, lipophilic compounds such as dimethyl mercury. Borrowing from Wood, the most important chemical and biochemical reactions are shown in a Hg cycle, whereby the process of methylation stands at the fore because of the high bioaccumulation of the organic Hg species. In the atmosphere, methylation can occur photochemically if CH_3 radicals and Hg vapour are present. It is assumed that microbial methylation takes place under both anaerobic (via methylcobolamine) and aerobic conditions (in cells that produce methionine). As a result of microbial processes, Hg(II) ions are reduced to form elemental Hg or are biomethylated. An additional methylation, or else the process of disproportionation, leads to dimethyl mercury, which can accumulate in food chains and which can also end up in the atmosphere due to the high vapour pressure. In sediments, Hg can temporarily be removed from the cycle as slightly soluble sulphide, but it can be remobilised by bacteria – even as $CH_3S–Hg–CH_3$, which is also bioaccumulated.

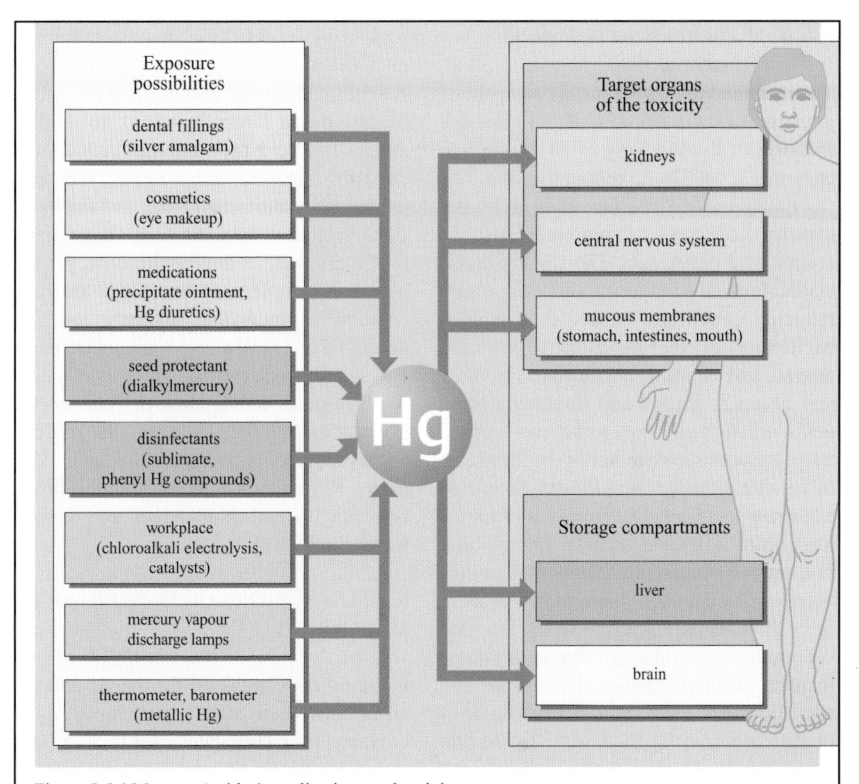

Figure 5.6.4 Mercury 'spider': applications and activity

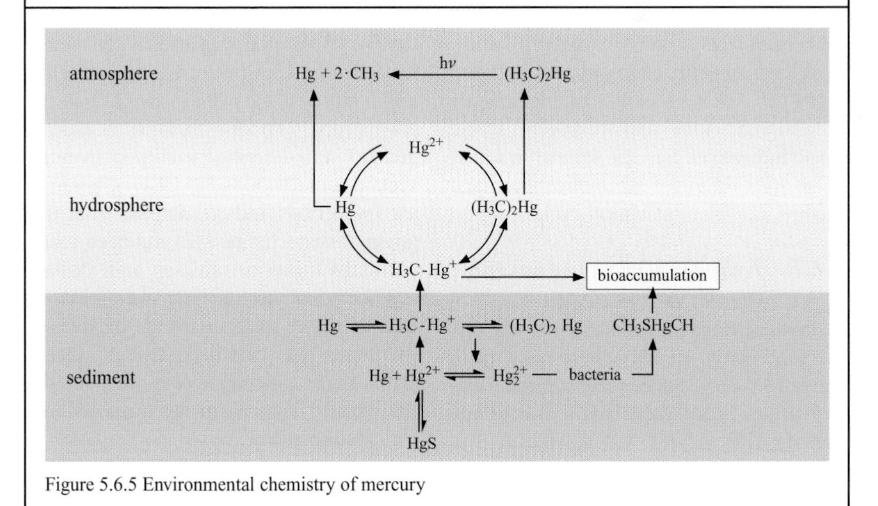

Figure 5.6.5 Environmental chemistry of mercury

5.6.6 *Ecochemical reactions of arsenic*

The metalloid arsenic occurs in compounds in oxidation states of –3 (AsH_3), +3 and +5. Because of the blocking of SH groups in enzymes, As(III) compounds, e.g. as arsenous acid or arsenite, have a higher toxicity than As(V) compounds such as arsenic acid or arsenate. The slightly dissociated form $HASO_2$(aq) is formed under reducing conditions. Arsenate is reduced bacterially in the pedosphere to form arsenite and can then be methylated by fungi and bacteria: methyl and dimethylarsenic acids are formed. Anaerobic and aerobic transformations make possible the formation of volatile dimethyl- and trimethylarsine as extremely toxic organic derivatives of arsine. Very little is known thus far about these mechanisms under environmental conditions, e.g. in the pedosphere or hydrosphere or in the boundary layers. The transfer into the atmosphere is also possible in this same manner, although due to oxidation, the less toxic dimethyl arsinic acid (cacodylic acid) or photochemically also trimethylarsenic oxide can be formed. They return to the arsenic cycle in water and in sediments. In digested sludge, arsenic can be immobilised initially as slightly soluble arsenic sulphide. With high sulphide concentrations and also high pH levels, readily soluble arsenic tetrathionate ions and relatively stable monothioarsenate ions are formed in water; these ions return to the hydrosphere and thereby into the ecochemical cycle.

5.6.7 *Transformations of arsenic species in the ocean*

With reference to the behaviour of arsenic in the food chain, we show here once again (Figure 5.6.7) a detailed picture of the transformations in sea water and sediment and then to fish. At pH 7.2, arsenic acid is present as $HAsO_4^{2-}$ ion; it is metabolised microbially via steps 1–6 to form arsenous acid (2), to methylarse(o)nic acid (3 – after oxidation and biomethylation), to methyl arsinic(III) acid (4 – another reduction), dimethyl arsonic acid (5 – repeated oxidation and biomethylation), and finally to dimethylarsinic acid (6). Overall, primary producers such as microorganisms, phytoplankton, zooplankton and algae are in a position to methylate arsenate ions as described and furthermore to convert them into slightly toxic compounds such as water-soluble arsenic carbohydrate derivatives – as glycosides (7–9) and as fat-soluble arsenophospholipids. Glycosides with side-chains (R) up to 6 C atoms and phosphate residues in this chain have been found. Compound 9 with

$$R = -CH_2-CH(OH)-CH_2O-P(O_2H)-OCH_2-CH(OH)-CH_2OH$$

is the parent compound for the fat-soluble arsenic species in algae and mussels. Algae are able to accumulate arsenic from sea water. Arsenic levels in brown algae up to 30 mg kg^{-1} dry weight have been detected. Organisms of higher trophic stages are not capable of transforming arsenate or methylated arsenic acids into arsenobetaines (13), which represent the main species in fish and crustaceans. The term 'betaine' is used in general to describe trialkylammonium acetate ions ($R_3N^+-CH_2-COO^-$); betaines are zwitterions and are derived from this group. Arsenocholine (12) has been found in crabs. Ballin, Kruse and Rüssel (1994) postulate the transformation of arsenic sugars to form arsenic betaines in sediments via dimethylarsenoyl ethanol (10). These arsenic species are rapidly excreted by humans in the urine without biotransformation.

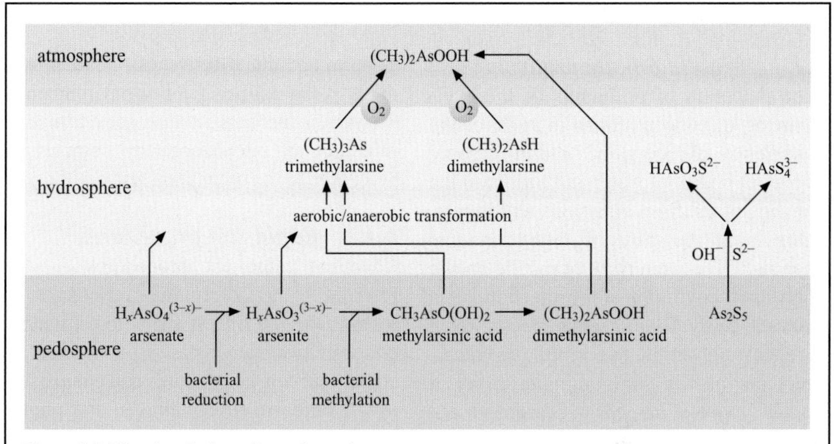

Figure 5.6.6 Ecochemical reactions of arsenic

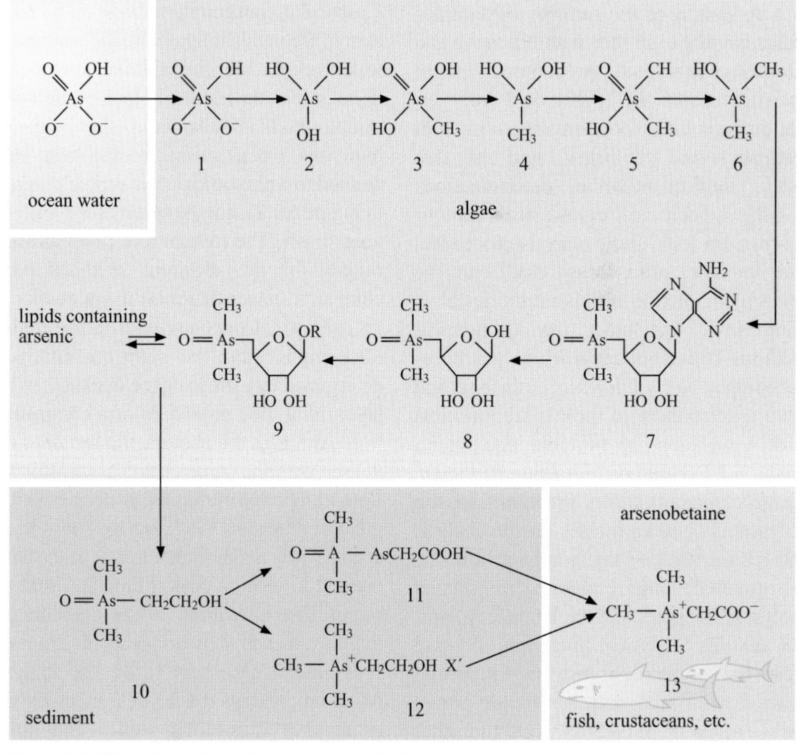

Figure 5.6.7 Transformations of arsenic species in the ocean

6 Problem- and Activity-Orientated Environmental Analysis

6.1 Environmental analysis – strategies and concepts

6.1.1 Mobile on-site analysis

'The detection of pollutants in the environment, of concentrations in soils, of the occurrence of inorganic and/or organic impact in bodies of water and further to the contents of air from emissions and immissions is at the core of mobile on-site analysis... The main tasks of mobile on-site analysis in environmental protection lie in two areas today: the utilization of tests which are quick and simple to perform, and measurements at the sampling site make it possible to make immediate decisions about protective actions which must be taken. The preliminary investigation of environmental samples which are strategically taken leads to a reduction of the number of samples which must be examined with expensive and laborious analytical procedures in an analytical laboratory' (Schwedt, 1996). On-site analysis includes orientation tests with test papers and test strips, rapid tests that make semi-quantitative determinations possible, such as colorimetric colour comparison tests, field analysis procedures and mobile procedures that can be performed outside a laboratory such as photometric and other physicochemical methods (from voltammetry to gas chromatography, X-ray fluorescence analysis and mass spectrometry in mobile laboratories) with hand tools or portable equipment. Figure 6.1.1 (based on a drawing by the Dr. Lange company) shows the characteristic differences between mobile on-site analysis and routine analysis using apparatus or the instrumental, largely automated special analyses. The costs and the labour involved, but also the analytical precision, increase from the orientation test up to special analysis with recognised reference procedures (based on ISO, DIN, VDI and other standards). On the other hand, the number of samples and the effectiveness of the procedures in the defined field of environmental protection increase as one goes from the reference or inter-laboratory comparison procedure to the orientation test.

6.1.2 Rapid-test procedures

Chemical rapid-test procedures – also referred to as alternative procedures – include the use of test strips to estimate a concentration range based on a selective colour reaction, colour cards and comparator tests as colorimetric methods, and photometric tests with field (pocket) photometers and ready-to-use reagents. In the colour card tests, one compares the colour designated for a particular concentration range on a colour card in the incident light with the colour of a water sample after the addition of reagents. In a comparator test the colour of the water sample itself is considered. The precision increases when going from a test strip method to a photometric test (upper diagram in Figure 6.1.2), but the speed of the analysis goes down. The measurable concentration ranges for the different methods were compared (lower diagram) using copper as an example. Concentrations in the region ≥ 0.1 mg L^{-1} can be semi-quantitatively determined by increasing the thickness of the layer (long tube procedure of a comparator test). Although the photometric test does not detect (with the same chemical transformations) low concentrations, it does have the greatest precision. Gas detector tubes in air analysis play a role similar to that of the latter method in water analysis; they are used for monitoring maximum workplace concentration values in the workplace, for example. It is always necessary to do the analysis based on prescribed analytical methods and procedures if limiting values and guide values established by law must be examined.

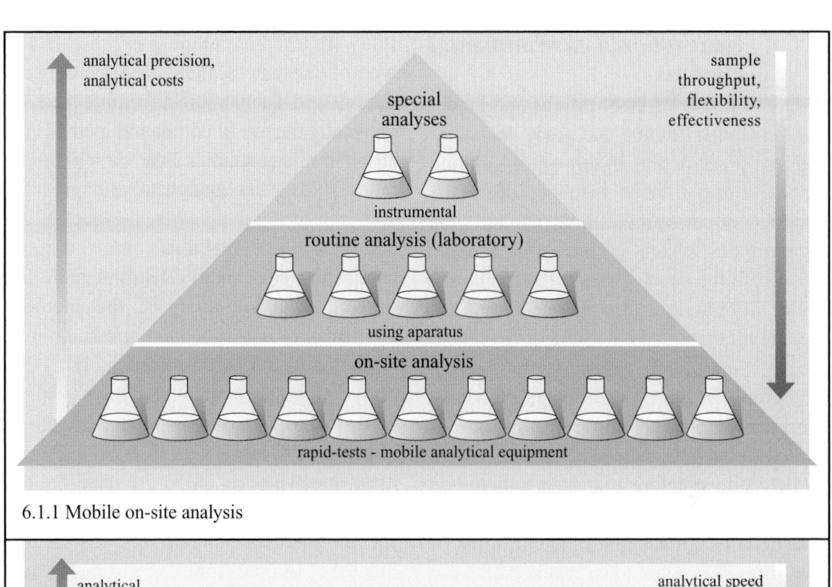

6.1.1 Mobile on-site analysis

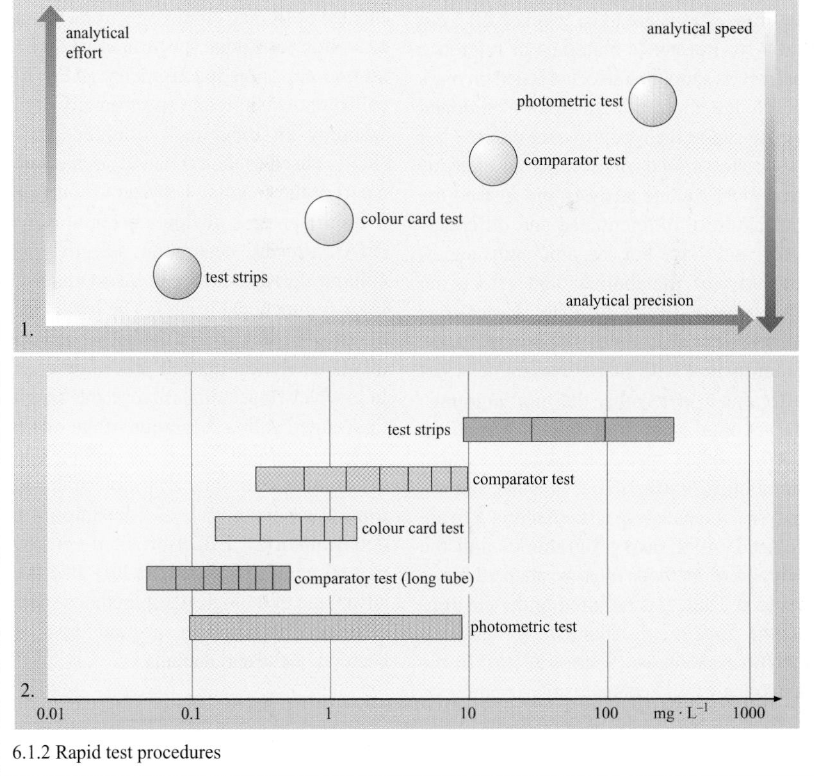

6.1.2 Rapid test procedures

6.1.3 *Conceptional environmental analysis*

A large number of laws, rulings and guidelines establish limiting and guide values for numerous substances in air, water, soil and refuse. Long lists of data are often the product of inappropriately planned test programmes, which in spite of or perhaps because of the flood of data in the end do not make it possible to make a realistic statement about danger to the environment (see Section 6.1.4 on estimating the danger). Therefore, the environmental analyst who is working in a problem-oriented mode is forced to limit the number of parameters to be determined before a study is to begin on the basis of a concrete formulation of the question. This means that the entire list of limiting values is not to be 'worked off', but that a few reference materials or reference parameters should be selected based on one's knowledge about a landfill, an abandoned storage site or the type of waste water.

A conceptional environmental analysis means less routine analysis and instead the application of differentiated and differentiating methods. For organic substances, processes of metabolism and effects on organisms (with the aid of biological test methods) are determined. To evaluate areas contaminated with heavy metals, it is not sufficient to determine the total amounts. The results from sequential extractions (Section 4.3.11) provide indications about the mobility. Furthermore, in water studies, analyses of element species (Section 5.6) are required. Analytical programmes and the selection of methods must be adapted to the problem. Thus, it is required of the environmental analytical chemist to develop analytical strategies (Section 6.1.6). In the first stage of an ecochemically meaningful study, the test programme is established. The type of substances, their expected concentrations and the matrix itself require methodologically the use of compound methods (as an optimal adaptation of the sample preparation steps to the analytical method) or of coupling techniques, where separation and determination methods are directly connected to one another instrumentally. For organic pollutant analysis, the methodological emphasis is on gas chromatography (GC) and liquid chromatography (LC) and on high-performance methods, high-performance capillary GC or HPLC. Fourier transform infrared spectrometry (FTIR) and mass spectrometry (MS) are used for the identification of substances. For element species analyses, suitable methods include linkage of LC and atomic spectrometry such as atomic absorption spectrometry (AAS) or atomic emission spectrometry (AES, also called optical emission spectrometry, OES), whereby an inductively coupled plasma (ICP) is used as the excitation source and as a part of the coupling between LC and AES. With the process of flow injection analysis (FIA), after LC separation, selective post-column derivatisation can be conducted to identify functional groups. The partial steps of an efficient and informative environmental analytical general procedure consist in a selective accumulation of the analytes (associated with a separation of the matrix), a chromatographic separation and the application of both atomic and molecular spectrometric methods of detection and determination. Effect-oriented environmental analysis (Section 6.1.7) also takes advantage of biological test methods such as enzyme inhibition tests and tests for bacteria, algae and daphnia.

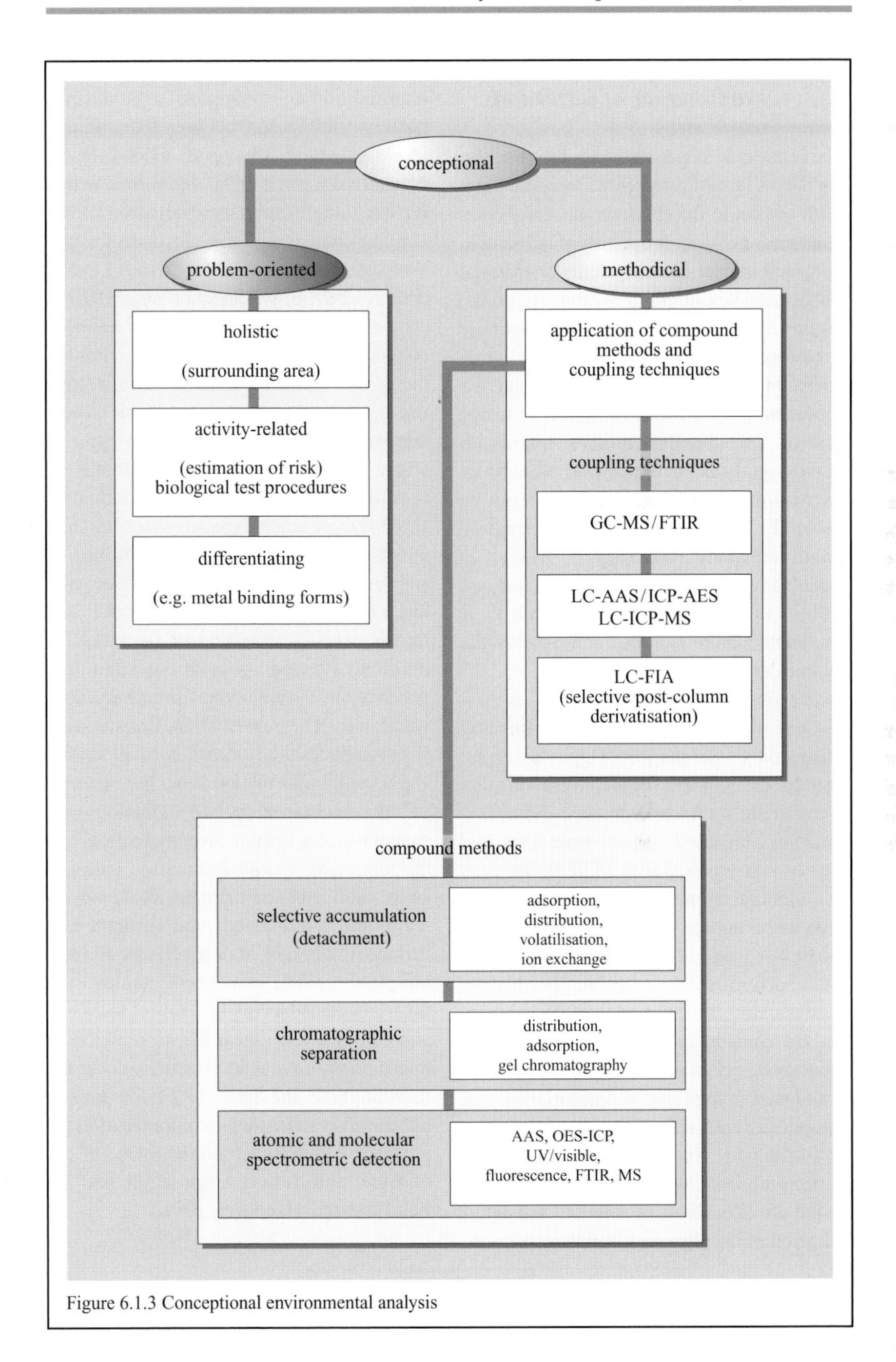

Figure 6.1.3 Conceptional environmental analysis

6.1.4 Risk assessment during examination of hazardous waste sites

The term 'risk' is generally used to describe the risk or hazard represented by something. With respect to the environment, risk represents the sum of all properties of substances by which impact on ecosystems or material damage can be caused. 'Substances or preparations are described as environmentally hazardous if they themselves or their transformation products are suited to change the condition of the balance of nature, of water, soil or air, climate, animals, plants or microorganisms such that immediately or later on, risks for the environment can be induced' (Chemical Law of Germany). Environmentally hazardous properties include low degradability (high persistence), ability to accumulate, and mobility. A hazardous nature can also appear after the biotransformation of a substance.

The first step along the way to an estimation of the risk lies in performing surrounding area analysis. This is a general term for the collection of all information that is important for a more detailed characterisation of a localised suspected site. This first part of surrounding area analysis includes the identification of the location of an abandoned storage site (a storage site before the Refuse Law of 1972 was enacted, such as unauthorised refuse dumps, production residues, etc.), the description of the location, and the compilation of base data and documentation. Hazardous waste sites are abandoned storage sites and abandoned sites (properties with operations that have been closed down), from which environmental impacts originate for the soil, a body of water or the air. 'Localisation' refers to the determination of the physical location of the area.

Depending on the extent of the base data required for an evaluation with defined goals, we differentiate between four different levels of proof. The term 'evaluation' is defined very generally as the formal action for the interdisciplinary evaluation of the existing data set and the subsequent establishment of the action items. Proof Level 1 (PL1) is part of the detection stage. PL2 is attained when the orientation study has been completed. After initial systematic measurements and studies, the abandoned storage site is investigated to the point where confirmed knowledge about the type of already existing or expected contamination is available. To do so, sufficient information about the presence, release, spreading and effects of contaminants must be obtained on site. Detailed studies take place at PL3. After this a risk evaluation can be performed as a target-specific evaluation of the risks for defined affected populations that are generated by the emissions from a hazardous waste site. The remediation investigation (PL4) includes the summation of all studies and provides information about the transport of the contaminant (considering the processes of sorption, decomposition and mobilisation), the contamination pathway (water, soil, air) and the entry of a contaminant into a protected area (ground and surface waters, soil, atmospheric air and soil atmosphere). The goal of all studies is to secure or remediate as the sum of technical and administrative measures to produce an interruption or reduction (securing) over the mid-range or the removal of the impact (clean-up) over the long term for the affected areas to be protected. (*Source:* State Offices of Water and Refuse, Hildesheim, and for Soil Research, Hannover, 1989.)

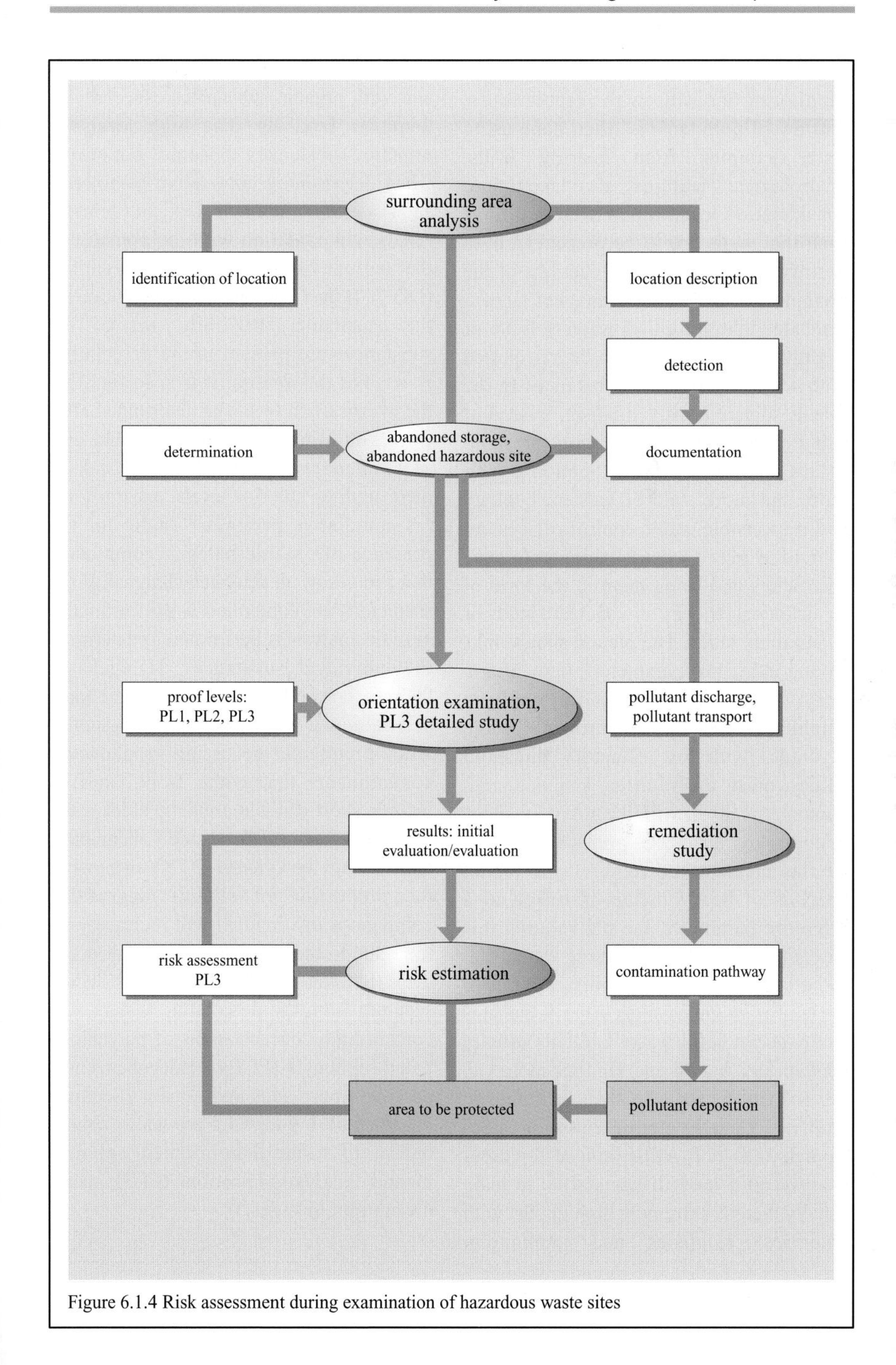

Figure 6.1.4 Risk assessment during examination of hazardous waste sites

6.1.5 Water studies near landfills

Monitoring landfills with respect to the outflow of pollutants is often done using water samples from bearing wells. Hydrological features are taken into consideration when a new well is drilled, such that wells are to be present in both upstream and downstream sections of the groundwater. Substances can enter landfill leachates that are either water-soluble or slightly soluble but that can be transported with water via organic substances in the role as solvent or as solubiliser. Instead of using a list of parameters with threshold values and guide values (Section 6.1.3), more and more state offices for ecology and responsible water control offices are switching over to establishing reference parameters and using them in the form of an analytical strategy as a standard study or an auxiliary study. The picture shows links among individual parameters. Only after a positive sensory smell test for hydrogen sulphide (sulphide) or hydrogen cyanide (cyanide) with low olfactory threshold values must quantitative analysis, e.g. photometrically or with the aid of ion chromatography, also be performed. In a similar manner, the requirement for an analysis of heavy metals is linked to a reference parameter, the pH value in this case. Only in acidic percolating waters is it to be expected that a mobilisation out of the landfill might have occurred. Leachates from special solid waste landfills usually contain less As, Pb and Hg, but more Cd, Fe, Ni, Cu, Cr and Zn than in domestic landfills. Usually atomic emission spectrometry (as ICP-AES) is used for rapid analysis. If a landfill has started to leak, usually there is extremely high impact with chlorides, sulphates and ammonium compounds in special solid waste leachates and with organic substances especially in domestic landfills. The high levels of organic substances become noticeable when determining cumulative parameters such as the chemical oxygen demand (COD; via oxidation with dichromate) or the amount of dissolved organic carbon (DOC). If the COD levels are high, it raises the suspicion that oily or tar-like substances are present in the percolating water. For this reason, it is specified that the hydrocarbon levels be determined after extraction using IR spectrometry (likewise as a cumulative parameter) as a check. If there are high chloride levels, one must test (as cumulative parameter AOX) for the presence of organic halogen compounds that can adsorb to activated charcoal. If the suspicions are confirmed, a gas chromatographic analysis is performed to determine the individual substances. Halogenated hydrocarbons (HHC) are some of the most persistent and readily accumulating pollutants. Frequently occurring groundwater contaminants that come from landfills include short-chained aliphatic chlorinated hydrocarbons, chlorinated benzenes, toluene, and the xylenes (BTX). In general, the proportion of slightly degradable compounds in a landfill will increase with time due to the transformations that occur there (Section 4.5). Percolating oils, which separate from the percolating waters, can contain high concentrations of polychlorinated biphenyls (PCBs), chlorobenzenes, persistent pesticides and dioxins (Sections 5.4 and 5.5). Figure 6.1.5 provides a contribution to a problem-oriented environmental analysis (Section 6.1.3) using meaningful linkages.

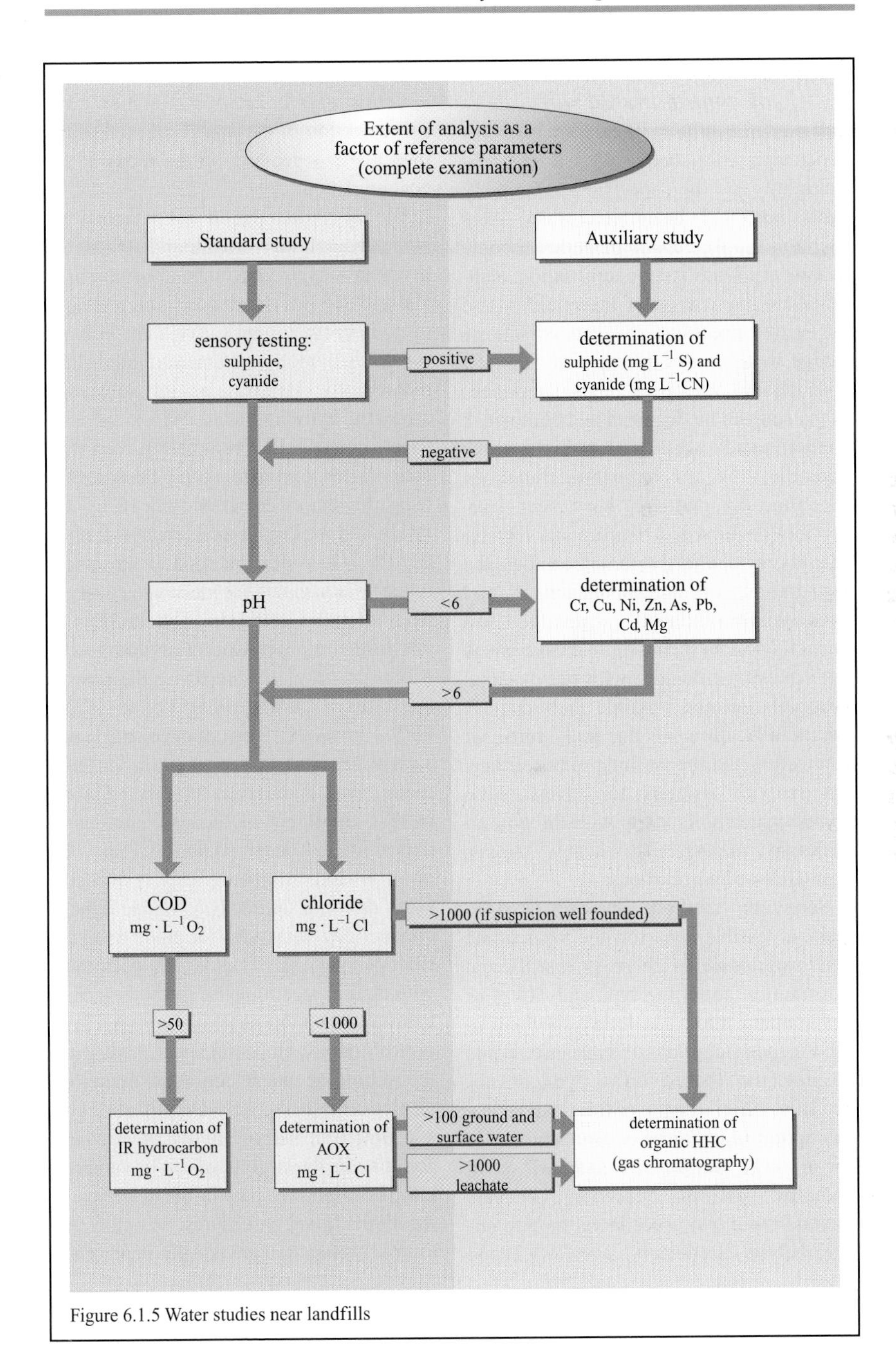

Figure 6.1.5 Water studies near landfills

6.1.6 Strategies for the analysis of contaminated soils

Soil contamination in general includes the deposition of substances that did not originate in the specific soil itself. Emissions and hazardous waste sites (Section 6.1.4), results of anthropogenic dealings such as extreme fertilisation, inappropriate application of insecticides, and the use of excessive amounts of sewage sludge are all part of the deposits. Even if a considerable proportion of the substances in the soil can be degraded and eliminated, contaminated soil represents a risk potential, or its essential functions (Sections 4.1 and 4.2) have been irreversibly destroyed. A soil impact register should provide initial information about the distribution of anthropogenously and geogenically induced, wide-area soil impact. The data should be able to be linked with questions dealing with heavy metal accumulation and possible mobilisation, the acidification of the soil, fertiliser application and the washing out of nitrate. An analysis strategy for investigating contaminated soils starts with the type of pollutant impact, e.g. heavy metals, pesticides or hydrocarbons.

Screening methods are supposed to make it possible to narrow the focus of the test programme if there is insufficient information about the type and extent of soil contamination. The term 'screening' is used if specific groups of substances are to be detected (based on a 'yes or no' decision) either using tests from a statistical standpoint (e.g. for large-area studies) or for a large number of samples or in complex samples. Screening methods include rapid test procedures, mobile on-site analysis (Sections 6.1.1 and 6.1.2) and biochemical or biological tests such as immunoassays or enzyme inhibition tests. The selection of the analytical methods is then made depending on the focus of the contamination.

For oil contamination or contamination by hydrocarbons (including halogenated hydrocarbons) in general, methods are used that include the chromatographic methods of gas chromatography (GC; for volatile substances), high-performance liquid chromatography (HPLC: e.g., for polycyclic aromatic hydrocarbons, PAHs), infrared spectrometry (IR; summation determination after joint extraction), fluorescence measurement as detection with HPLC for PAHs, and the Dräger air extraction method (DAEM), in which the volatile substances in aqueous samples are transferred into gas detector tubes using air. The analysis of pesticides requires extraction and purification processes in preparing the sample (Sections 6.1.8 and 6.1.9) before GC or HPLC analysis. Inorganic contaminants such as heavy metals or inorganic ions also require extraction or, in the case of heavy metals, treatment of the soil samples to determine the total contents. Today the powerful ion chromatography is available (with direct or indirect UV/visible light or conductivity detection) for anion analysis. Heavy-metal analysis is largely performed with atomic spectrometric methods such as atomic absorption spectrometry (AAS) or optical/atomic emission spectrometry for the simultaneous detection of numerous elements (as ICP-AES). Questions of cost and time (i.e. the possibility of extensive automation for large numbers of samples) play an important part in the selection of the methods and procedures. Sampling and sample preparation are equally important.

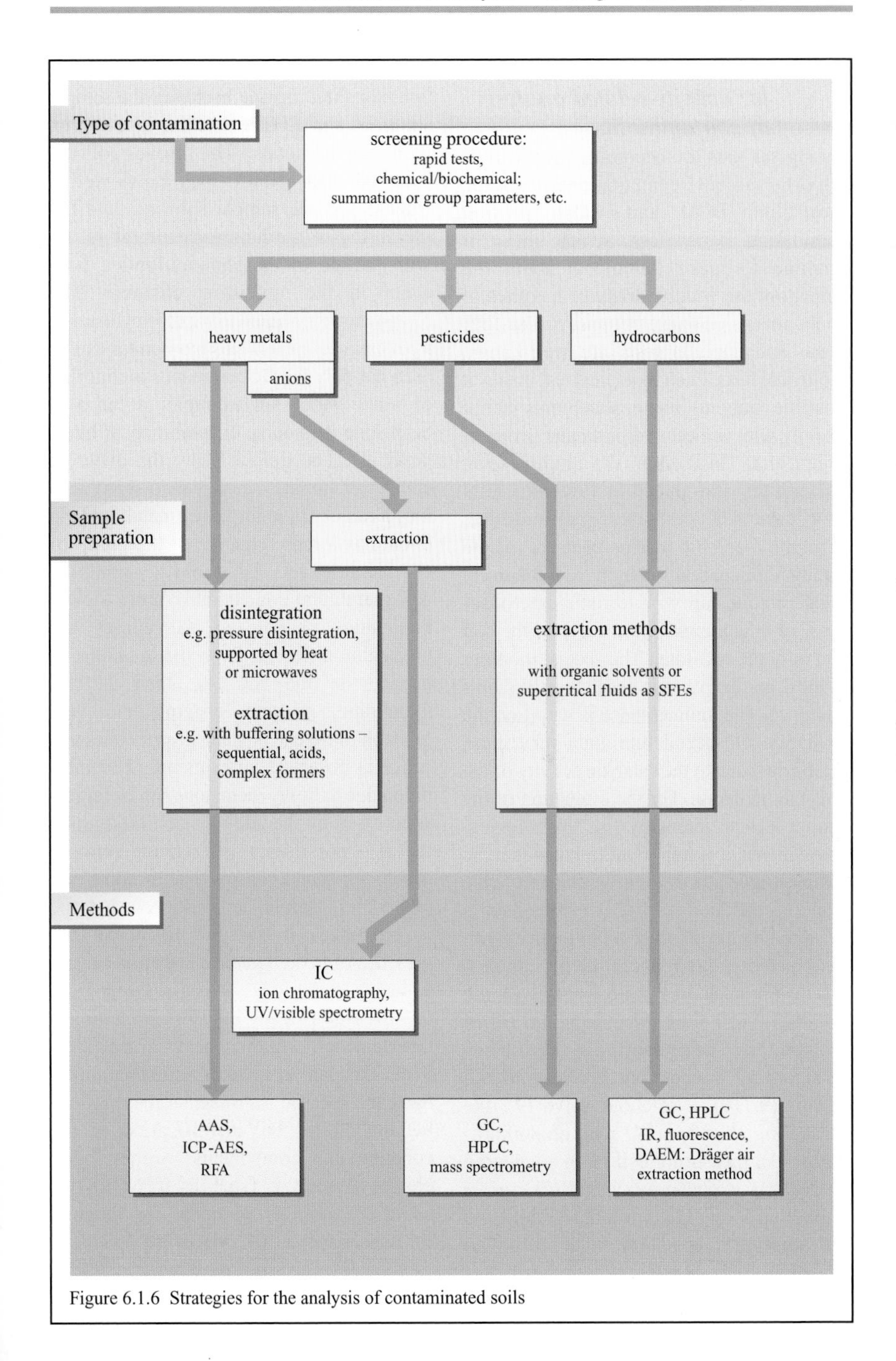

Figure 6.1.6 Strategies for the analysis of contaminated soils

6.1.7 Luminescent bacteria test for activity-related analysis of contaminants

Biological tests are becoming increasingly important in environmental analysis since they, like BOD and COD, provide summation parameters, which make it possible to make statements about the activity of the widest variety of mixtures of contaminants on living organisms. The fish, algae and daphnia tests are well-known biological tests. Each biological test makes it possible only to make statements about toxicity with respect to a particular group of organisms, including the luminescent bacteria test introduced in 1989 as a DIN procedure. This test uses gram-negative, facultative aerobic marine bacteria (of the family Vibrionaceae): they have a relatively close relationship with terristic enterobacteria, of which most species live in the soil and in bodies of water. The special property of these marine bacteria lies in their bioluminescence. The luminescence is based on the oxidation of special luminous substances (luciferins) due to the catalytic activity of the enzyme luciferase. For the evaluation of the luminescence intensity for toxicological assessments, it is important to know that the luminescence process represents a part of the bacterial metabolism. If this metabolism is impaired by the presence of substances with toxic activity, this reduces the bioluminescence. Automated biological tests with the possibility of online monitoring are called biomonitors. The measuring system shown in Figure 6.1.7 (upper part) consists of two pumps (9, 10), a solenoid valve (9), the measuring chamber (14) with photomultiplier (13), measuring cell (12), a solenoid valve (8), a four-way valve (7) and a tempering loop (11). Various solutions are fed through tubes 1–6, such as saline solution (to create the optimal living conditions for these marine bacteria), the sample solution, and EDTA solution to complex with heavy metals. The mixing of the bacterial suspension (which stores well if frozen) and the sample solution, and the measuring of the luminescence (at 15 °C) with the aid of the photomultiplier, both occur in the measuring chamber. The biomonitor is operated using the keyboard of a personal computer. This biomonitor can be used not only for the continuous monitoring of waste water, for example; it can also determine the course of inhibition of luminescence. The graphs show the different courses of luminescence reduction via lead ions as an example for heavy metals and via phenol as a representative of the group of organic pollutants. Lead ions penetrate the cell membrane considerably more slowly than does phenol, so the curves are noticeably different. Even the locations of damage to the bacteria, such as cell membrane, enzyme systems, etc., are probably different. A steep drop in luminescence is observed with phenol. The inhibition due to heavy metal ions can be largely suppressed by the addition of EDTA, such that it is possible to differentiate between activities of contaminants. The example of a real water sample demonstrates that the curve makes it possible to detect the presence of toxic organic substances as well as heavy metal ions. After the steep drop, there is still a steady reduction in luminescence, which points to heavy metal ions. With this activity-related information in hand, the specific instrumental methods can be applied for the identification of the contaminant components, either with elemental analysis (such as ICP-OES) or using separation methods for organic substances such as GC-MS or LC-MS.

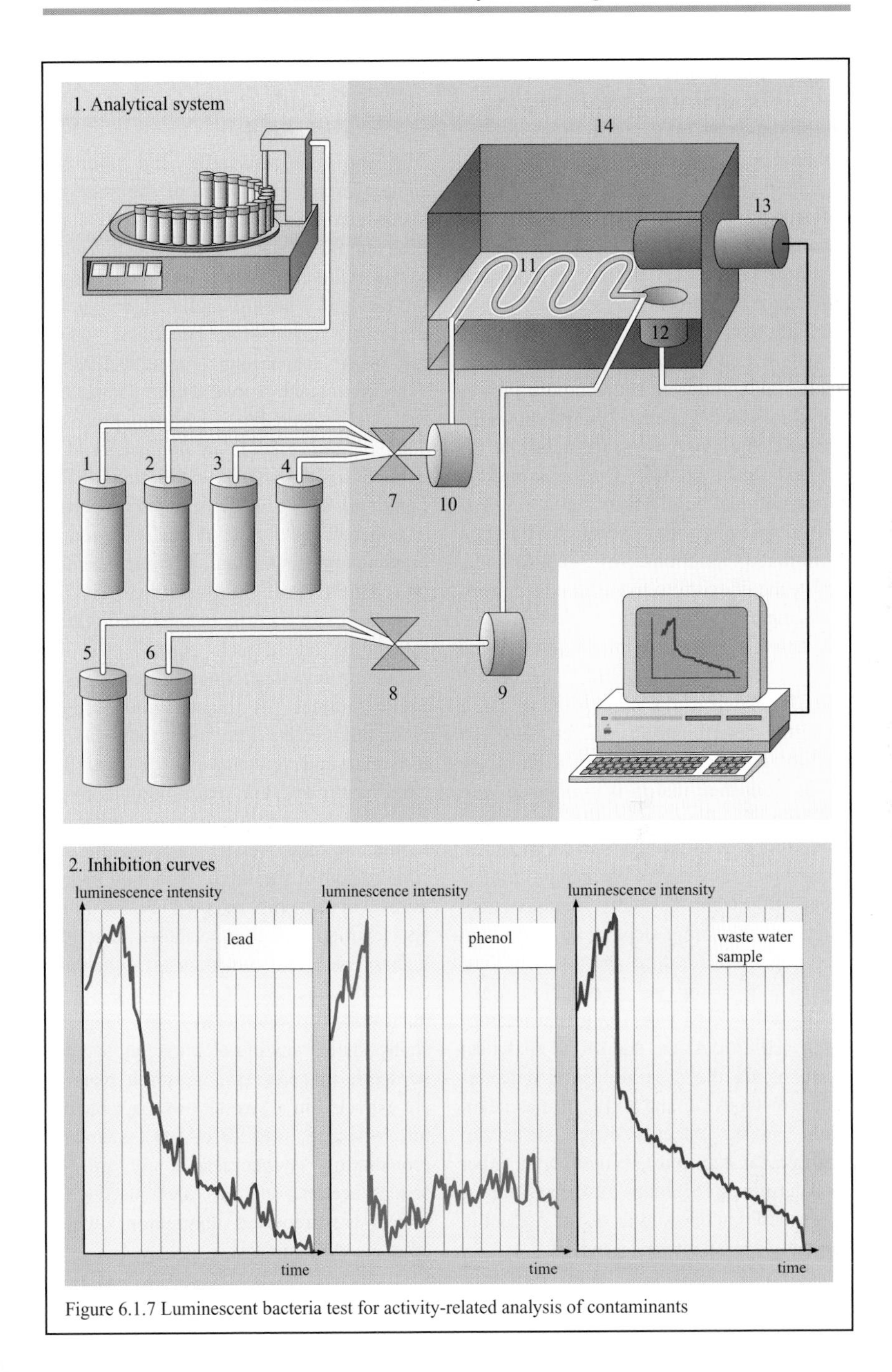

Figure 6.1.7 Luminescent bacteria test for activity-related analysis of contaminants

6.1.8 Outline for pesticide analysis in environmental samples

In addition to the sampling, in the analysis of contaminants especially in the trace range, sample preparation plays a decisive role in influencing the result. With the high degree of precision afforded by modern analytical instruments, the total error of an analytical procedure is determined especially by the sample preparation. The simplified outline (Figure 6.1.8) showing an example for the analysis of chlorinated hydrocarbons such as polychlorinated biphenyls (PCBs) shows the protocol, the most important steps and the methods to be applied. During sampling, attention must be given to the representativity (especially with respect to the question formulated: Sections 6.1.3–6.1.6) and, before the study starts, to the homogeneity of the sample (examples here would be human fat, animals or their organs, and activated sludge). Since the analytical results are usually related to the percentage of fat, a separate determination of fat must be performed. Then a first separation of the PCBs from their matrix is performed with the help of column or Soxhlet extractions using different solvents or solvent mixtures. In the block showing the cleanup procedures (separation from the matrix components), there are alternating steps of concentration (e.g. by concentrating the solvent) and purification using gel permeation chromatography (GPC) or liquid chromatography (HPLC). At the end of the analytical protocol are the separation and determination methods GC and HPLC in association with mass spectrometric detection. Purification steps must be monitored, either by determining recurrence of the analytes or by the addition of a substance not present in the sample as an internal standard.

6.1.9 Modern sample preparation for soils contaminated with PCBs

The very time-consuming and laborious sample preparation steps of the standard procedure outlined in Section 6.1.8 have been simplified since the 1990s with the aid of new techniques. Effective extractions with mixtures of *n*-hexane and acetone can be performed in PTFE containers using microwave technology (Figure 6.1.9, 1), whereby acetone represents the component that is excitable by microwaves. Temperatures of 80 °C are optimal for PCB extraction. Superfluid extraction (SFE, Figure 6.1.9, 2) uses the solution properties of supercritical fluids: by the selection of pressure and temperature, different densities of a supercritical fluid and therefore the solution capacity can be adjusted. Here, toluene is added as a modifier to the frequently used supercritical carbon dioxide. Solid samples are placed into airtight extraction cells. The liquefied gas is loosened and removed in the restrictor (capillary tube). The main modules of an SFE device are those for the generation, regulation and relief of pressure. A comparison of the three extraction procedures for several selected PCBs (with Ballschmiter numbers) shows that the highest levels in a contaminated soil can be detected using GC-ECD with the aid of microwave-supported extraction. A significantly smaller amount of extracting agent is needed in this process: 2 g sample material are extracted in a Soxhlet process with 100 mL *n*-hexane over 90 cycles (secondary remediation is necessary); 30 mL *n*-hexane/acetone (3:1) are used for microwave-supported extraction, which takes about 30 minutes.

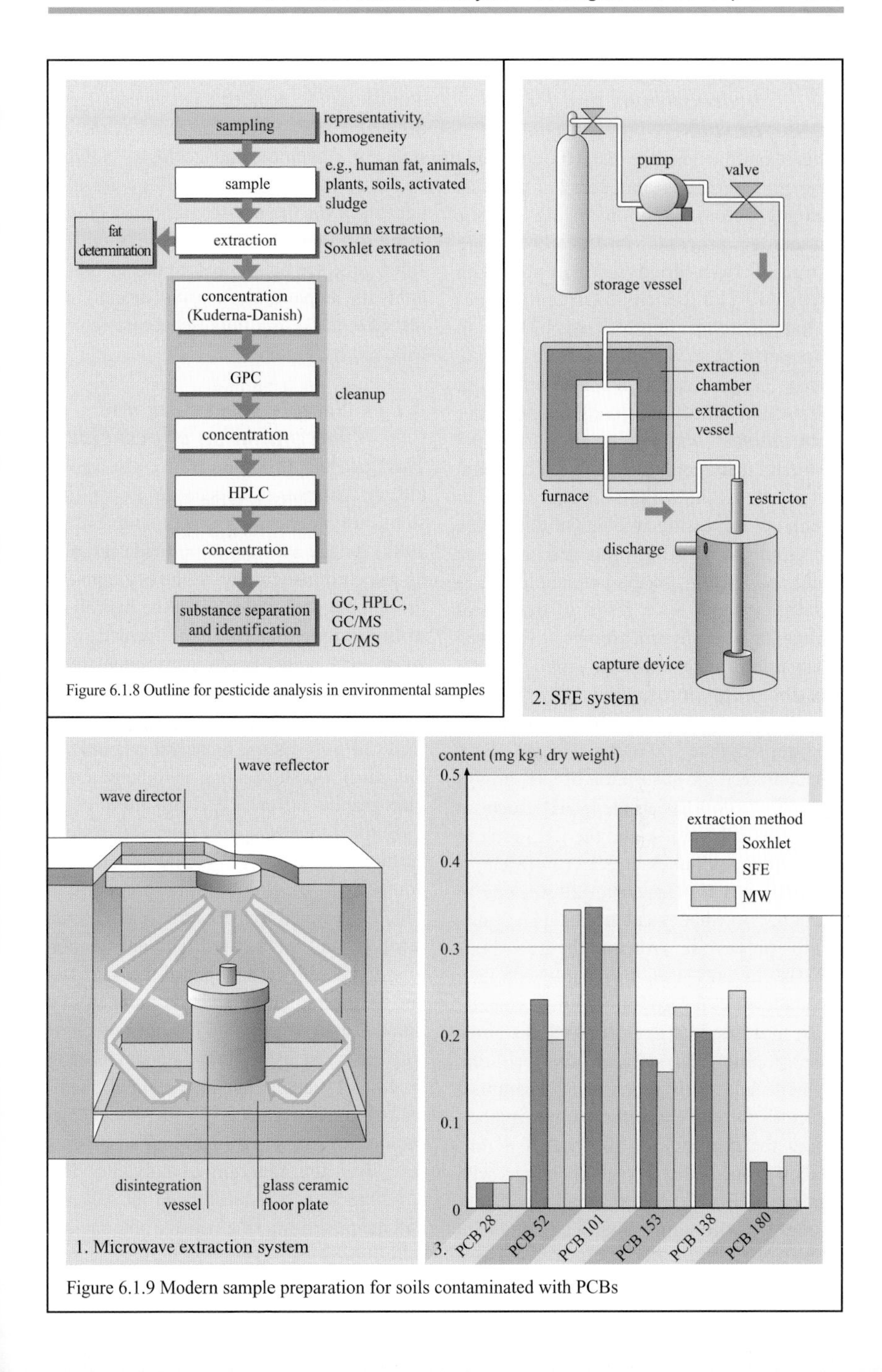

Figure 6.1.8 Outline for pesticide analysis in environmental samples

Figure 6.1.9 Modern sample preparation for soils contaminated with PCBs

6.1.10 Analytical methods for hydrocarbons and PAHs in soils

This example (Figure 6.1.10) compares screening methods and detailed analysis straight across (Section 6.1.4). In the screening process, hydrocarbons are extracted from dried soil samples with Freon 113 (1,1,2-trichlorotrifluoroethane) or tetrachlorohexafluorobutane S 316 in an ultrasound bath. Drying is necessary to avoid the formation of emulsion. Ultrasound extraction is considerably faster than solvent extraction in a Soxhlet extractor (Section 6.1.9). The ultrasound action also causes the soil aggregates to break up; increasing amounts of turbulence occur in the limiting layer area between solid and solvent, so the extraction time is relatively short. As a result of the use of halogenated hydrocarbons, infrared absorption measurements are directly possible after centrifugation, filtering and drying of the extract using anhydrous sodium sulphate. IR measurements detect the characteristic absorption of CH_3 groups at 3.38 μm ($\nu = 2958\ cm^{-1}$), CH_2 groups at 3.42 μm (2924 cm^{-1}), and the CH group of aromatic compounds at 3.30 μm (3030^{-1}). Recording a UV spectrum allows one to recognise to what extent aromatic and especially polycyclic aromatic hydrocarbons (PAHs) are present in the sample with their characteristic absorption spectra, typically with multiple maxima. A rapid gas chromatographic analysis in a short column (which yields only incomplete separation) provides further characteristic information about the presence or absence of hydrocarbon groups: aliphatic hydrocarbons with *n*-alkanes, *iso*-alkanes, cycloalkanes/naphthenes or aromatic hydrocarbons with alkylated 1- to 5-nuclear and naphthene aromatics. The detailed analysis uses either Soxhlet extraction or ultrasound extraction with a cyclohexane–acetone solvent mixture, whereby cleanup steps are also necessary for the pre-separation of groups using solid-phase extraction (with modified silica gels; see below) before GC or HPLC analysis. A photodiode array detector (to record spectra) or a fluorescence detector is used for PAH analysis.

6.1.11 Sample processing and determination of pesticides

The flowchart in Figure 6.1.11 also shows clearly that numerous sample preparation steps are necessary (Specht and Tilkes, 1981) for the analysis of pesticide residues by gas chromatography. At the beginning of the sample preparation, alkaline hydrolysis of the sample material is performed, with subsequent liquid–liquid distribution steps. The extracts are then cleaned up with the aid of gel chromatography (using Bio Beads made of polystyrene or polyacrylamide in this case) and short silica gel column chromatography (after a derivatisation step). Even the derivatisation of pesticides (esterification with methanol/sulphuric acid) is a component of this sample preparation, which is adapted to the matrix, the material (analyte) and the method. The purification of extracts, the enrichment of the analytes, and the separation of matrix components by liquid–liquid distribution has been replaced by solid-phase extraction more and more in the last 10 years. In the latter method, chemically modified silica gels with phenyl or *n*-octadecane groups as non-polar adsorbents take over the function of solvents. The advantages are a reduced need for solvents and the possibility of extensive automation.

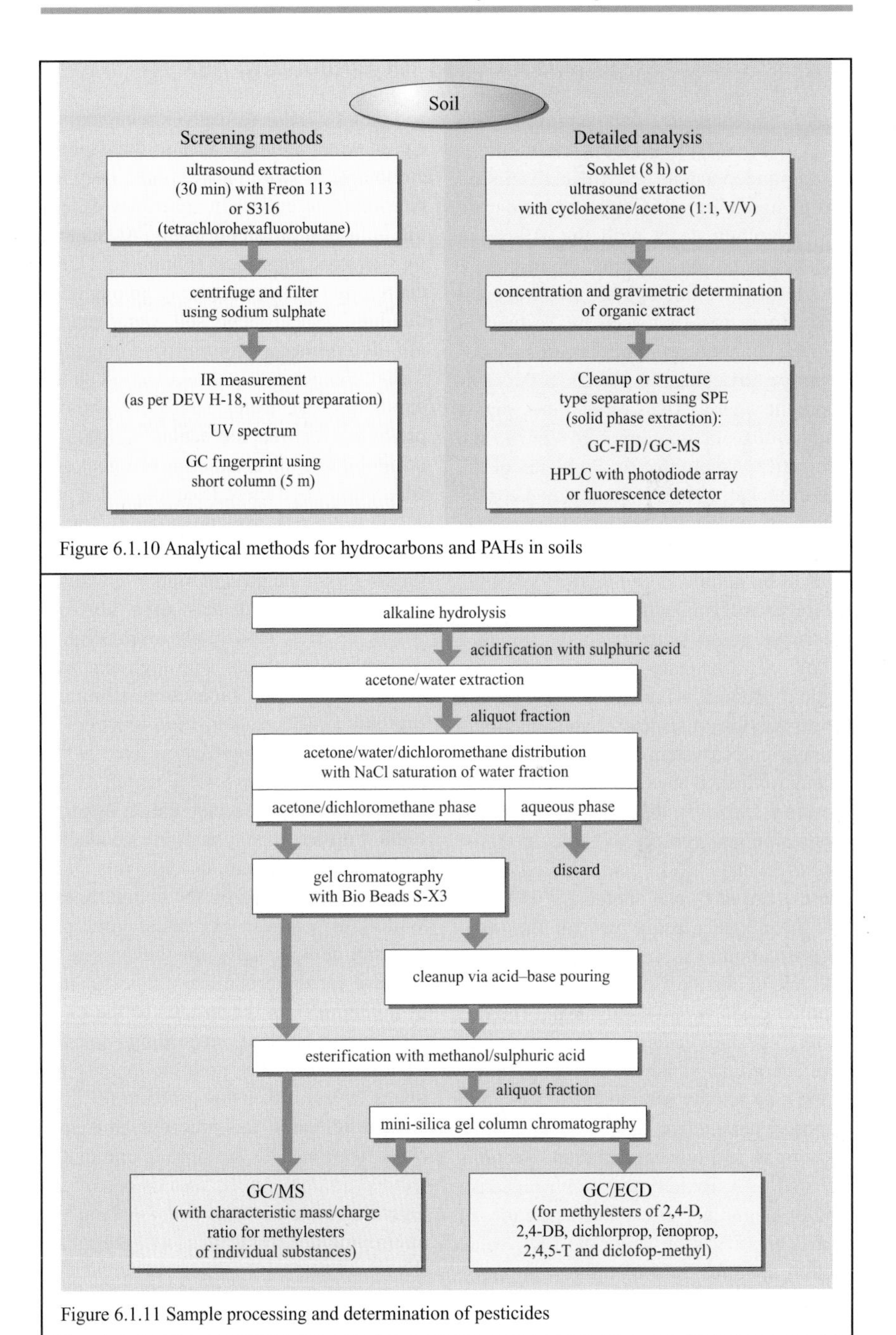

Figure 6.1.10 Analytical methods for hydrocarbons and PAHs in soils

Figure 6.1.11 Sample processing and determination of pesticides

6.2 Ecotoxicological concepts for evaluating risk

6.2.1 Flowchart for rapidly classifying chemicals

As an independent scientific discipline (from toxicology, biology and ecology), ecotoxicology deals with the effects of substances on the inhabited environment. The term ecotoxicology as environmental toxicology was first used by Truhaut in 1969 as an application of the principles of chemical toxicology to objects in the environment such as fish, water fleas, plants and entire ecosystems. In 1978 the 'Scientific Committee on Problems of the Environment' (SCOPE) formulated a definition as follows: 'A branch of toxicology dealing with the study of toxic effects, caused by natural or synthetic pollutants, on the animal (including human), plant and microbial components of ecosystems as a whole.' The primary goal of ecotoxicological studies is considered to be the determination of structural and functional changes in ecosystems under the influence of environmental chemicals (Römpp, 1993: *Lexikon Umwelt*). The results form the basis for an ecological risk analysis. Ecochemistry and ecotoxicology are closely linked to one another. In 1985 the EC developed a procedure for the rapid identification of a possible environmental risk due to chemical substances in order to monitor environmental chemicals. The two most important criteria of this procedure are, according to Parlar and Angerhöfer (1991), an activity parameter (toxicity) and exposure parameters, such as degradability, dispersion and bioaccumulation (Sections 1.7 and 5.1). Toxicity tests with rats, fish and/or daphnia (water fleas) provide an initial classification using the LD_{50} value (lethal dose for 50% of the test animals) and the EC_{50} value (effective concentration, e.g. in water at which 50% of the daphnia are unable to swim or are dead). They are considered to be very toxic if they exceed the indicated threshold levels. At Stage 2 for less toxic substances (Figure 6.2.1), the distribution and degradability among or in the three compartments soil, water and air are determined. A simplified fugacity model (Section 6.2.3) is used, based on the parameters of water solubility, vapour pressure, relative molecular weight, 1-octanol–water distribution coefficient and adsorption coefficient (Sections 5.1.6 and 5.1.7). A risk is postulated for a compartment if more than 1% of the substance can be present in it. Even with a biodegradability of less than 60–70% within 28 days, moderately toxic, slightly degradable substances with high exposure are tested for their bioaccumulation and ecotoxicity (Stage 3 in the Figure). The measure of bioaccumulation used is the P_{OW} value (Section 5.1.6), which at 3.5 corresponds to a bioaccumulation factor of >300. Further toxicity tests are conducted with different animal species (rats, fish, daphnia), depending on the compartment. With two types of tests for one compartment, usually the lower one is decisive for the preliminary classification of a chemical as 'hazardous to the environment' (labelled according to the Chemical Law). The possible toxicity for plants or soil organisms (Section 6.1.7) is not considered in this procedure for rapid classification. The example is one of the rapid procedures in the area of ecotoxicological profile analysis for checking the environmental relevance of chemicals (Korte, 1992).

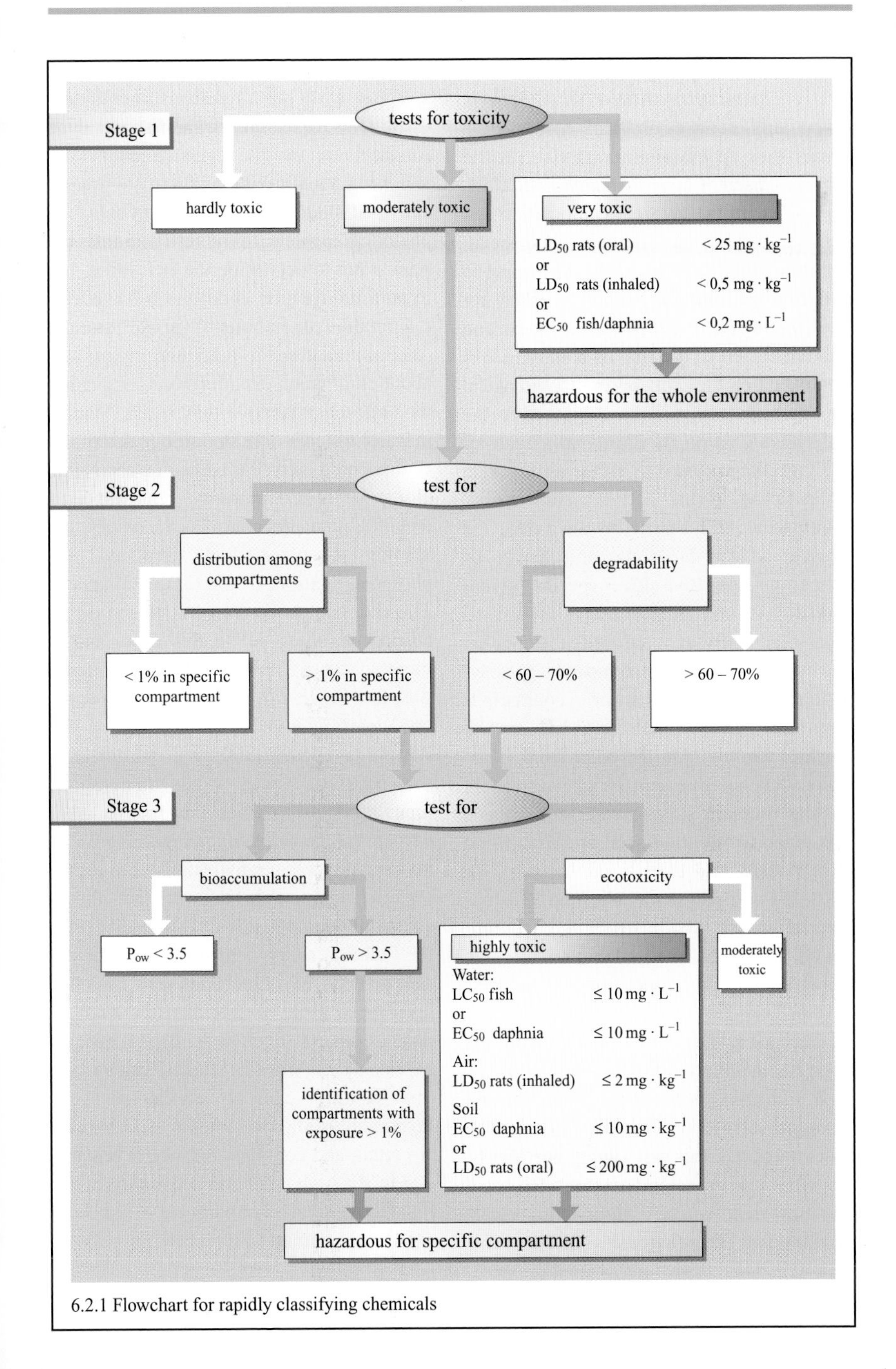

6.2.1 Flowchart for rapidly classifying chemicals

6.2.2 Developing an environmental risk profile

Even this ecotoxicological test system establishes an environmental risk profile (from Bahadir, *Chemie unserer Zeit*, 1991, Issue 5) with the help of accumulation and degradation, using coumarin and hexachlorobenzene as an example. According to a definition from the Federal Ministry for the Environment (see Bahadir, Parlar and Spiteller, 1995), 'the risk of a substance is dependent on the probability that biological or other systems will be subjected to the substance and on its ability to impair or damage these systems, either alone or in cooperation with other substances.' Coumarin (2H-1-benzopyran-2-one) is present in the blossoms and leaves of numerous grass and clover species (glycosidically bound in part), and is formed biosynthetically as a natural substance by the hydroxylation of cinnamic acid, glycosylation and cyclisation. However, coumarin is toxic for higher animals (oral LD_{50} = 680 mg kg^{-1} for rats). On the other hand, hexachlorobenzene does not occur naturally in the environment; it is a pesticide (used as a fungicide for treating seed and as a wood preservative; oral LD_{50} = 1.5–10 g kg^{-1} for rats). In this test system, called the 'environmental hazard profile test', 'the accumulation of radioactively labelled, organic compounds (^{14}C) at environmentally relevant concentrations (1 ppm–50 ppb) is determined at different trophic levels of the food chain (microorganisms, green algae, fish and rats), as are their microbial and photodegradation' (Bahadir, 1991). The assessment assumes that a substance poses a potential risk to the environment if it accumulates readily but does not easily decompose. The relationship of chemicals to the environment is determined using test systems with the organisms mentioned, usually rats for mammals, and fish and algae for the lower trophic levels (Section 5.1), and using transformations (in waste water – activated sludge or physicochemically via photodegradation). In the test with rats, the goal is not to determine the LD_{50} dose, but to gain information about the behaviour in mammalian metabolism: about storage (bioaccumulation) in particular organs and about elimination overall (monitoring using the radioactive label) within 5 days. Studies in waste water or activated sludge determine ecotoxicologically the course of chemicals along the refuse pathway. This test area includes biotic processes. On the other hand, abiotic processes are simulated via photodegradation, or photomineralisation. The chemical to be tested is placed on an adsorbent (silica gel in this case) and is radiated for 17 hours with light having wavelengths less than 290 nm. Afterwards one determines what percentage of the starting compound (100 ng g^{-1} adsorbent in this case) has been totally mineralised (converted to carbon dioxide), what percentage remains unchanged, or what percentage was transformed into other organic compounds due to the photochemical decomposition. The results from the four test areas yield the environmental risk profile: hexachlorobenzene accumulates much more readily than the strongly polar coumarin (Section 5.1.6) and decomposes only slightly. The studies which led to this finding can be conducted very favourably using the radiotracer technique. A certain and complete balance is possible due to the high detection sensitivity of the radioactively labelled carbon (^{14}C) in both organic substances.

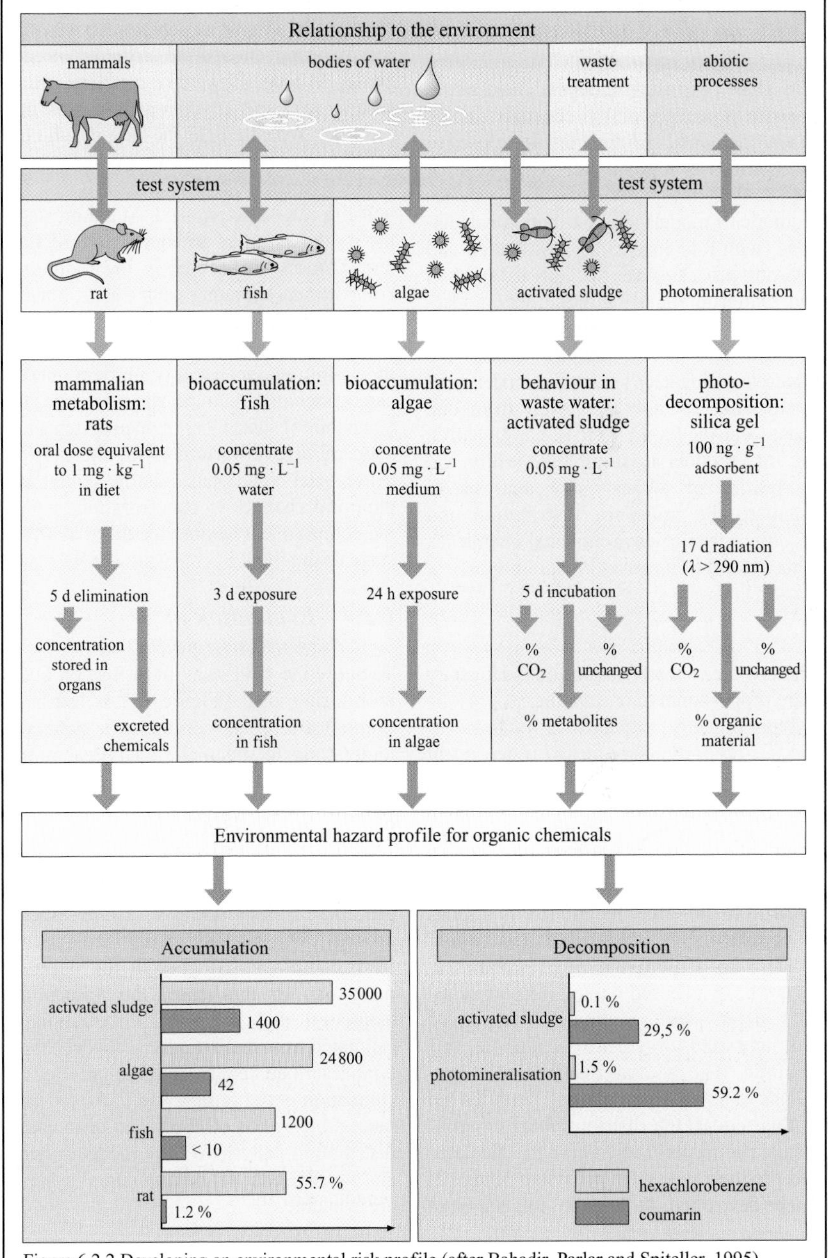

Figure 6.2.2 Developing an environmental risk profile (after Bahadir, Parlar and Spiteller, 1995)

6.2.3 Exposure analysis based on the 'Unit World' fugacity model

The term 'exposure analysis' is used to describe prognoses on the concentration of an environmental chemical that is exposed to an organism or a population. The bases for the calculations (approximations) to generate (numerical) models are distribution calculations (with reference to the distribution and transport processes, see Sections 1.7 and 5.1) and environmental simulation models (taking into account the processes of decomposition and accumulation). Using the principle of fugacity (f) (as the tendency of a substance to distribute itself from one phase over the entire system, i.e. globally), fugacity models are completed within the framework of an exposure analysis to calculate the transport, distribution and decomposition of environmental chemicals. For a system of n phases in equilibrium,

$$f_1 = f_2 = \dots f_n$$

whereby the concentration c and the fugacity f are proportional to one another: $c = f \times z$.

The fugacity capacity z (substance-specific proportionality factor) is dependent on the water solubility, the vapour pressure, the 1-octanol–water distribution coefficient, the type of phase and the interactions between the chemical and the phase. The fugacity capacity can be calculated separately from the values mentioned for a series of compartments (Parlar and Angerhöfer, 1991). The 'Unit World' system shown in Figure 6.2.3 is based on six compartments: air, water, soil, sediment, suspended sediment and organisms – fish in this example. The spatial distribution corresponds to actual volume ratios. With the aid of this model, the distribution of environmental chemicals (Stage 1) can be calculated and prognoses on their behaviour (Stage 2) can be developed. At Stage 1 – calculation of the distribution of a substance among the phases – we assume an unreactive compound that is homogeneously distributed in each phase. At Stage 2, we take reactions (transformations) and advection processes into account. We calculate the rate at which a substance is removed from a compartment using first-order kinetics, from which the half-life can be determined. At a third stage, the model assumes an equilibrium of flow and differing fugacities in the individual compartments. Among other things, transfer coefficients are determined. Non-diffusive transfer processes (such as deposition, sedimentation, resuspension or nutrient uptake) can be included. A fourth stage describes the behaviour of substances in open systems in a state of disequilibrium (with the help of differential equations). Finally, spatial and temporal changes in concentrations within the compartment can be calculated at a fifth stage of the model.

6.2.4 Evaluation of monitoring data

To apply the fifth stage from Section 6.2.3, monitoring data (Figure 6.2.4, left-hand graphs) are used (monitoring is generally understood to mean the analytical monitoring of substance concentrations in the environment). They are plotted against a known total volume. The total amount M, the concentration c_{m} and the scattering σ are calculated and displayed graphically. This procedure is described as a forward-directed process. The backwards-directed process (right-hand graphs) goes in the opposite direction. In this case, the values are estimated and expected distribution is estimated from them. Fugacity models of the type described here can also be used for the simulation or the comparison of the environmental behaviour of chemicals or to create distribution patterns if only model systems and no real data or environmental systems are available.

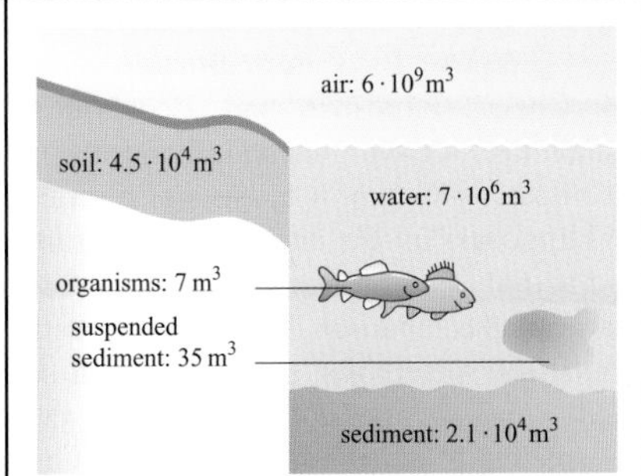

Figure 6.2.3 Exposure analysis based on the 'Unit World' fugacity model

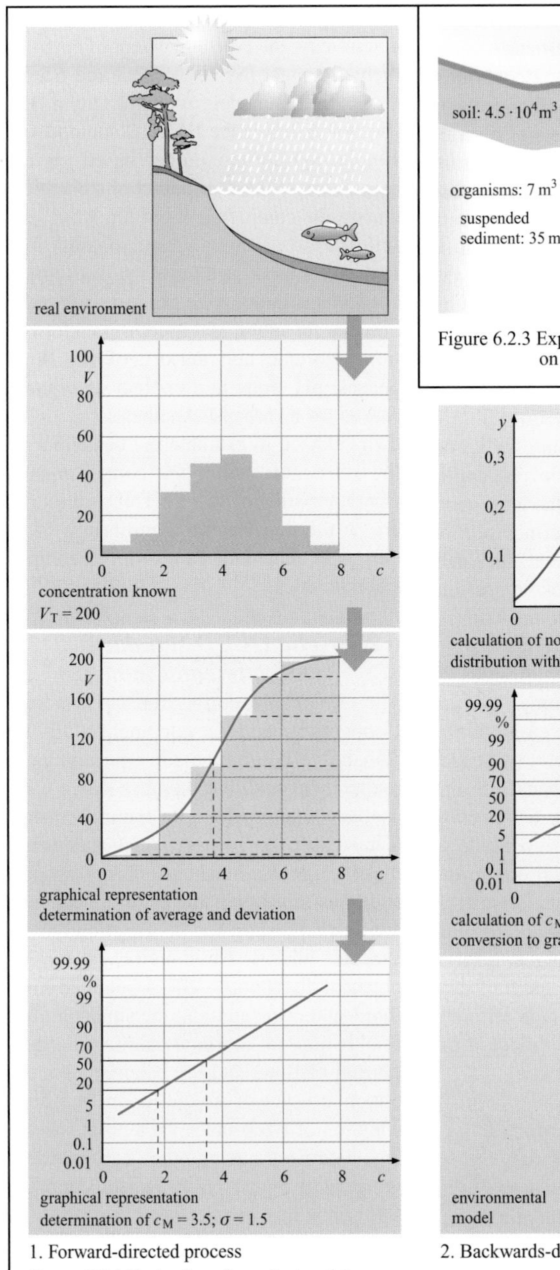

1. Forward-directed process

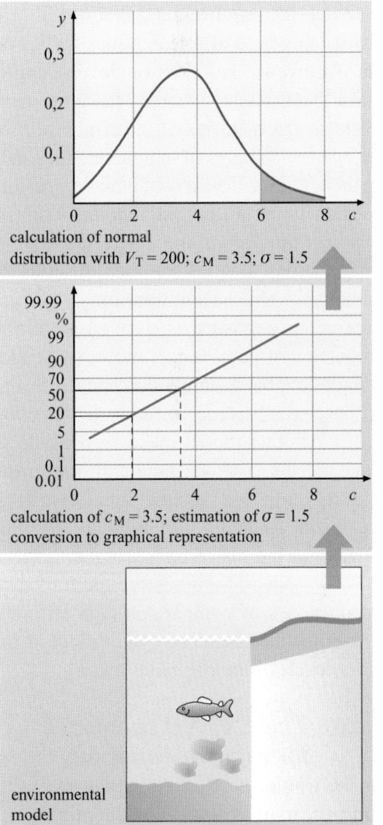

2. Backwards-directed process

Figure 6.2.4 Evaluation of monitoring data

6.2.5 *Risk evaluation according to the E4 Chem model*

E4 Chem (Exposure and Ecotoxicity Estimation for Environmental Chemicals) is a computer program. It consists of a series of submodels for the analysis of exposure and activity and of a system for transformation and comparison of the acquired data (Parlar and Angerhöfer, 1991). One of these models makes it possible to determine the release rate of environmental chemicals by means of the available information via the application pattern. According to Korte, one of the main goals of ecological chemistry is to establish environmental impact. He divides this field of research into studies on the occurrence, production levels, application pattern, etc., whereby the latter term refers to the quantitative determination of the operative ranges of individual chemicals by the end user. These data make it possible to calculate or at least estimate the material primary impact in the operative ranges (living areas and environmental areas of humans). The accumulation potential relative to exposure is derived from the properties of persistence, mobility, geoaccumulation, bioaccumulation and the extent of the contamination of the water (Section 5.1.6). Another model makes it possible to calculate the ecotoxic potential due to the toxicity data for aquatic and terrestrial organisms. In the E4 Chem model, the individual data are scaled up to a higher level of risk assessment; the transformation of data for purposes of comparison (for different environmental chemicals) takes place at the lower levels of the computer model.

6.2.6 *The EXTND submodel for exposure analysis*

On the whole, there are two areas for evaluating the risk potential of chemicals: (i) checking the spectrum of activity and (ii) determining the type and extent of environmental contact (Sections 6.2.3 and 6.2.5). One of the submodels of the fugacity model (Section 6.2.3) is the EXTND submodel, a simple equilibrium model based on the calculation of the distribution of a chemical among the compartments of air, water and soil (the three thermodynamic phases). The foundation takes as a basis the distribution coefficients (among the phases), first-order kinetics for biotic and abiotic transformations, volume and density of the three phases, pH value and content of organic carbon (as a sorbent). Another submodel is then selected to estimate the behaviour of the substance within the compartments. Characteristics of the EXTND submodel are a thermodynamic equilibrium in a closed system with first order decomposition kinetics.

6.2.7 *The OECD standard environmental model*

The Organization for Economic Cooperation and Development (OECD) also has a committee for environmental matters. In 1982 the member states were challenged to provide data about chemicals using a list of parameters, which data would make a risk assessment possible. Transport and transformation pathways of a substance are normally shown as its function during exposure analysis based on these data. The potential distribution (PEC = potential environmental concentration; the situation after establishing a thermodynamic or equilibrium of flow) and the potential concentration in a state of disequilibrium (e.g. after emission of a substance via a point source) are estimated using the compartment volumes in Figure 6.2.7. Fugacity models are used in turn in calculating the PEC.

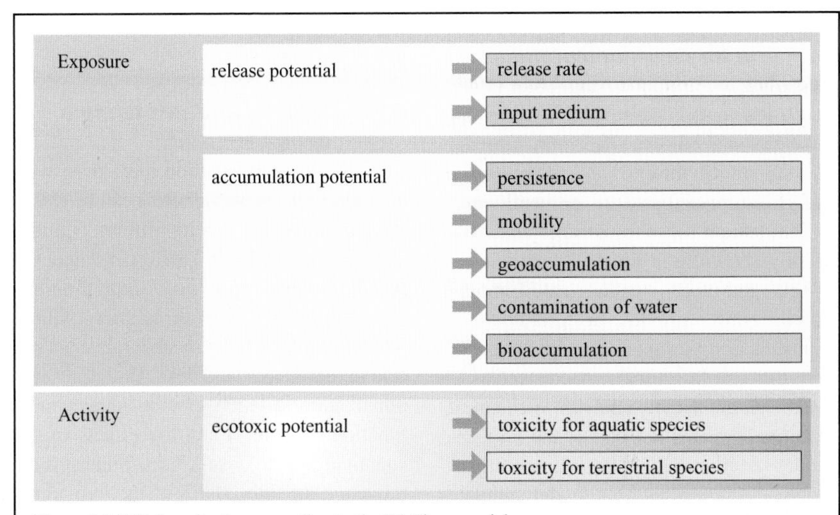

Figure 6.2.5 Risk evaluation according to the E4 Chem model

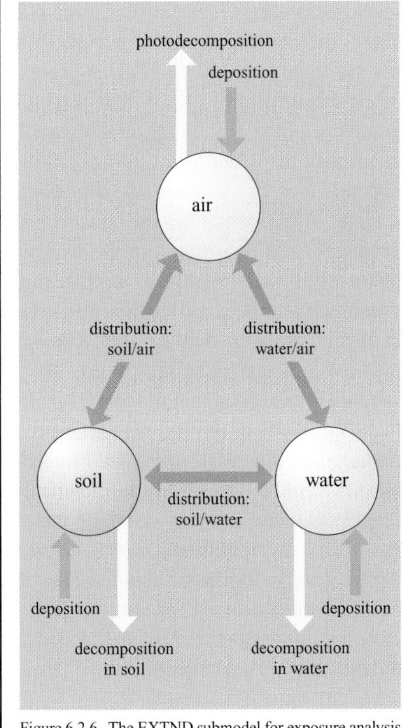

Figure 6.2.6 The EXTND submodel for exposure analysis

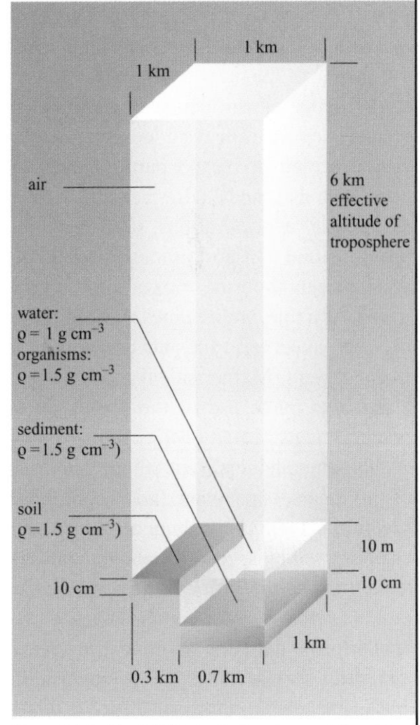

Figure 6.2.7 The OECD standard environmental model

6.2.8 *Compartments and processes in an environment model*

According to Parlar and Angerhöfer (1991), only a few models are suitable for providing information about the effect of a substance on a system. On the one hand, the substance- and system-related data are frequently insufficient, and on the other hand ecotoxicity is usually shown as a function of toxic test effects and even system-related models usually only provide a prognosis of exposure. Long-term effects in particular are very difficult to estimate. A concept which deals with the activity analysis of chemicals in natural systems is a part of the E4 Chem model (Section 6.2.5). Based on compartments and processes, this model simulates the most important dynamic processes. Nutrient cycles (Sections 5.1.3–5.1.5) are taken into special consideration. This even makes it possible to make statements about the change of the mass equilibrium (Section 6.2.9) under the influence of an environmental chemical. The term 'compartment' is used in ecology to mean a part of a complex ecosystem that can be described and defined by material concentrations, transformation processes and transport processes via border areas to neighbouring 'boxes' or compartments. As a rule, one assumes an equilibrium of the observed substances within a compartment. Mathematical models especially also make use of the formation of compartments. Three of the compartments of this submodel represent the trophic levels of producers, consumers and decomposers (Section 5.1.4). These three areas are each characterised by special metabolic pathways and by a specific form of nutrition. The two other material-oriented compartments correspond to the domains 'inorganic nutrients' and 'dead organic matter'. They connect the trophic-oriented compartments via the transfer of material.

6.2.9 *Activity analysis using a computer study*

Changes in the population mass even with long-term effects of chemicals can be calculated using the compartment model described in Section 6.2.8. To determine the population dynamics, we assume the individual to be in one of three states: viable (able to reproduce); non-viable (alive but not able to reproduce); and dead. Both the nutritional status (which can be shown as a function of satiety) and the effects of the chemical (as a dose–effect relationship) influence the three states of the individual. The contribution of viable individuals to the population mass is positive, that of non-viable individuals is negative. Figure 6.2.9 shows the development of the mass in the compartments during and after an exposure of 2000 units of time (dose 0.5 units) (Parlar and Angerhöfer, 1991). The graphs make it clear that the reaction of the population depends on the location in the trophic system. A reduction in the mass of the producers and consumers in the exposure phase stands in contrast to an increase in the decomposers and, by association, an increase in dead organic matter and inorganic nutrients. Afterwards the mass of the producers increases again slightly, whereas that of consumers (with a higher degree of damage) continues to decrease. From this mass-oriented model for the development of ecosystems, one can also conclude that the increased activity of the decomposers causes the increase in inorganic mass and that the risk of washing out, of an outflow of material, thereby increases due to the low productivity of the producers.

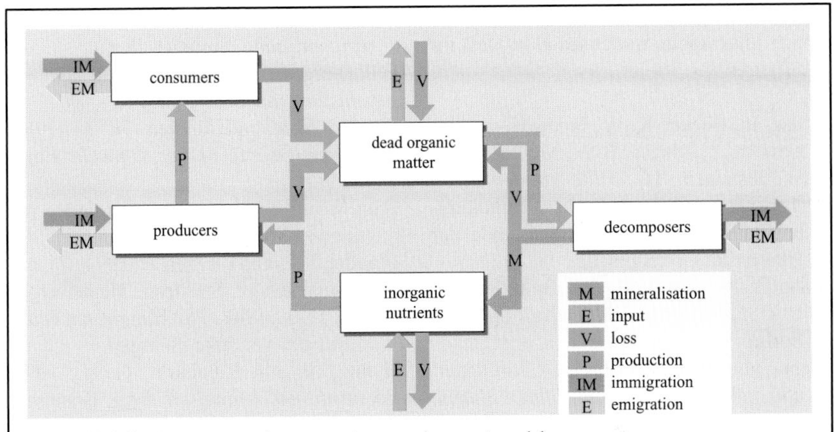

Figure 6.2.8 Compartments and processes in an environment model

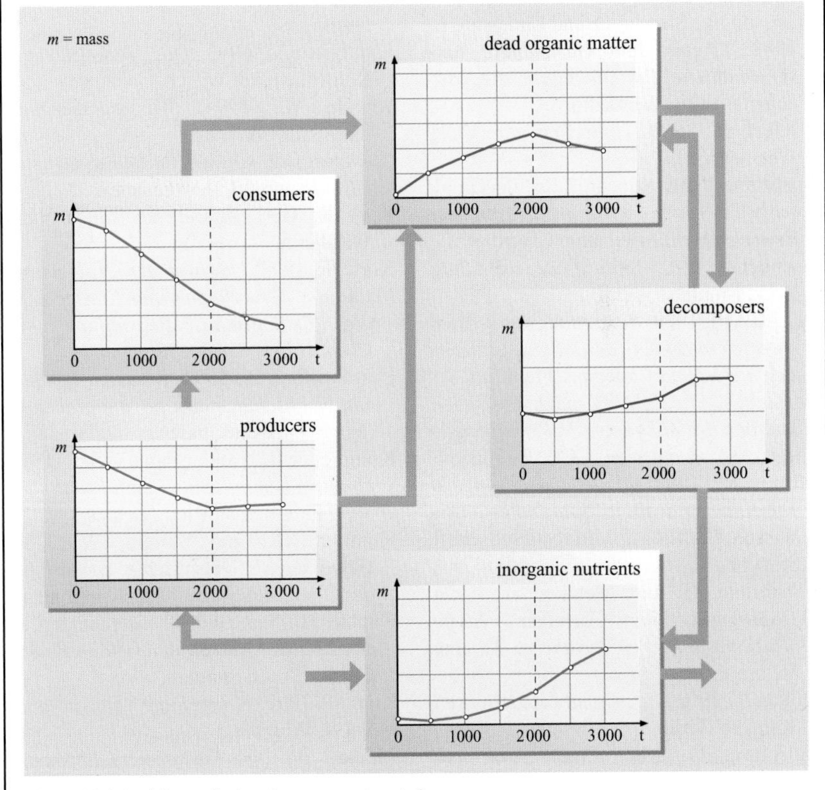

Figure 6.2.9 Activity analysis using a computer study

Bibliography

The following books served as sources from which illustrations were used as models for the figures.

Aldag, R., Becker, K.-W., Frede, H.-G., Hugenroth, P., Klages, F.-W., Meyer, B. and Wildeshagen, H., 1982. *Bodenkunde. Aspekte und Grundlagen.* Grundkursleitfaden: Institut für Bodenkunde der Universität Göttingen.

Bahadir, M., Parar, H. and Spiteller, M., 1995. *Springer Umweltlexikon.* Springer, Berlin.

Bank, M., 1994. *Basiswissen Umwelttechnik. Wasser, Luft, Abfall, Lärm, Umweltrecht.* Vogel, Würzburg.

Berndt, J., 1995. *Umweltbiochemie.* Fischer (UTB 1838), Stuttgart.

Birgerson, B., Sterner, O. and Zimerson, E., 1988. *Chemie und Gesundheit. Eine verständliche Einführung in die Toxikologie.* VCH, Weinheim.

Bliefert, C., 1994. *Umweltchemie.* VCH, Weinheim.

Clausthaler Umwelttechnik-Institut GmbH (ed.), 1990. *Analytik - Messen und Bewerten von Emissionen.* Clausthal.

Daunderer, I.M., 1995. *Gute im Alltag.* Beck, Munich.

DECHEMA, 1989. *Beurteilung von schwermetallen in Böden von Ballungsgebieten: Arsen, Blei und Cadmium.* Frankfurt.

Domsch, K H., 1992. *Pestizide im Boden. Mikrobieller Abbau und Nebenwirkungen auf Mikroorganismen.* VCH, Weinheim.

Dörner, K., 1993. *Akute und chronische Toxizität von Spurenelementen.* Wissenschaftliche Verlagsgesellschaft, Stuttgart.

Eisenbrand, G. and Metzler, M., 1994. *Toxikologie für Chemiker. Stoffe, Mechanismen, Prüfverfahren.* Thieme, Stuttgart.

Fabian, P., 1992. *Atmosphäre und Umwelt.* Springer, Berlin.

Fellenberg, G., 1977. *Umweltforschung. Einführung in die Probleme der Umweltverschmutzung.* Springer, Berlin.

Fellenberg, C., 1992. *Chemie der Umweltbelastung.* Teubner, Stuttgart.

Fiedler, H.J. and Rösler, H. J., 1988. *Spurenelemente in der Umwelt.* Enke, Stuttgart.

Fritz, W. and Kern. H., 1990. *Reinigung von Abgasen.* Vogel, Würzburg.

Graedel, T. E. and Crutzen, P., 1994. *Chemie der Atmosphäre.* Spektrum, Heidelberg.

Hapke, H.-J., 1988. *Toxikologie für Veterinärmediziner.* Enke, Stuttgart.

Heintz, A. and Reinhardt, G. A., 1993. *Chemie und Umwelt.* Vieweg, Braunschweig.

Hütter, L. A., 1994. *Wasser und Wasseruntersuchung.* Salle + Sauerländer, Frankfurt.

Katalyse e.V., 1993. *Das Umweltlexikon.* Kiepenheuer & Witsch, Cologne.

Knoch, W., 1994. *Wasserversorgung, Abwasserreinigung und Abfallentsorgung. Chemische und analytische Grundlagen.* VCH, Weinheim.

Koch, R., 1991. *Umweltchemikalien.* VCH, Weinheim.

Korte, F., 1992. *Lehrbuch der Ökologischen Chemie. Grundlagen und Konzepte für die ökologische Beurteilung von Chemikalien.* Thieme, Stuttgart.

Kowaleski, J. B., 1993. *Altlasten-Lexikon. Ein Nachschlagewerk für die praktische Arbeit.* Glückauf, Essen.

Kümmel, R. and Papp, S., 1990. *Umweltchemie. Eine Einführung.* Dt. Verl. für Grundstoffind., Leipzig.

Kummert, R. and Stumm, W., 1992. *Gewässer als Ökosysteme. Grundlagen des Gewässerschutzes.* Teubner, Stuttgart.

Kuntze, H., Niemann, J., Roeschmann, G. and Schwerdtfeger, G., 1983. *Bodenkunde.* Ulmer, Stuttgart.

Kunz, P, 1992. *Behandlung von Abwasser.* Vogel, Würzburg.

Markert, B., 1994. *Environmental Sampling*

for Trace Analysis. VCH, Weinheim.

Matschullat, J., Heinrichs, H., Schneider, J. and Ulrich, B., 1994. *Gefahr für Ökosysteme und Wasserqualität. Ergebnisse interdisziplinärer Forschung im Harz*. Springer, Berlin.

Matschullat, J. and Müller, G., 1994. *Geowissenschaften und Umwelt*. Springer, Berlin.

Merian, E., 1991. *Metals and Their Compounds in the Environment. Occurrence, Analysis, and Biological Relevance*. VCH, Weinheim.

Müller, H. J., 1991. *Ökologie*. Fischer, Jena.

Parlar, H. and Angerhöfer, D., 1991. *Chemische Ökotoxikologie*. Springer, Heidelberg.

Philipp, B., 1993. *Einführung in die Umwelttechnik*. Vieweg, Braunschweig.

Pietsch. J. and Kamieth, H., 1991. *Stadtböden. Entwicklungen, Belastungen, Bewertung und Planung*. Blottner, Taunusstein.

Römpp, 1993. *Lexikon Umwelt*. Thieme, Stuttgart.

Sattler, K. and Emberger, J., 1990. *Behandlung fester Abfälle*. Vogel, Würzburg.

Schachtschabel, R., Blume, H.-P., Hartge, K.-H. and Schwertmann, U., 1984. *Lehrbuch der Bodenkunde*, Enke, Stuttgart.

Schultheiß, S. and Coos, W., 1993. *Altlasten. Eine Einführung für Naturwissenschaftler, Ingenieure und Planer*. Clausthaler Tektonische Hefte 28: Sven von Loga, Cologne.

Schwedt, G., 1995. *Mobile Umweltanalytik*. Vogel, Würzburg.

Schwedt, G., 1996. *Toxikologisches Lexikon zum Umweltchemikalienrecht (ChemG, GefStoffV und ChemverbotsV)*. Vogel, Würzburg.

Sigg, L. and Stumm, W., 1994. *Aquatische Chemie. Eine Einführung in die Chemie wässriger Lösungen und in die Chemie natürlicher Gewässer*. Teubner, Stuttgart.

Stegmann, L., 1993. *Bodenreinigung. Biologische und chemisch-physikalische Verfahrensentwicklung unter Berücksichtigung der bodenkundlichen, analytischen und rechtlichen Bewertung*. Econornia, Bonn.

Trudinger, P.A. and Swaine, D.J., 1979. *Biogeochemical Cycling of Mineral-Forming Elements*. Elsevier, Amsterdam.

VDI Analytik bei Abfallentsorgung und Altlasten, 1990. VDI, Düsseldorf.

Wild, A., 1995. *Umweltorientierte Bodenkunde*. Spektrum, Heidelberg.

Wille, F., 1993. *Bodensanierungsverfahren*. Vogel, Würzburg.

Ziechmann, W. and Müller-Wegener, U., 1990. *Bodenchemie*. BI-Wiss.-Verl., Mannheim.

Other references

Ballin, U., Kruse, R. and Rüssel, H.-A., 1994. Determination of total arsenic and speciation of arseno-betaine in marine fish by means of reaction – headspace gas chromatography utilising flame-ionisation detection and element specific spectrometric detection. *Fresenius J. Anal. Chem.*, **350**(1/2): 54–61.

Garrels, R.M. and Mackenzie, F.T., 1971. *Evolution of Sedimentary Rocks*. W.W. Norton, New York.

Specht, W. and Tilkes, M., 1981. Gas-chromatographische Bestimmung von Rückständen von Pflanzenbehandlungsmitteln nach Clean-up über Gel-Chromatographie und Mini-Kieselgel-Säulen-Chromatographie; 4. Mitteilung: Gas-chromatographische Bestimmung von 11 herbiciden Phenoxyalkan-carbonsäuren und ihren Estern in Pflanzenmaterial. *Fresenius Z. Anal. Chem.*, **307**: 257–64.

Index